AF877566

BRAKES

HBJ

Harcourt Brace Jovanovich, Publishers
Technology Publications
San Diego New York Chicago Austin Atlanta Washington, D.C.
London Sydney Tokyo Toronto

BRAKES

James G. Hughes
Rio Hondo College

Copyright © 1987 by Harcourt Brace Jovanovich, Inc.

All rights reserved. No part of this publication may be
reproduced or transmitted in any form or by any means, electronic or mechanical,
including photocopy, recording, or any information storage and retrieval system,
without permission in writing from the publisher.

Requests for permissions to make copies of any part of the work should be mailed to:
Permissions, Harcourt Brace Jovanovich, Publishers, Orlando, Florida 32887.

ISBN: 0-15-505545-3

Library of Congress Catalog Card Number: 86-82886

Printed in the United States of America

Cover design: Bill Agee
Production supervision and interior design: Linda Zuk, WordCrafters Editorial Services, Inc.

The information contained herein has been developed by the author
for publication by Harcourt Brace Jovanovich, Inc.
Inquiries regarding the content of this work should be addressed to the author and/or publisher.
The National Institute for AUTOMOTIVE SERVICE EXCELLENCE
is neither the author nor the publisher of this work
and has no responsibility for the material contained herein.

The information contained herein has been developed from the best available sources.
Technology Publications and the author cannot assume responsibility
for the accuracy of resource data or consequences of its application.
Persons using this book are cautioned to refer to vehicle and
equipment manufacturers' manuals for specific service information.
Also be advised to observe all safety precautions herein and to
refer to manufacturers' procedures for specific vehicle and personnel safety directions.

ASBESTOS DUST HAZARD WARNING
Brake shoe and pad lining can contain from 30% to 70% asbestos.
Studies have indicated that breathing asbestos dust can cause asbestosis and cancer.
During brake servicing, the technician should wear an approved air-purifying respirator.
It should be worn during all procedures starting with the removal of the wheels
and including reassembly.

CONTENTS

PREFACE

An essential part of any successful automotive training program is a solid foundation on which to develop the information to be presented. To save time and effort and ensure that technicians receive the training they need to prepare them to enter into the automotive service field, the author has built this book around the task list prepared by the National Institute for AUTOMOTIVE SERVICE EXCELLENCE (ASE).

This comprehensive list of job-related diagnostic and repair tasks was initially prepared by representatives of the automotive service industry. ASE updates each of the individual certification areas by assembling a group of industry specialists from domestic and import manufacturers, aftermarket suppliers, working technicians, and educators. This group reviews the task list, adding new and eliminating old technology. Each proposed change must be approved by consensus of the entire group. The end result is a task list that represents the body of knowledge each working technician should possess.

Because this book was developed from the task list on which the ASE certification tests are based, using the book for initial or refresher training on automotive brake systems should help prepare a technician for the tests.

Facts About ASE Certification

ASE is a nonprofit organization dedicated to improving the quality of automotive service and repair. It is governed by a board of directors representing all sectors of the automotive industry.

ASE's voluntary certification program offers experienced technicians the opportunity to take any of the eight automotive certification tests during the May or November testing periods. Each test consists of either 40 or 80 questions on diagnosis and repair techniques. To be certified, the technician must pass one or more tests and be able to prove a minimum of two years' related education and/or work experience. Successfully completing all eight tests qualifies a technician as a Certified Master Automobile Technician (CMAT).

The eight ASE test areas are:

1. Engine Repair (80 questions)
2. Automatic Transmission/Transaxle (40 questions)
3. Manual Drive Train and Axles (40 questions)
4. Suspension and Steering (40 questions)
5. Brakes (40 questions)
6. Electrical Systems (40 questions)

7. Heating and Air Conditioning (40 questions)

8. Engine Performance (80 questions)

Applicants must sign up in advance to take these tests. For complete information about ASE certification and test registration, contact:

National Institute for AUTOMOTIVE SERVICE EXCELLENCE
1920 Association Drive
Reston, VA 22091-1502
(703) 648-3838

Certification for Automobile Technician Training Programs

Future generations of automotive service professionals are in the hands of today's vocational educators. Automotive instructors have the opportunity not only to train technicians for entry-level jobs, but also to encourage and foster their professionalism. To further this effort, ASE offers voluntary certification for automobile technician training programs upon the recommendation of the National Automotive Technicians Education Foundation (NATEF).

NATEF's goals are to develop and encourage automotive technical education. The foundation examines the structure and resources of automotive technician training programs and evaluates them in relation to certain nationally accepted standards. Among these standards is a task list nearly identical to the ASE task list.

To make effective use of this book, the training program should include:

1. Appropriate theory, safety, and support instruction will be required for the technician to perform each task. This instruction must include identification and use of tools and equipment required to accomplish the task. It must also include training in how to locate and use current reference materials from accepted industry publications that present the manufacturer's recommendations regarding the task (see Appendix A).

2. The training program should have performance standards for each task covered in the curriculum. For example: *Working alone or with one other person, you will remove and replace a minimum of four wheel cylinders on two instructor-assigned automobiles. The manufacturer's recommended procedure and specifications will be followed. Time limit will be twice the flat-rate time allowed in the Motors or Chilton flat-rate manual.*

Each technician's progress should be monitored and evaluated against these performance standards.

3. A feedback system should be in place that informs the technician which tasks have been completed successfully.

Special Features of This Book

1. True case history examples

2. Shop tips

3. Safety cautions

4. More than 120 questions similar to those on the actual ASE Brake test

5. Addresses for obtaining manufacturers' service manuals

6. Addresses for obtaining the most commonly used comprehensive shop manuals

ACKNOWLEDGMENTS

I am grateful for the comments and suggestions of the individuals who contributed to the development of this book. I extend special thanks to William J. deKryger, Central Michigan University; Carl E. Dinkel, Tarrant County Junior College; Earl McCartney, Northern Virginia Community College; Tom Orr, Rio Hondo College; Don Neman, Automotive Training Institute; and Bill J. Steen, Citrus College.

Grateful acknowledgment is also made to the following companies, which provided reference literature used in the preparation of this book:

American Brakeblock Division, Abex Corporation; Allied Aftermarket Division, Allied-Signal Corporation; American Honda Motor Company, Inc.; American Motors Corporation; Ammco Tools, Inc.; Automotive Caliper Exchange, Inc.; Barrett Brake Service Equipment; John Bean Division, FMC Corporation; Bear Manufacturing Corporation; Brush Research Manufacturing Company, Inc.; Buick Motor Division, General Motors Corporation; Cadillac Motor Division, General Motors Corporation; Chevrolet Motor Division, General Motors Corporation; Chrysler Corporation; CR Industries; E. Edelmann and Company; EIS Division, Parker Hannifin Corporation; Ford Parts and Service Division; General Motors Corporation, Product Service Training Division; Gibson Products, Division of Rolero, Inc.; Heli-Coil Products, Division of Mite Corporation; Isuzu Motors Limited; K-D Tools, Inc.; K&W Products, Inc.; Kelsey Products Division, Kelsey-Hayes Company; Lee Manufacturing Company, Division of Rolero, Inc.; Lisle Corporation; Lucas Industries, Inc.; Mac Tools, Inc.; Master Parts Division, Airtex Products, Inc.; Mighty Distributing System; Neward Enterprises Inc.; Nissan Motor Corporation; Oldsmobile Motor Division, General Motors Corporation; Pontiac Motor Division, General Motors Corporation; Quaker State Oil Company; Raybestos-Manhattan, Inc.; Raymark Industries, Inc.; SD Specialty Distributors; S.K. Wellman Corporation; Snap-on Tools Corporation; Specialty Products Company; Star Machine and Tool Company; Stuart-Western, Inc.; Toyota Motor Sales, U.S.A., Inc.; TRW Replacement Parts Division, Automotive Aftermarket Group; Wagner Division, Cooper Industries, Inc.; Wagner Electric Sales Corporation; Western Brake Company.

The author gratefully acknowledges the National Institute for AUTOMOTIVE SERVICE EXCELLENCE for granting special permission to reproduce its Brake task list.

James G. Hughes
CMAT

ABOUT THE AUTHOR

James G. Hughes has been a full-time automotive training instructor at Rio Hondo College in Whittier, California for eighteen years, and was the national service training instructor for Toyota Motor Sales, U.S.A., Inc. He has written several textbooks and conducted numerous workshops for state- and nationwide audiences. Mr. Hughes is a Certified Master Automobile Technician and a member of the Society of Automotive Engineers. He was selected as the outstanding community college automotive instructor in California for 1985.

BRAKES

Section A

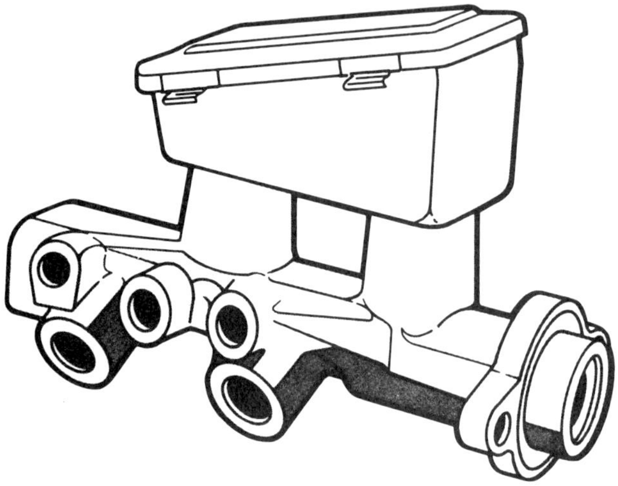

After studying this section, and with hands-on practice on a "live" vehicle, you should be able to perform the following tasks:

- Diagnose poor stopping, pulling, or dragging caused by problems in the master cylinder; determine needed repairs.
- Diagnose poor stopping, pulling, dragging, high or low pedal, or hard pedal caused by problems in the step bore master cylinder and internal valves (e.g., volume control devices, quick take-up valve, fast-fill valve, pressure regulating valve); determine needed repairs.
- Measure and adjust pedal push rod length.
- Check master cylinder for defects by depressing brake pedal and observing brake action to determine whether there is a leak in the hydraulic system and/or whether the primary cups are defective.
- Check master cylinder for secondary cup defects by inspecting for fluid leakage at the connection between power booster and cylinder or boot.

HYDRAULIC SYSTEM DIAGNOSIS AND REPAIR

- Remove master cylinder from vehicle.
- Disassemble, clean, inspect master cylinder and reusable parts, and clean or hone cylinder bore (if necessary and recommended by manufacturer).
- Reassemble master cylinder.
- Bench bleed (check for function and remove air) master cylinder.
- Install master cylinder in vehicle; test operation of the hydraulic system.
- Diagnose poor stopping, pulling, or dragging caused by problems in the brake fluid, lines, and hoses; determine needed repairs.
- Inspect brake lines and fittings for leaks, dents, kinks, rust, cracks, or wear; tighten loose fittings and supports.
- Inspect flexible brake hoses for leaks, kinks, cracks, bulging, or wear; tighten loose fittings and supports.
- Replace brake lines (double flare and ISO types), hoses, fittings, and supports.
- Select, handle, store, and install brake fluids (including silicone fluids).
- Diagnose poor stopping, pulling, or dragging caused by problems in the hydraulic system valve(s); determine needed repairs.
- Inspect, test, and replace metering (hold-off), proportioning (balance), pressure differential, and combination valves.
- Inspect, test, replace, and adjust load or height sensing-type proportioning valve(s).
- Inspect, test, and replace brake warning light system switch and wiring.
- Reset brake pressure differential valve (if necessary).
- Bleed (manual, pressure, vacuum, or surge) and/or flush hydraulic system.
- Pressure test brake hydraulic system by applying heavy force to pedal; hold for 15 seconds and inspect for leaks.

HYDRAULIC SYSTEM OPERATING PRINCIPLES

The hydraulic system controls an automobile's braking action. The system includes a master cylinder, fluid lines and hoses, control valves, and calipers and/or wheel cylinders (Figure 1).

When the driver steps on the brake pedal, foot force creates hydraulic pressure in the master cylinder. Braking occurs when this pressure is transmitted through lines and hoses to calipers and/or wheel cylinder pistons.

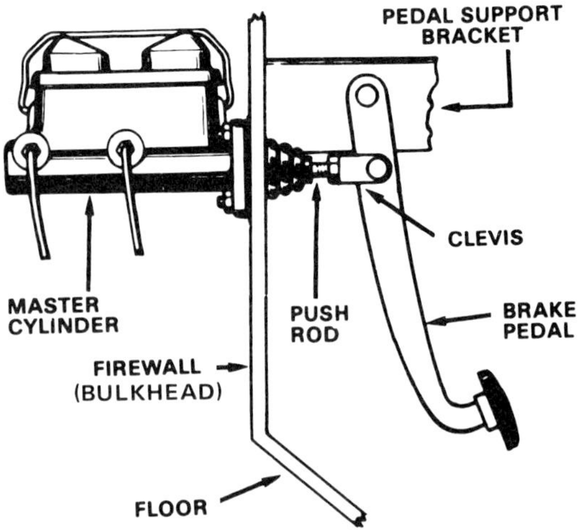

FIGURE 2
Suspended pedal—manual brakes. (Courtesy of Allied Aftermarket Division, Allied-Signal Corporation)

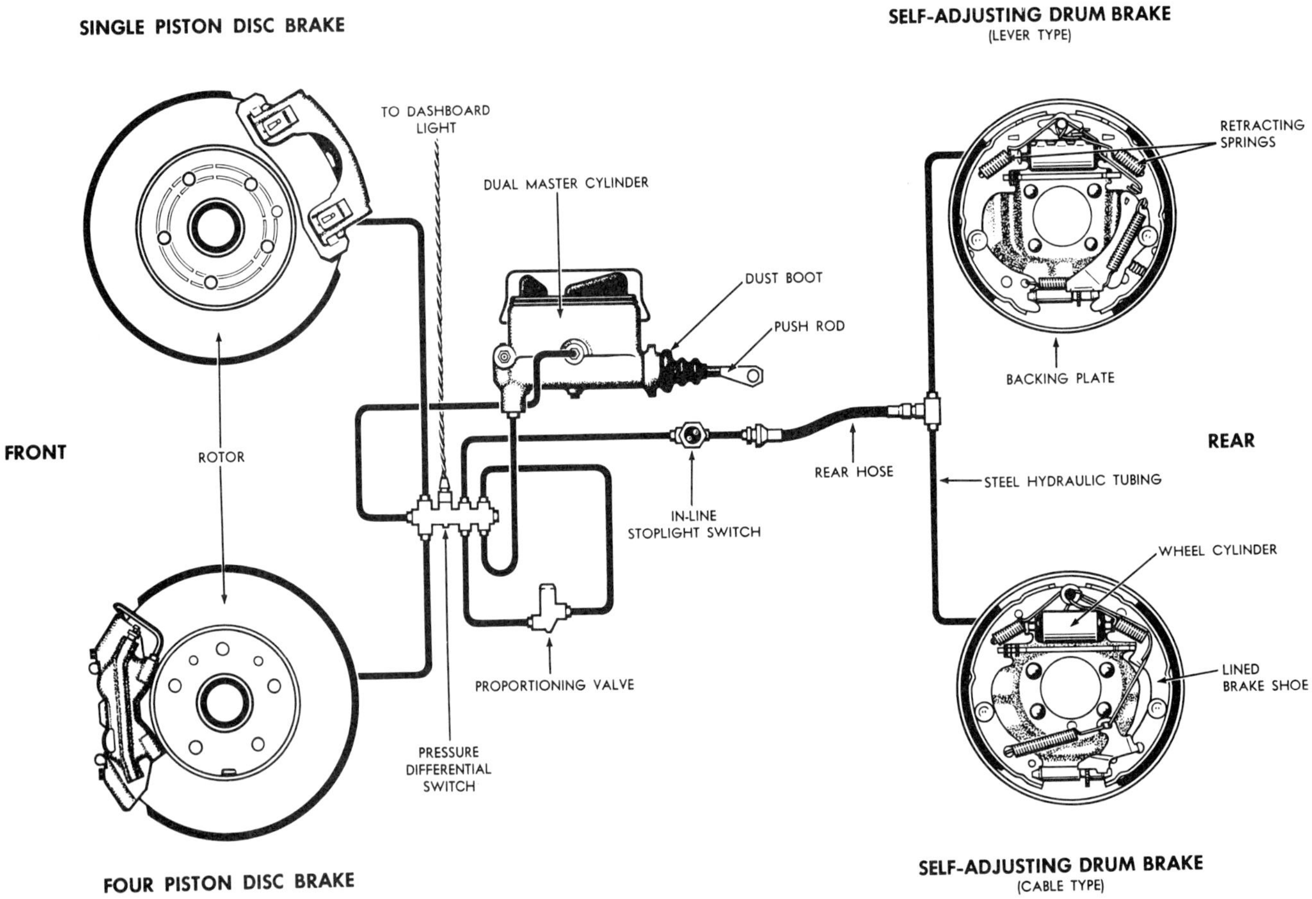

FIGURE 1
An example of parts you may find in the hydraulic braking system on older vehicles, as well as in some newer vehicles. Four different brake designs are shown. In practice, brake designs are paired, front or rear. (Courtesy of Raybestos-Manhattan, Inc.)

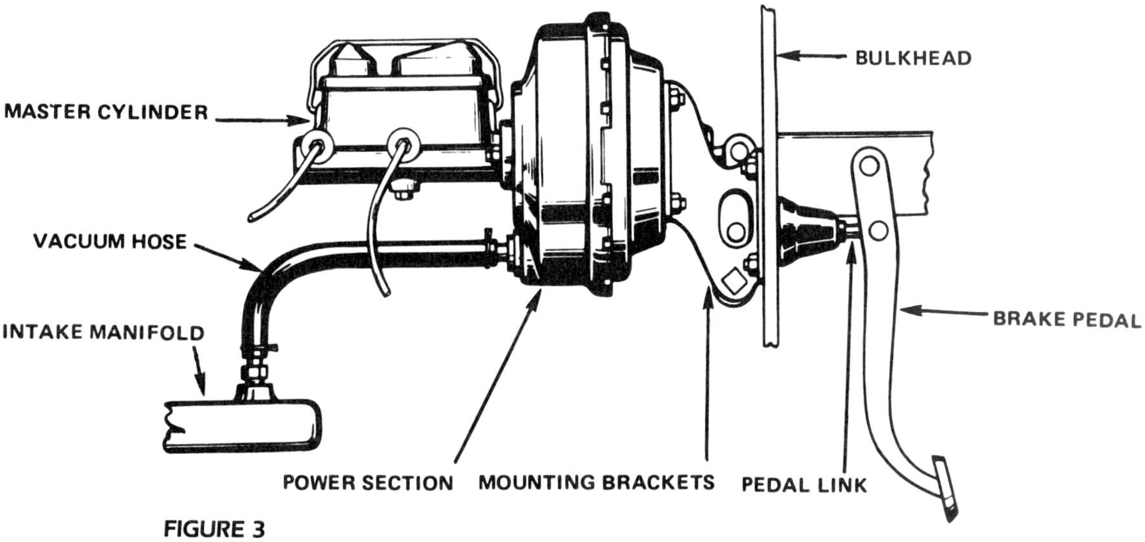

FIGURE 3
Suspended pedal—power brakes. (Courtesy of Allied Aftermarket Division, Allied-Signal Corporation)

MASTER CYLINDER

The master cylinder is a type of pump that is controlled by a push rod attached to the brake pedal. Most passenger cars use a suspended pedal (Figures 2 and 3).

The master cylinder is generally made of cast iron with the fluid reservoir built into it, or of aluminum with a moulded plastic reservoir (Figures 4 and 5).

Disc brake master cylinders are much like drum brake master cylinders. The main

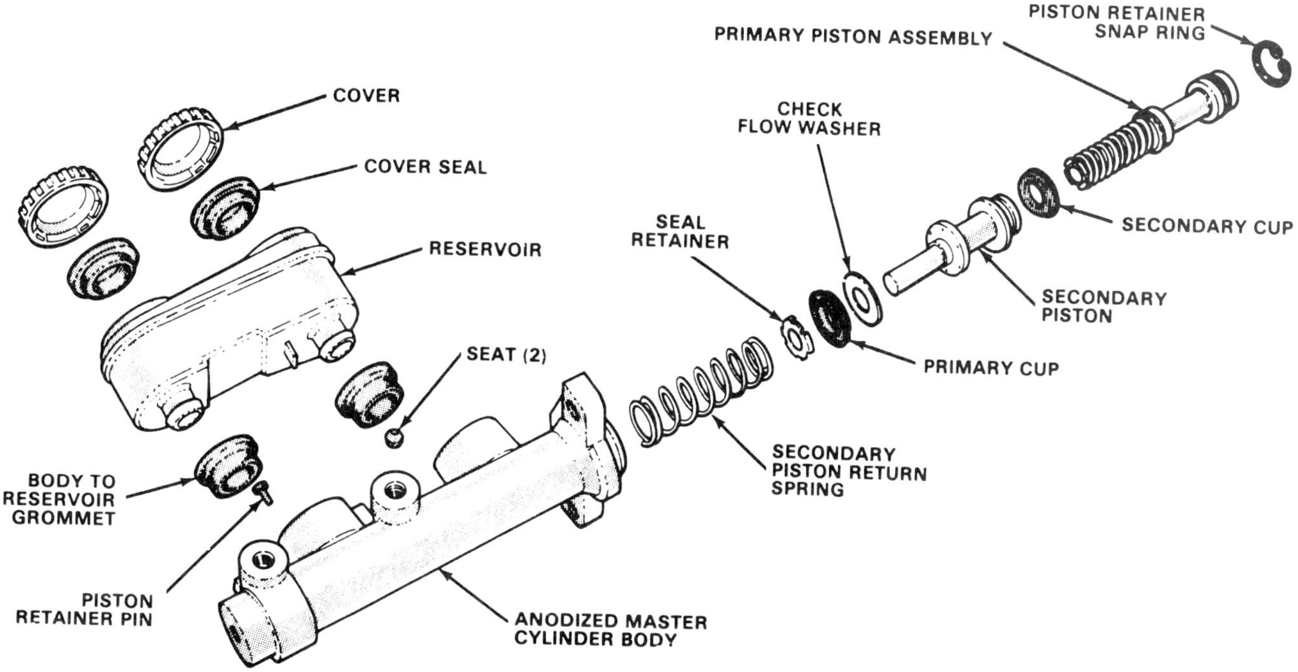

FIGURE 4
Dual system aluminum master cylinder with plastic reservoir. (Courtesy of Chrysler Corporation)

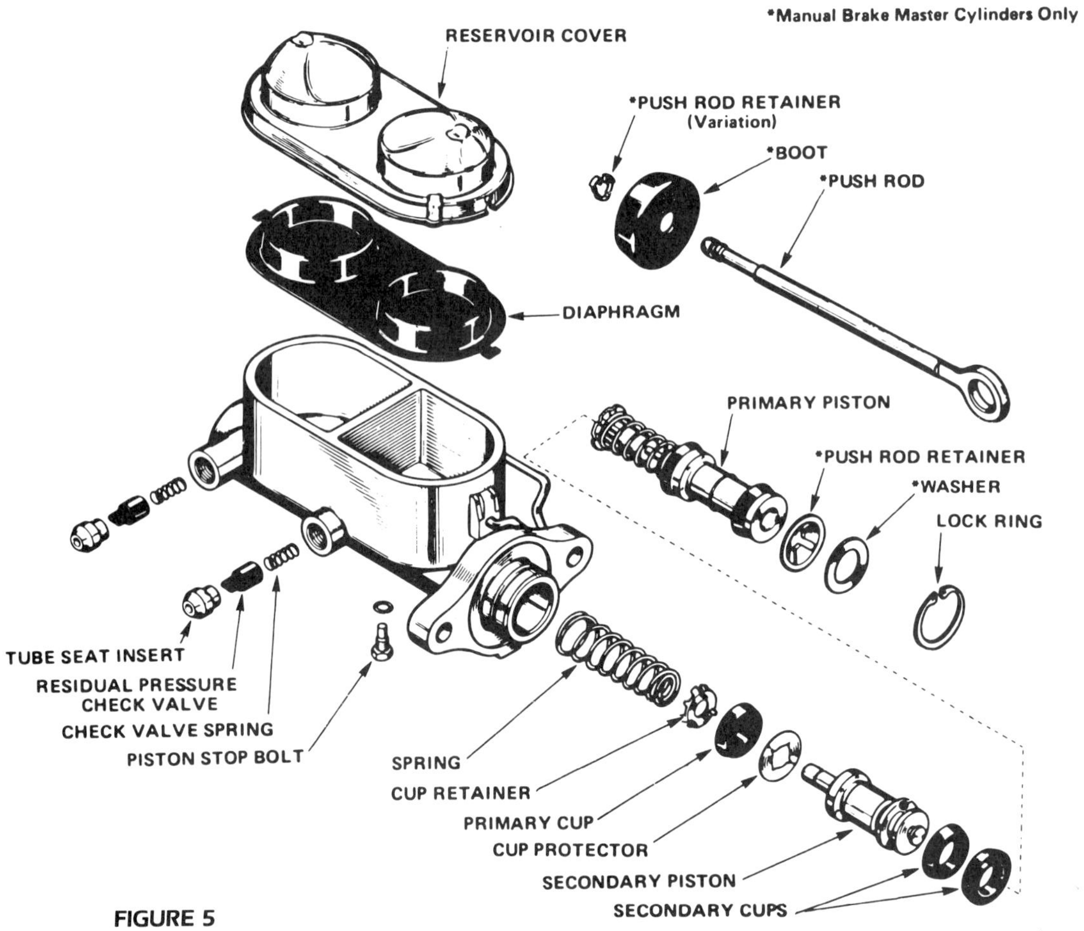

FIGURE 5

A dual system drum brake master cylinder. Note the equal size of the two fluid reservoirs and the use of residual pressure check valves at both outlets. These valves maintain 8 to 16 pounds per square inch (psi) pressure in the hydraulic lines at all times to prevent air from entering the system when the brakes are released. (Courtesy of Allied Aftermarket Division, Allied-Signal Corporation)

difference is in the size of their brake fluid reservoirs. Disc brakes are connected to a larger reservoir. Another difference is in the use of residual pressure check valves (refer back to Figure 5). Outlets connected to disc brakes do not contain a check valve.

COMMENT: *Residual pressure check valves have been done away with on many newer cars. To prevent air from entering past the wheel cylinder cups, expanders are being used (Figure 6).*

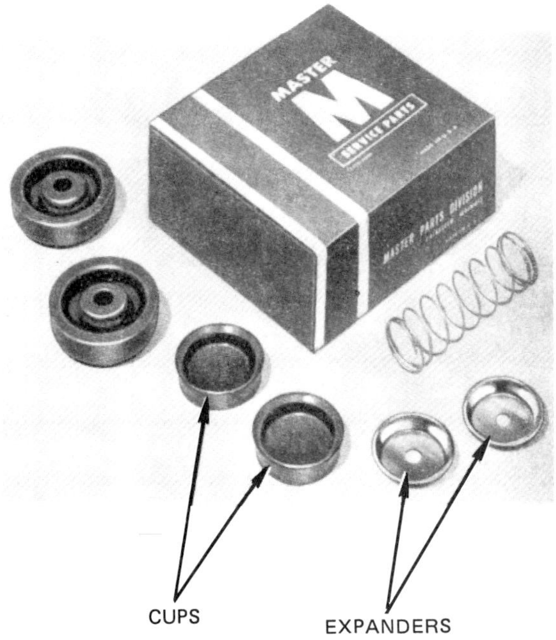

FIGURE 6

Expanders are designed to keep cup sealing lips in constant contact with the wheel cylinder bore. This prevents air from entering past the cups. (Courtesy of Master Parts Division, Airtex Products, Inc.)

Dual and Diagonally Split Systems

Beginning with the 1967 model year, all passenger cars are required to be equipped with a dual system master cylinder. The front brakes connect to one piston and reservoir, and the rear brakes to the other. If either the front or rear brake system fails, the other system remains operational. This axle-to-axle arrangement is common on front engine, rear wheel drive vehicles.

Most front wheel drive vehicles use a diagonally split brake system (Figure 7) instead of the axle-to-axle system. With this arrangement, if either half of the system fails, the other half has equal stopping power.

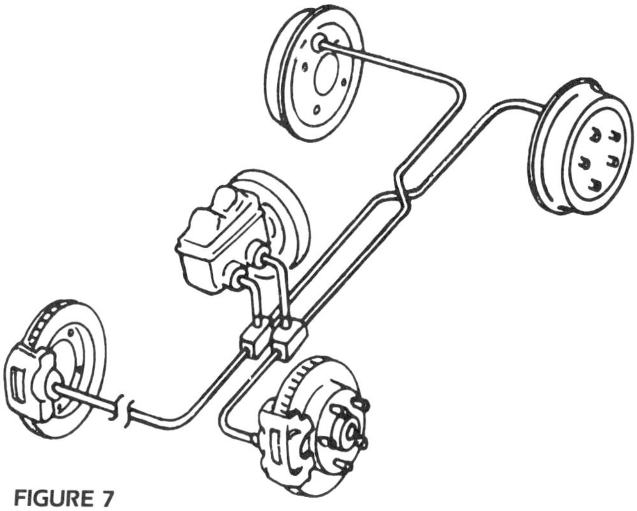

FIGURE 7
Diagonally split system. Since both rear brakes and both front brakes are on separate circuits, a proportioning or metering valve must be used in each circuit if required in the system. (Courtesy of Wagner Division, Cooper Industries, Inc.)

Step Bore Master Cylinder

A step bore master cylinder is shown in Figure 8. This cylinder is the same as other cylinders except that the forward part of the cylinder bore is slightly smaller in diameter than the rear part. This difference contributes to better brake pressure balance on certain vehicles.

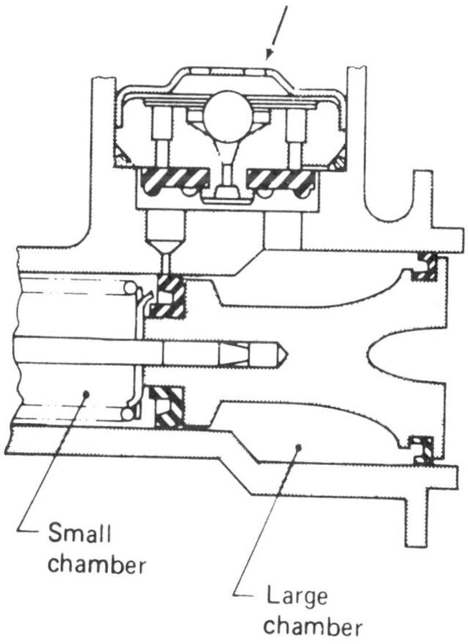

FIGURE 8
In a step bore master cylinder, the forward part of the bore is slightly smaller in diameter than the rear part. This particular step bore master cylinder includes a fast-fill valve (often called a quick take-up valve). (Courtesy of Nissan Motor Corporation)

Quick Take-Up Master Cylinder

In 1980, General Motors introduced the quick take-up master cylinder on its "X" body cars (Figure 9). This master cylinder provides a large volume of fluid at low pressure with only a small amount of pedal movement.

When the brakes are first applied, more fluid is displaced in the primary low pressure chamber than in the primary high pressure chamber. This extra fluid passes around the primary piston lip seal into the primary high pressure chamber and on to two wheels. Since the pressure and displacement are equal in both the primary and secondary systems, the secondary piston must move farther because its bore is smaller. Flow and pressure to the two wheels it serves are the same as they are to the two wheels served by the primary piston. Low pressure has now been applied throughout the system.

The quick take-up valve holds pressure in the primary low pressure chamber up to a certain psi value. When the pressure goes beyond this amount, the valve opens and fluid

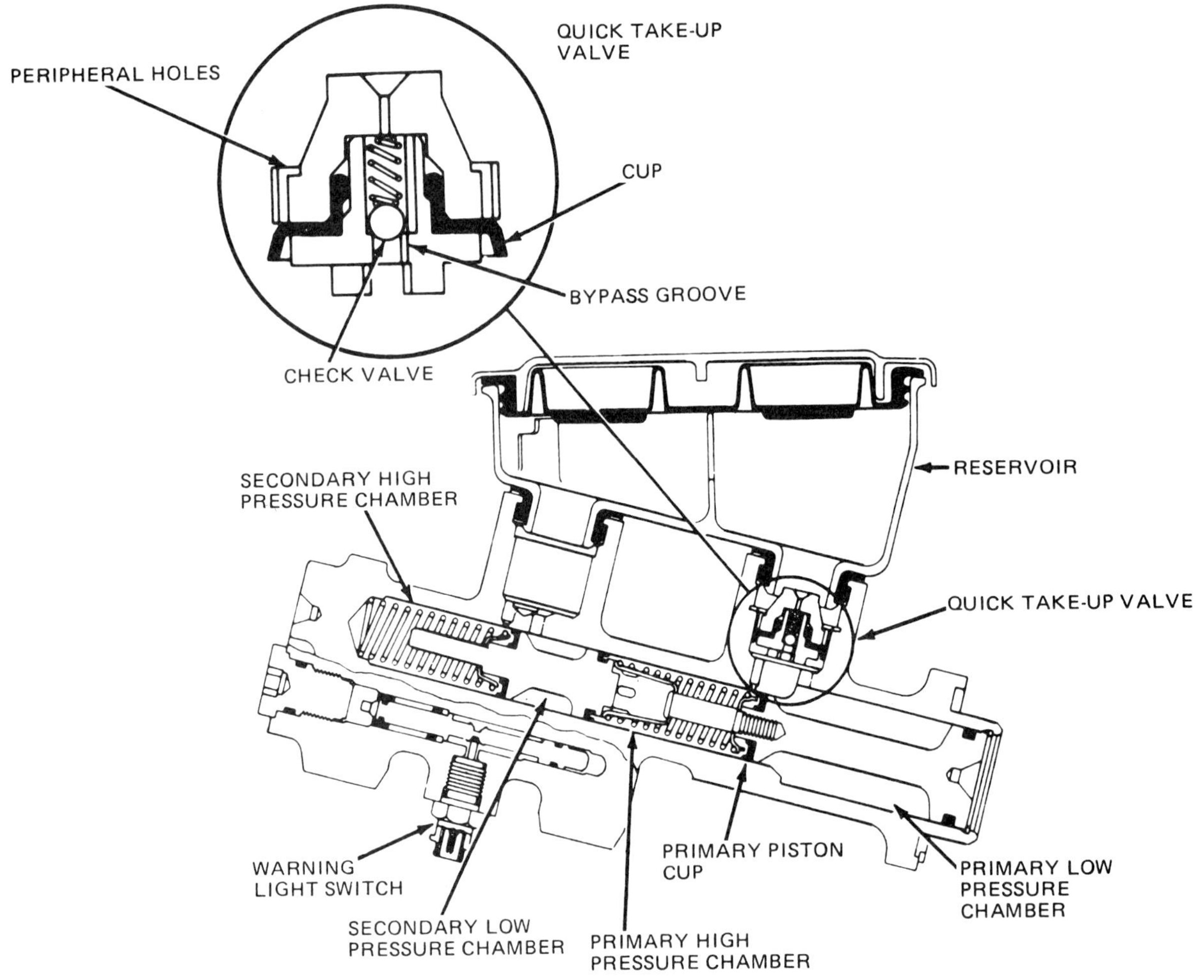

FIGURE 9

The General Motors quick take-up master cylinder. The brake warning light switch and a quick take-up valve are inside the aluminum housing. (Courtesy of Wagner Division, Cooper Industries, Inc.)

passes to the reservoir. The primary and secondary pistons now function as a normal dual master cylinder. Higher braking pressure will develop because these pistons operate in a smaller bore.

When the brakes are released, fluid makeup for the primary piston occurs through the holes in the outside edge of the quick take-up valve.

The quick take-up master cylinder can be used with a system that uses low drag calipers. These special calipers use piston seals that pull the pads completely away from the rotor when the brake pedal is in the released position. This gives better fuel mileage.

Pedal Push Rod

Some push rods are adjustable. The two general methods used to make an adjustment are:

1. By an eccentric bolt at the brake pedal (Figure 10).

2. By an adjustable clevis or eye bolt at the end of the push rod (Figure 11).

With either method, brake pedal free play can be increased or decreased to obtain the correct adjustment. *Pedal free play is the free*

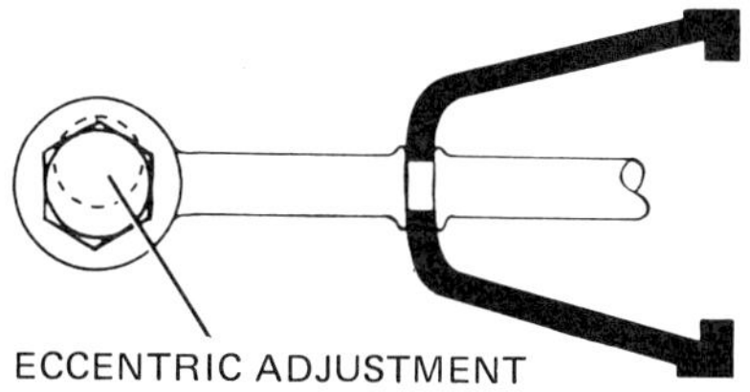

FIGURE 10
Some brake pedal push rods are adjusted by turning an eccentric bolt.

travel of the pedal before any movement of the master cylinder piston occurs. A lack of free play can cause the brakes to drag and eventually lock up. Too much free play can result in a low pedal.

For most vehicles, the correct adjustment will provide 1/16″ to 1/4″ travel of the foot pad from the fully released position to close the master cylinder compensating port(s) and build up system pressure (Figure 12).

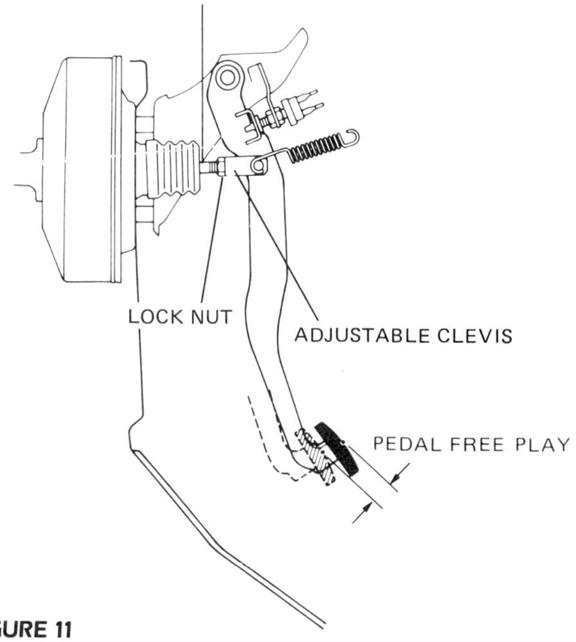

FIGURE 11
This pedal push rod is adjusted by loosening a lock nut, then turning a clevis. (Courtesy of Toyota Motor Sales, U.S.A., Inc.)

MASTER CYLINDER SERVICING

The following inspection procedures will help the technician determine the condition of the master cylinder and its need for service:

1. Inspect the exterior of the master cylinder. The presence of brake fluid indicates an external leak. Check for signs of paint removal on the power brake booster unit or on the bulkhead (firewall). Refer to Figure 13. This indicates that the master cylinder primary piston assembly has a defective secondary cup.

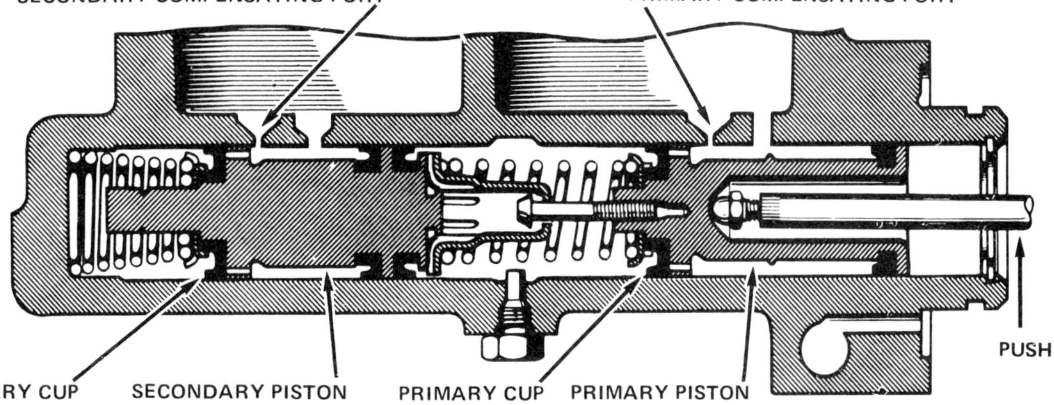

FIGURE 12
Master cylinder in the brakes-applied position. (Courtesy of Allied Aftermarket Division, Allied-Signal Corporation)

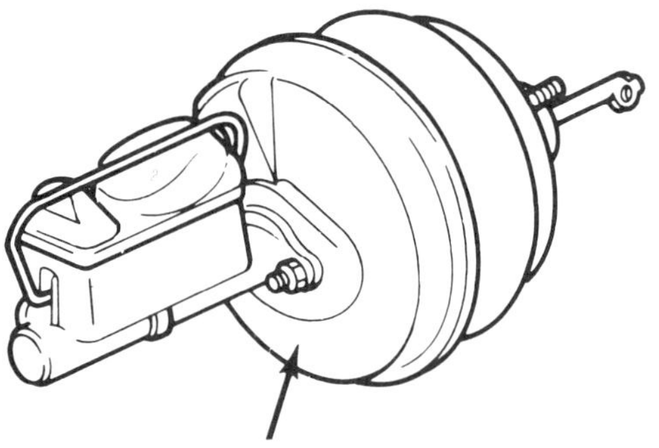

FIGURE 13
Paint removal at this point indicates a defective primary piston secondary cup. (Courtesy of Ford Parts and Service Division)

2. Remove the reservoir cover or cap and check the brake fluid level. It should be within 1/4″ of the top of the reservoir. Also, the fluid should be clean, with no residue in the bottom of the reservoir or other evidence of contamination. If there is residue present, the hydraulic system should be flushed.

3. Inspect the vent holes in the reservoir cover or cap. The holes must be clean and open.

4. Check the reservoir diaphragm (if used). It must not have holes or other damage. This can allow entry of water (from moisture in the air) and dirt.

5. Watch the fluid in the reservoir while an assistant lightly "jabs" the brake pedal 1/4″ to 1/2″. The brake fluid should show agitation or spurting. This means that the compensating port(s) are closing and the push rod length is correct (Figure 14).

6. Apply heavy foot pressure to the pedal (with the engine running for power brakes) and check for sponginess and pedal reserve. The pedal should feel firm, and the foot pad should be a minimum of 2″ from the floor for manual brakes, 1″ from the floor for power brakes (Figure 15).

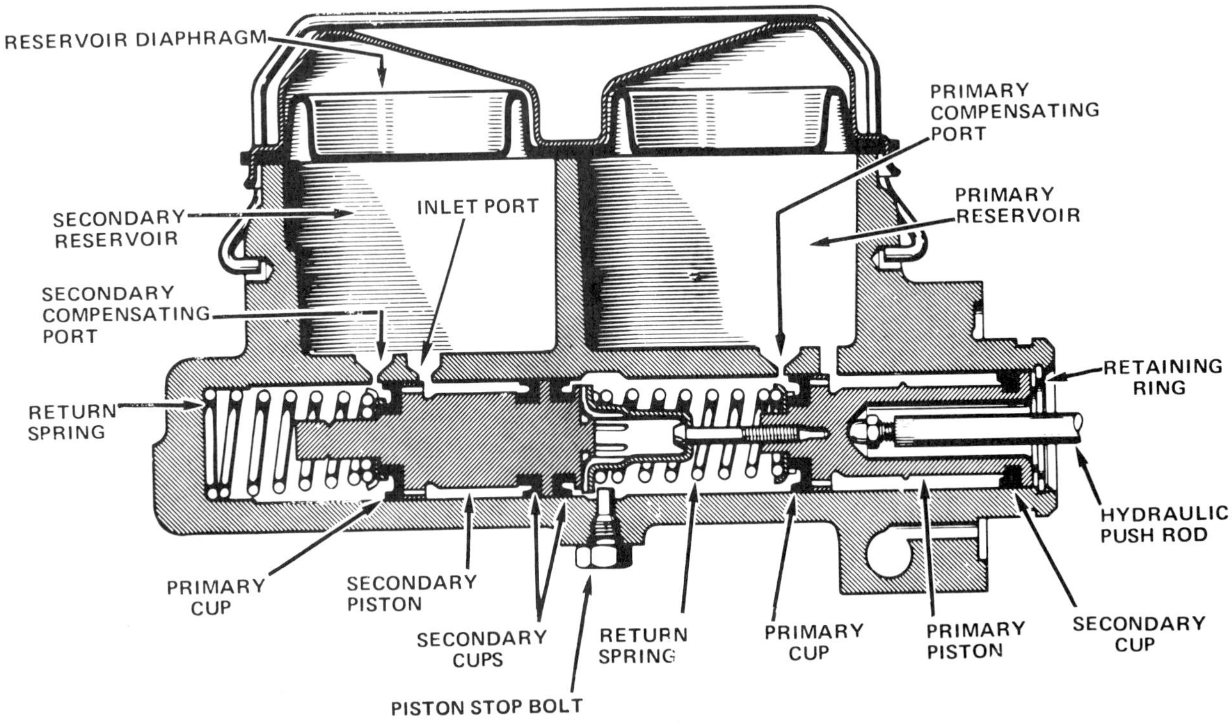

FIGURE 14
You can observe the movement of brake fluid in the reservoirs to see if the compensating ports are closing. (Courtesy of Allied Aftermarket Division, Allied-Signal Corporation)

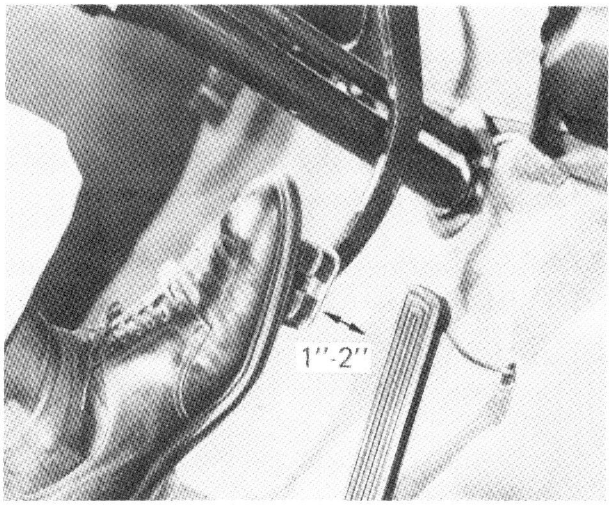

FIGURE 15
A brake pedal should be firm and have adequate reserve (distance from the floor).

7. Hold the brake pedal depressed with medium foot pressure (25 to 35 pounds) for 15 seconds. The pedal should not slowly sink or bottom out under steady pressure. This would indicate an internal master cylinder leak.

8. Depress the brake pedal several times and release it. Then quickly open and close a wheel cylinder bleeder screw. Check for a small spurt of brake fluid when the bleeder screw is opened. This indicates that some leftover (residual) pressure is being maintained in the line if the brake system is equipped with residual check valves. *Charts are available that show which vehicles use residual pressure check valves.*

The usual reason for a master cylinder failure is that the cups lose their ability to seal. Rubber parts wear with use or may deteriorate with age or fluid contamination. Corrosion or deposits can form in the cylinder bore due to moisture or dirt.

MASTER CYLINDER TROUBLESHOOTING

Refer to Chart 1 for some additional master cylinder problems.

MASTER CYLINDER REPLACEMENT

The typical step-by-step procedure for removing and replacing a master cylinder is as follows:

1. Before disconnecting or removing any part, use fender covers to protect the paint finish. *Brake fluid can remove paint.*

2. Carefully clear away or remove any parts that may get in the way.

3. Disconnect the brake lines from the cylinder. Use a flare nut wrench to prevent damage to the line nuts (Figure 16). Cover the ends of the lines and nuts to prevent dirt from entering their open ends.

4. Remove the nuts attaching the master cylinder to the power brake unit or to the bulkhead (firewall).

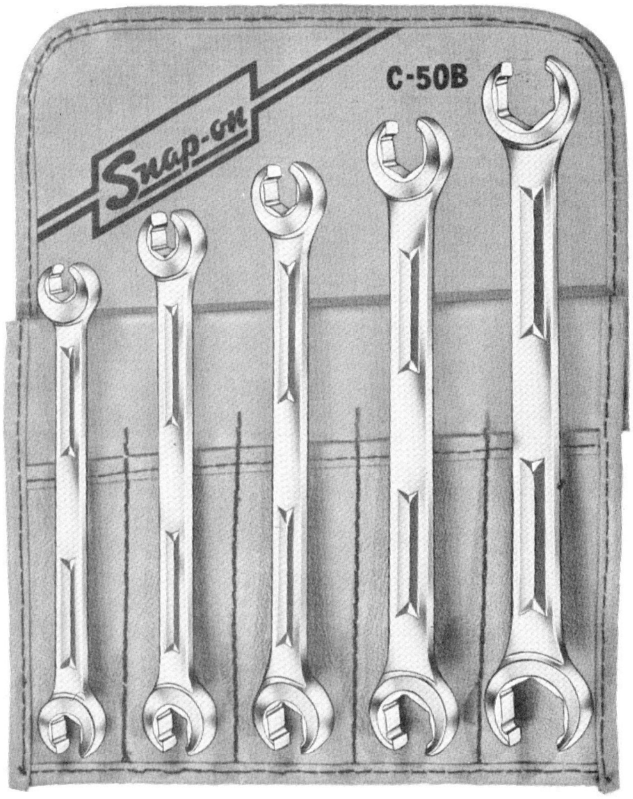

FIGURE 16
Metric flare nut wrenches. (Courtesy of Snap-on Tools Corporation)

CHART 1 TROUBLESHOOTING CHART FOR MASTER CYLINDERS, LINES, AND HOSES

Low Pedal
1. Low fluid level
2. Air in system
3. Soft or weak hoses (expanding under pressure)
4. Too much push rod free play
5. Internal leak in master cylinder
6. "Dirt" in master cylinder
7. Incorrect bleeding procedure
8. Defective quick take-up valve
9. Brakes out of adjustment

Spongy Pedal
1. Low fluid level
2. Air in system
3. Residual pressure check valve faulty
4. Plugged vent in master cylinder filler cover
5. Incorrect bleeding procedure
6. Defective or weak hydraulic hose
7. Drums too thin or cracked

Hard Pedal
1. Dirty, thick brake fluid
2. Cups swollen in master cylinder
3. Bore rough or corroded in master cylinder
4. Lines dented or clogged
5. Hoses kinked, collapsed, or clogged
6. Clogged or plugged fluid control valve (Ford) or quick take-up valve (GM) in master cylinder

Pulling
1. Crimped, dented, or plugged brake line
2. Defective chassis parts causing improper weight shift when brakes are applied

Fading Pedal (Falling Away Under Steady Foot Pressure)
1. Poor quality fluid (low boiling point)
2. Fluid leakage from loose hose and/or line connections
3. Cups damaged or worn in master cylinder
4. Drums turned too thin or cracked

Dragging (Slow or Incomplete Release of Brakes)
1. Cups swollen in master cylinder
2. Compensating port restricted in master cylinder
3. Too little push rod free play
4. Binding pedal linkage
5. Torn brake hose inner liner (acting as a one-way check valve)
6. Incorrect adjustment of cruise control vacuum release valve assembly, keeping the brake pedal from returning fully
7. Defective caliper hardware

Dragging (Right Front Caliper Only)
1. Faulty fluid control valve (Ford) or quick take-up valve in master cylinder (GM)

Rising Pedal on Successive Brake Applications
1. Too little push rod free play
2. Swollen cup in master cylinder obstructing relief (inlet) port
3. Brake fluid expanding from vaporization

NOTE: On some vehicles, the master cylinder push rod must be disconnected from the brake pedal before the master cylinder is removed.

5. Withdraw the master cylinder from the vehicle.

6. Before installing a new or repaired master cylinder, bench bleed it. This removes air from the cylinder bore and saves bleeding time on the vehicle.

Bleeder tubes can be made or purchased, and used as shown in Figure 17. Depressing and releasing the master cylinder piston cycles brake fluid from the cylinder bore into the reservoirs and removes air.

7. Install the master cylinder.

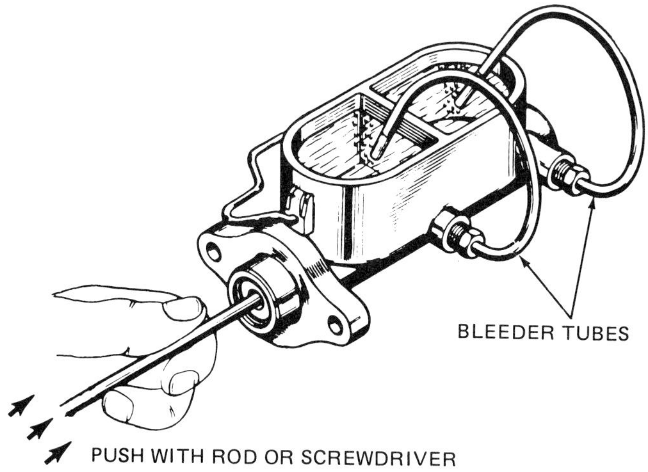

FIGURE 17
Bleeding the master cylinder before installing it on the vehicle. (Courtesy of American Motors Corporation)

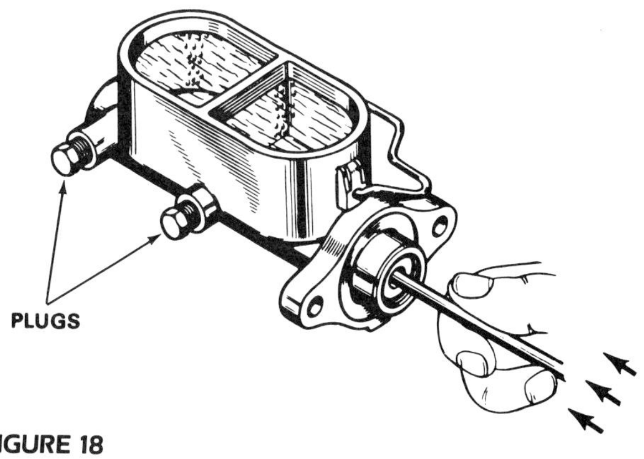

FIGURE 18
Bench bleeding the master cylinder without using bleeder tubes. Be careful to avoid spraying brake fluid. **Do not hold face directly above reservoirs.** (Courtesy of Allied Aftermarket Division, Allied-Signal Corporation)

NOTE: Bench bleeding also may be accomplished by plugging the outlet ports and depressing and releasing the piston (Figure 18).

SHOP TIP

A syringe can be used to bench bleed a master cylinder. Figure 19 shows one that can be purchased commercially for this purpose.

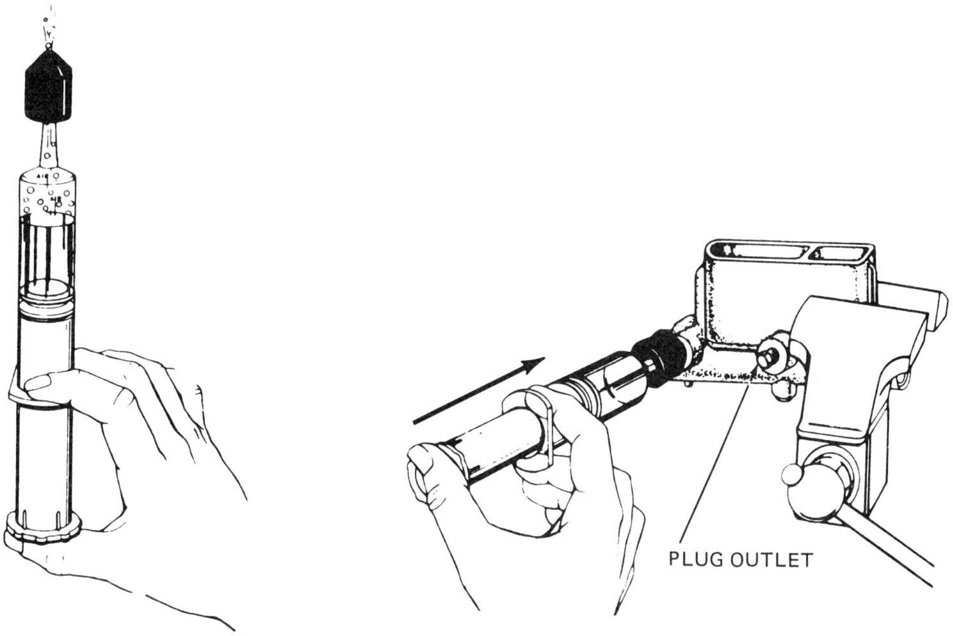

FIGURE 19
Bench bleeding the master cylinder with the EIS Sur-Bleed syringe. (Courtesy of EIS Division, Parker Hannifin Corporation)

SHOP TIP

When installing a master cylinder, you may find it easier to thread the brake line nuts into the outlet ports before tightening the master cylinder mounting nuts.

MASTER CYLINDER OVERHAUL

Today, many shops will not perform master cylinder overhaul. If a master cylinder is defective, they will replace it. This gives them higher profits and helps protect them from liability. In addition, overhaul may not be recommended by the manufacturer. However, master cylinder overhaul will be presented here because it is listed as an ASE task and is still performed by some shops.

Typical Tandem Master Cylinder Disassembly

1. Thoroughly clean the outside of the master cylinder, then remove the cover and diaphragm.

2. Remove the secondary or floating piston stop screw and gasket from the cylinder (Figure 20).

NOTE: Some late model master cylinders are not equipped with a piston stop screw.

3. Remove the piston lock ring at the end of the cylinder bore (if so equipped).

4. Turn the open end of the cylinder down and tamp it on a bench to remove the secondary or floating piston and spring. If the secondary piston sticks in the bore, use air

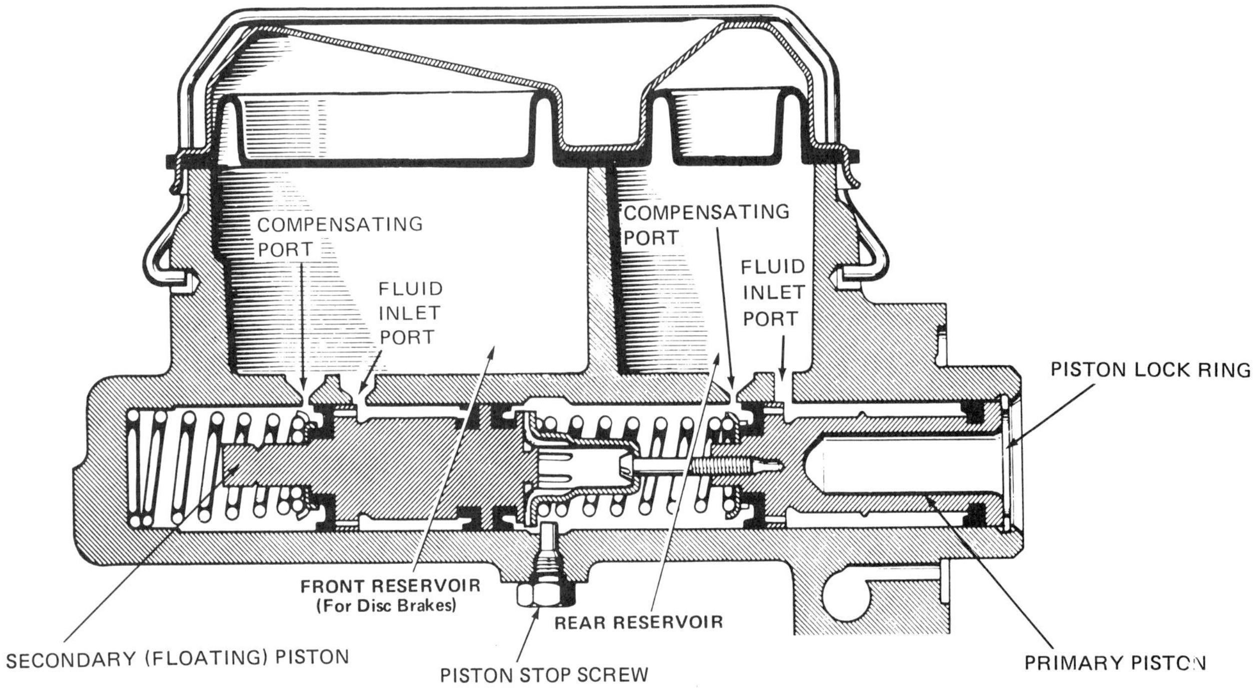

FIGURE 20
The piston stop screw is usually found on the outside of the master cylinder. However, in some cases it is located on the inside of the reservoir. (Courtesy of Allied Aftermarket Division, Allied-Signal Corporation)

pressure to force the piston out of the cylinder.

NOTE: New seals or cups must be installed when reassembling.

5. If any outlet port contains a residual pressure check valve, the tube seat insert must be removed so that the residual check valve can be replaced. If a port does not contain a check valve, do not remove the tube seat insert unless it is damaged.

To remove a tube seat insert, thread a self-tapping machine screw into the hole in the center of the insert and pull upward with pliers.

Cleaning and Inspection

Clean the master cylinder parts thoroughly, using denatured alcohol or brake fluid.

CAUTION: Do not clean parts with petroleum products (gasoline, kerosene, solvent, and so on).

Dry with compressed air. Inspect the cylinder bore for scoring or pitting. Cast iron cylinder bore walls that have light scratches or show signs of corrosion can usually be cleaned by honing (Figure 21). Hone lightly and use brake fluid as a lubricant while honing. Cylinder bores that have deep scratches or scoring should be replaced.

Clean and dry the master cylinder thoroughly after honing. Inspect the fluid inlet and compensating ports. These holes must be open for proper master cylinder operation.

NOTE: Under no conditions should an aluminum master cylinder bore be honed or cleaned with an abrasive material. This will remove the wear-resistant anodized surface. If the bore has defects, replace the master cylinder.

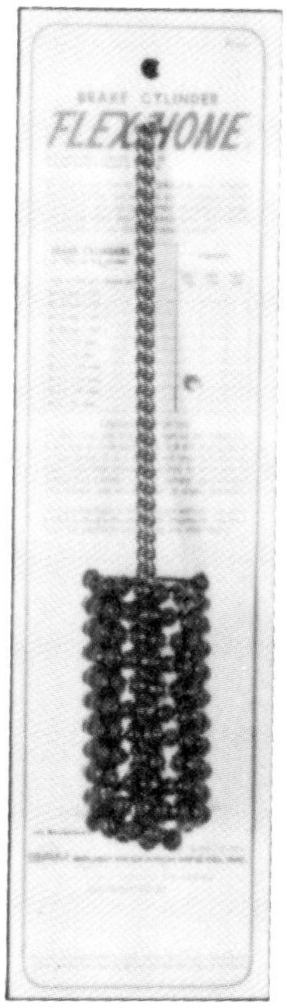

FIGURE 21
A Flex-Hone®. This is a widely used tool for honing master cylinders and wheel cylinders. (Courtesy of Brush Research Manufacturing Company, Inc.)

Reassembly

1. Carefully install the new cups or seals in the same positions and in reverse of the order in which they were removed.

2. Use brake fluid or assembly fluid very generously to keep the cups and seals lubricated. Assembling cups or seals dry can ruin them.

3. Put the pressure spring into the secondary piston retainer, then slide the assembly into the cylinder bore. Take care not to nick or gouge any rubber part.

4. Place the spring retainer of the primary piston assembly over the secondary piston shoulder and push both assemblies into the bore.

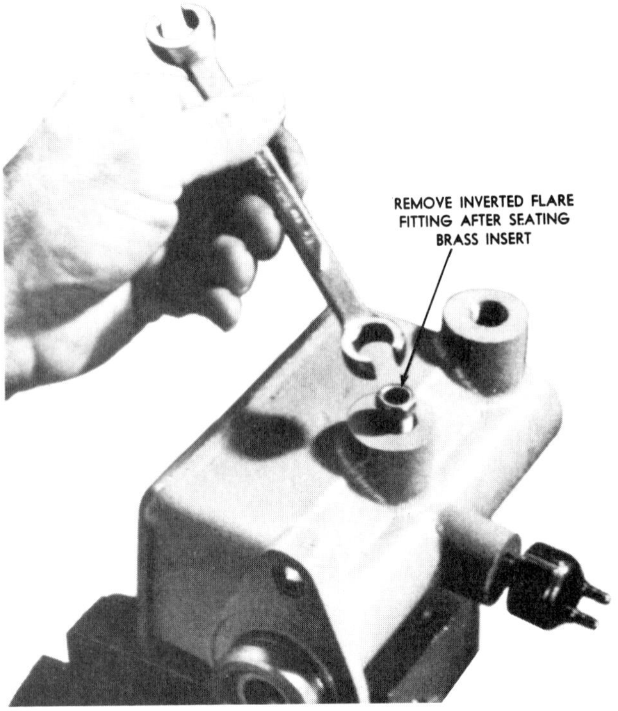

FIGURE 22
Using an inverted flare nut fitting to press in a tube seat insert. (Courtesy of Raybestos-Manhattan, Inc.)

5. Install and tighten the piston stop screw and gasket, while holding the pistons in their seated positions. Put the piston snap ring back on.

6. Install any residual check valves and springs.

7. Position new tube seat inserts in both pressure outlets. Using a flare nut, bottom each tube seat insert (Figure 22). Remove any brass chips that may have been sheared from the insert as it was pressed into the outlet.

BRAKE LINES, HOSES, AND FITTINGS

Hydraulic lines and hoses serve as the "arteries" and "veins" of the brake system (Figure 23). They transmit the hydraulic pressure developed in the master cylinder from the master cylinder to the brakes.

Lines

Brake lines are made of steel tubing, double flared at both ends to accommodate male fittings (Figure 24). They are available in different lengths and diameters, and can be bent to conform to the necessary configuration.

Most lines are double walled and coated to resist rust. The ends are double flared or have an International Standards Organization (ISO) type flare to prevent leakage (Figure 25 [a] and [b]).

Lines should be inspected periodically for leaks, kinks, cracks, rust, dents, and splits (Figure 26). *Never take a chance with a faulty line; replace it.*

SHOP TIP
Always start male fittings using fingers only. This will prevent cross-threading and stripping of threads.

Hoses

Hoses are the flexible links between the wheels and the car frame or body. They are furnished in different sizes, lengths, and end fittings. Figure 27 shows some typical brake hoses.

Hoses must be able to withstand high fluid pressures without expanding. They must be free to flex during spring deflection and wheel turn without being damaged.

Hoses should be inspected for cuts, chafing, cracks, twists, and loose supports (Figure 28 [a] and [b]). Two possible internal defects that can develop in a brake hose are pictured in Figure 29.

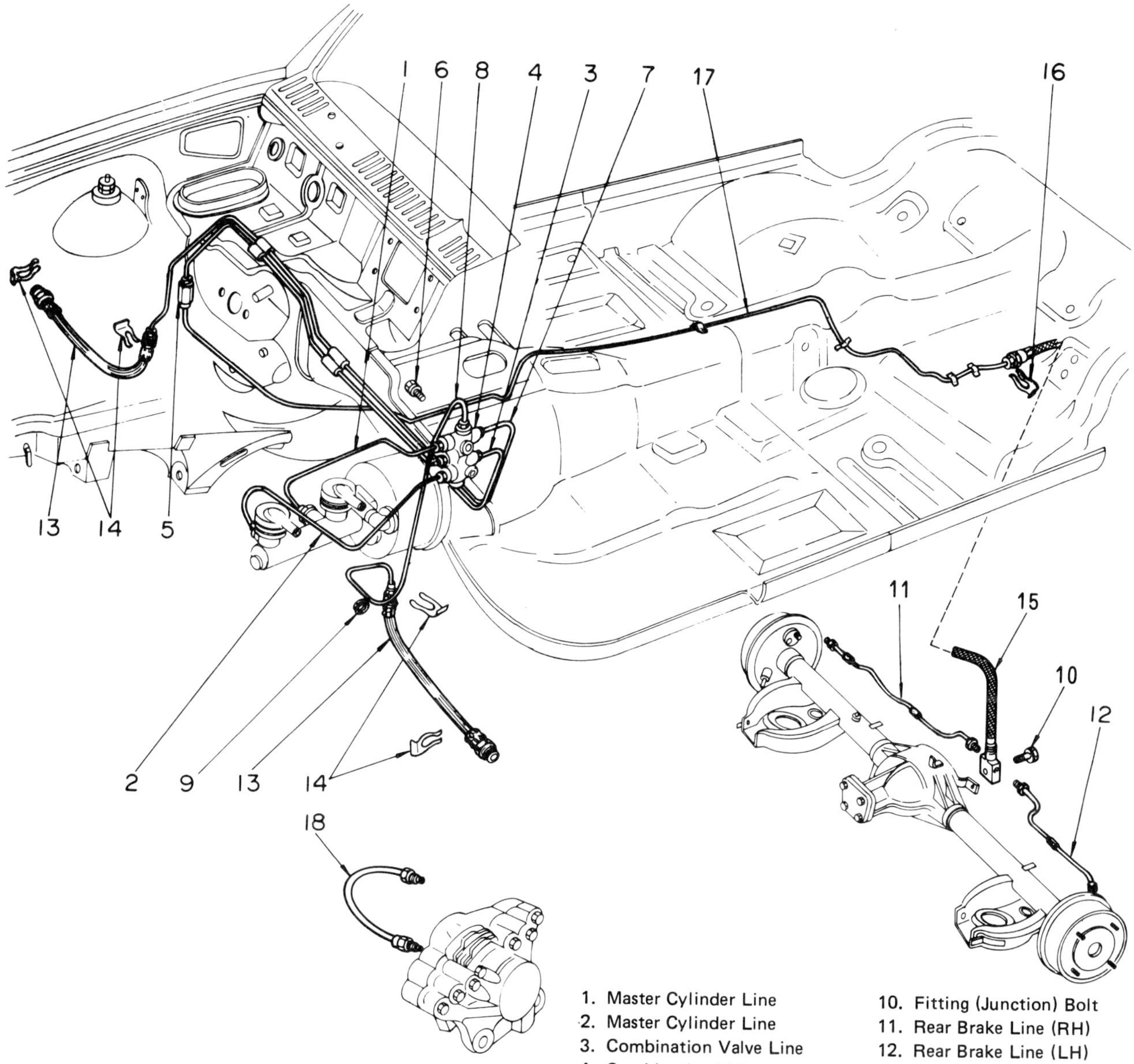

1. Master Cylinder Line
2. Master Cylinder Line
3. Combination Valve Line
4. Combination Valve
5. Connector
6. Combination Valve Mounting Bolt
7. Front Brake Line (RH)
8. Front Brake Line (LH)
9. Grommet
10. Fitting (Junction) Bolt
11. Rear Brake Line (RH)
12. Rear Brake Line (LH)
13. Flexible Hose (Front)
14. Hose Locks (Clips)
15. Flexible Hose (Rear)
16. Hose Lock (Clip)
17. Rear Brake Line
18. Hose-to-Caliper Line

FIGURE 23

A typical front wheel drive automobile brake line and hose layout. (Courtesy of Isuzu Motors Limited)

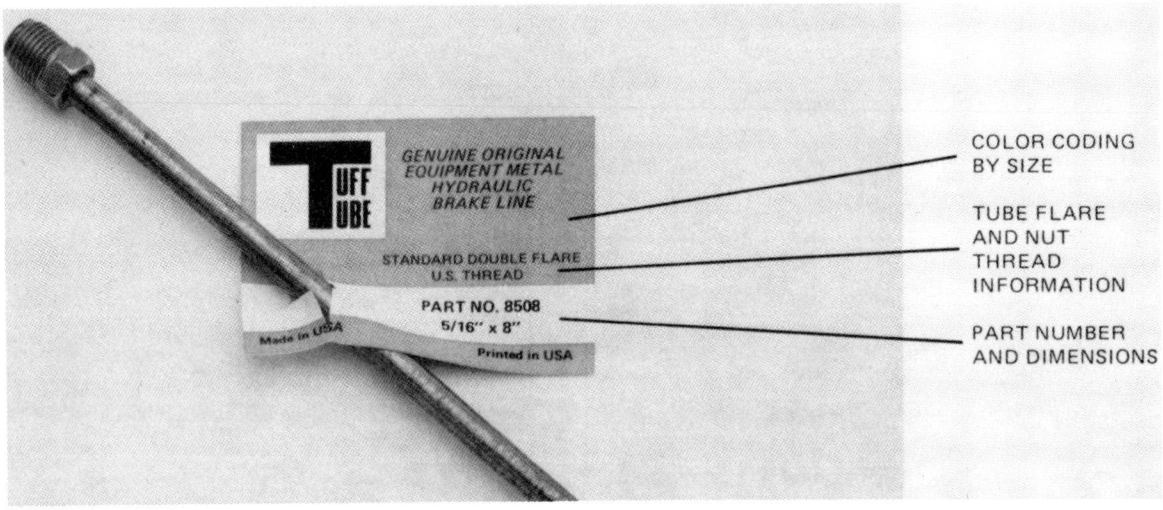

FIGURE 24
Brake lines are made of steel tubing, double flared at both ends. They come in different sizes. (Courtesy of Gibson Products, Division of Rolero, Inc.)

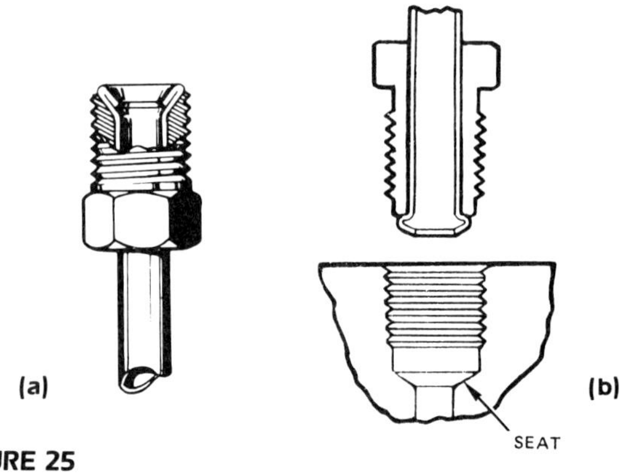

FIGURE 25
(a) Double flaring. (Courtesy of Allied Aftermarket Division, Allied-Signal Corporation) (b) ISO flare (sometimes called "bubble" flare) fittings are used by General Motors on their front wheel drive cars and on many imports. One advantage to this flare design is that tube seats are no longer needed. (Courtesy of Wagner Division, Cooper Industries, Inc.)

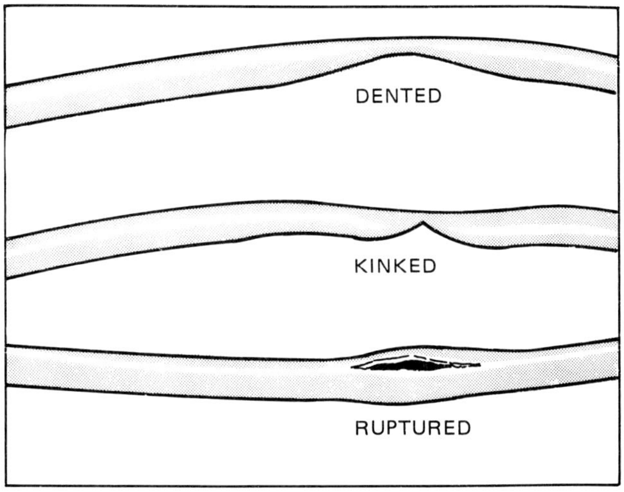

FIGURE 26
Brake lines should be inspected periodically for damage. (Courtesy of Allied Aftermarket Division, Allied-Signal Corporation)

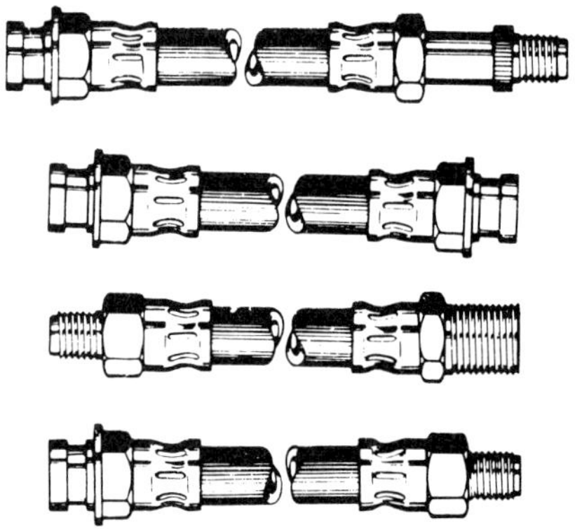

FIGURE 27
Brake hose ends. (Courtesy of Allied Aftermarket Division, Allied-Signal Corporation)

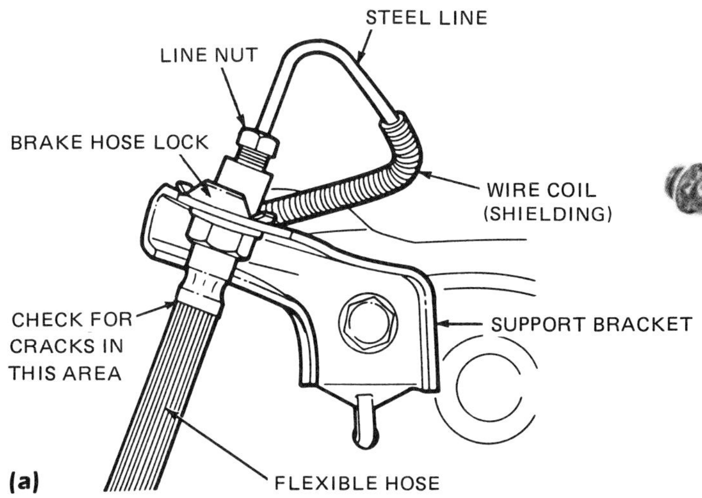

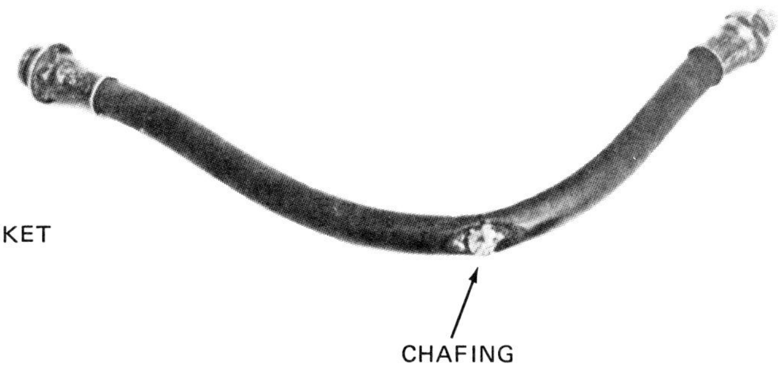

FIGURE 28
(a) Hoses should be inspected periodically for damage and loose supports. (b) A brake hose that is too long will rub against the chassis, the front end parts, or the tire. This can wear a hole through the cover and cause a hydraulic failure. (Courtesy of Raybestos-Manhattan, Inc.)

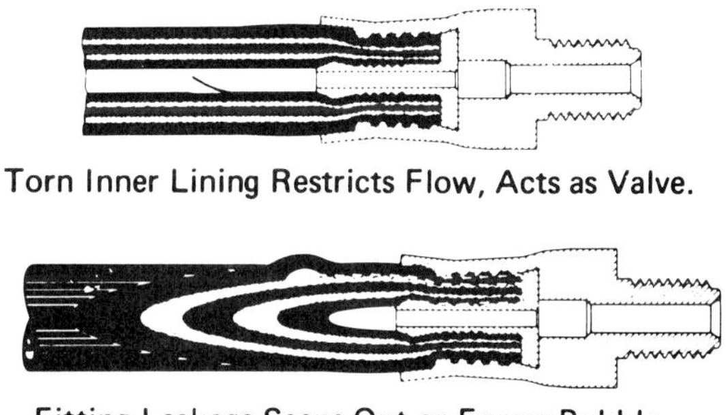

FIGURE 29
Internal defects can develop in a brake hose. (Courtesy of Wagner Division, Cooper Industries, Inc.)

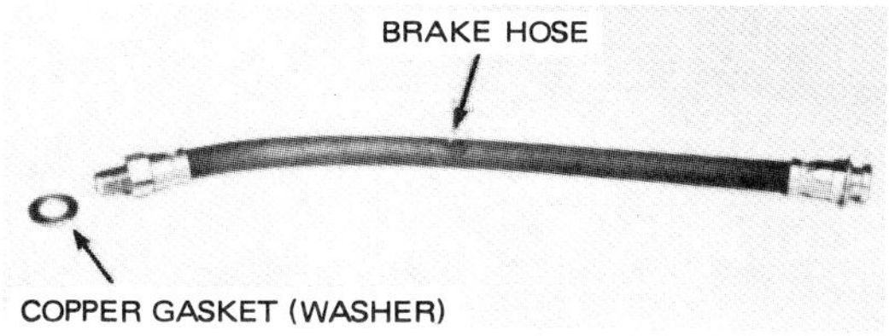

FIGURE 30
When replacing a brake hose, a new copper washer should be used.

When replacing a front wheel brake hose, attach the hose to the caliper and tighten securely.

NOTE: If a copper gasket is used, replace it with a new one (Figure 30).

Then attach the opposite end of the hose to the chassis bracket and brake line (Figure 31). Replace the lock clip that holds the hose to the chassis bracket.

Fittings

Brake line fittings include nuts, tees, unions, and adapters (Figure 32).

BRAKE FLUIDS

Brake fluid fills the parts of the hydraulic brake system and makes it into a complete operating unit. Some different brands of brake fluid are shown in Figure 33.

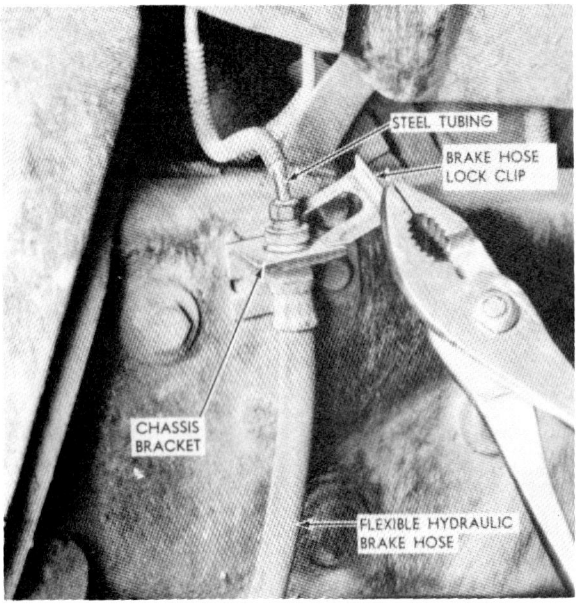

FIGURE 31
Installing a brake hose lock clip. (Courtesy of Raybestos-Manhattan, Inc.)

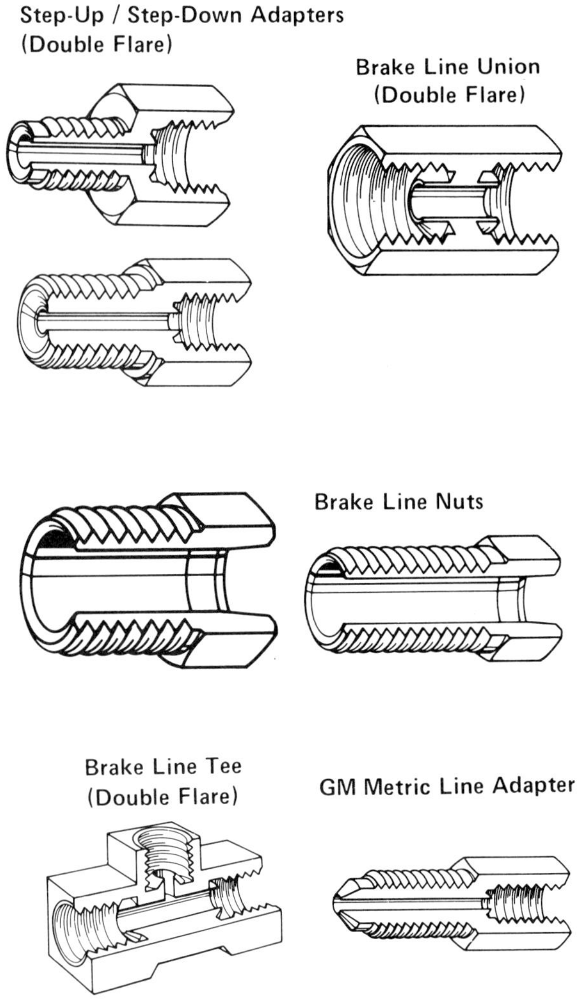

FIGURE 32
Various brake line fittings. (Courtesy of Gibson Products, Division of Rolero, Inc.)

Characteristics

Federal laws require that a brake fluid must meet certain standards and specifications before it is sold. Specifications have been set for three types of brake fluid for automotive use. They are DOT 3, DOT 4, and DOT 5. Each type must be a specific color if manufactured after September 1978. DOT 3 and DOT 4 are amber to clear in color. DOT 5 brake fluid is purple. Some major differences among these three fluids are listed below:

1. **DOT 3.** This is a non-petroleum-based fluid with a minimum dry boiling point of 401 degrees F.

2. **DOT 4.** This is a non-petroleum-based fluid with a minimum dry boiling point of 440 degrees F. It absorbs moisture more slowly than DOT 3 fluid.

3. **DOT 5.** This is a silicone-based fluid with a minimum dry boiling point of 500 degrees F. It will not attract moisture.

WARNING: *Some European original-equipment brake fluid is petroleum- (mineral oil)-based and is used in systems that are made with a special type of rubber. In order to conform to U.S. requirements, these fluids are color-coded green. Use a petroleum-based fluid only if it is recommended by the vehicle manufacturer. Do not mix a petroleum-based fluid with any other type of fluid. Failure to observe these precautions can result in brake failure; all the rubber components in the system (including hoses) can become soft and swollen.*

FIGURE 33
Different brands of brake fluid. Cans are marked to indicate specifications.

Silicone Fluids

Silicone brake fluids are *nonhygroscopic*. That is, they do not absorb moisture. Also, silicone fluids have a long-term shelf life and will not harm the paint finish. These are major problems with DOT 3 and DOT 4 fluids.

Changing Fluid

DOT 3 and DOT 4 brake fluid should be flushed annually because of moisture (water) contamination (Figure 34). According to fleet tests run by Dow-Corning, the average 1-year-old vehicle has about 2 percent water content in the brake fluid. If brake fluid absorbs just 3 percent moisture, its boiling point is lowered by about 25 percent. This can cause a spongy brake pedal or, under hard use, loss of brakes entirely.

***NOTE:** If a system is being converted to silicone fluid, the system should be flushed completely.*

Handling and Storage

1. Store brake fluid containers in a clean, dry place.

2. Always seal fluid containers tightly when they are not being used (Figure 35).

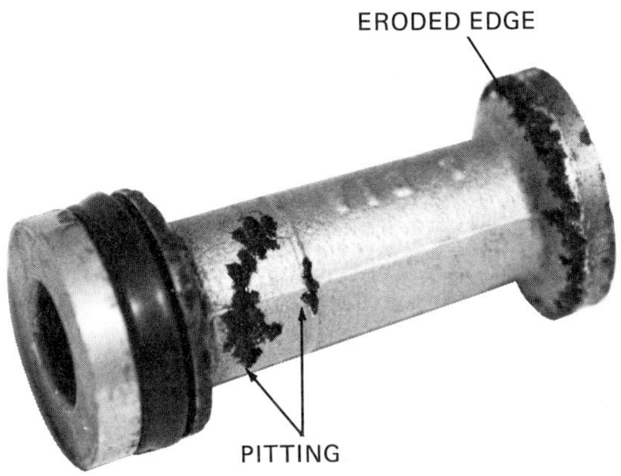

FIGURE 34
Internal brake parts can become corroded and damaged because of water in the brake fluid.

FIGURE 35
Make sure brake fluid containers are tightly sealed when not being used. If containers are left open, the fluid will absorb moisture from the air. Moisture contamination can result in "fluid boil" and braking loss.

3. Periodically flush pressure bleeders.

4. Never reuse old brake fluid.

5. DOT 3 and 4 fluids will mix with each other, but DOT 5 should not be mixed with DOT 3 or DOT 4.

HYDRAULIC SYSTEM VALVES

Automobile manufacturers use these valves in their disc brake systems:

1. Pressure differential valve/warning light switch.

2. Proportioning valve and/or

3. Metering valve.

These valves have many different designs. Often, a three-way combination valve is used (Figure 36). Do not attempt to service any of these valves (Figure 37). If a valve malfunctions, replace it with a new unit. The single exception is that the warning light switch can be replaced without replacing a valve assembly.

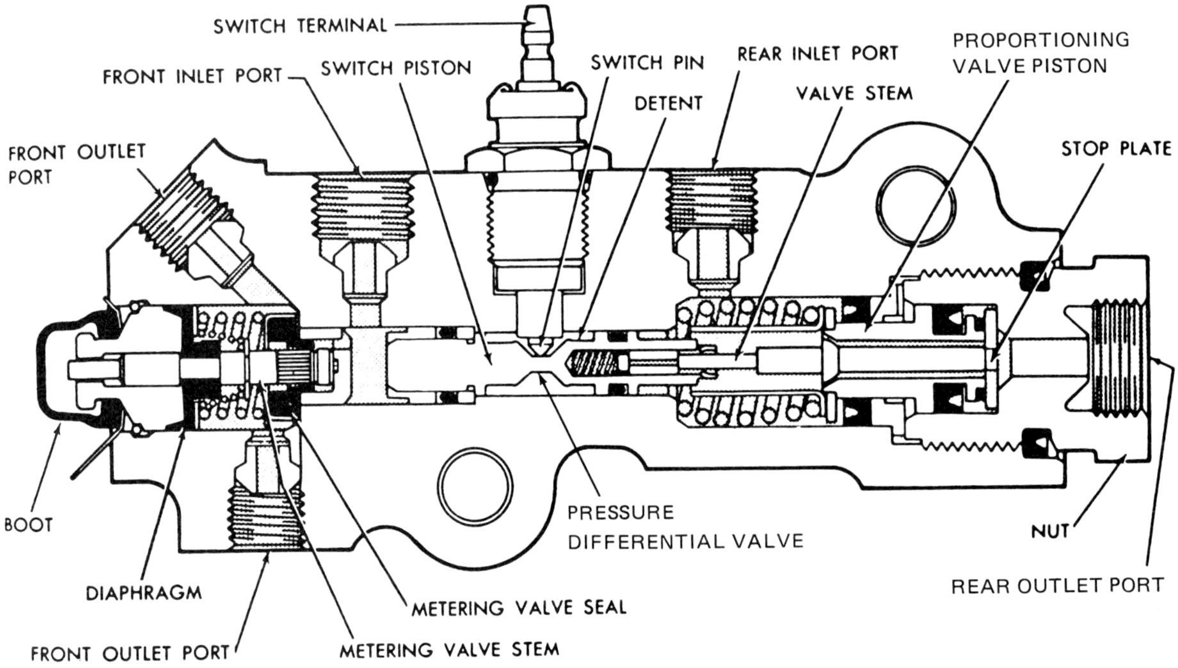

FIGURE 36
Sectional view of a late model three-way combination valve. Included are a proportioning valve, metering valve, and pressure differential valve. (Courtesy of General Motors Corporation, Product Service Training Division)

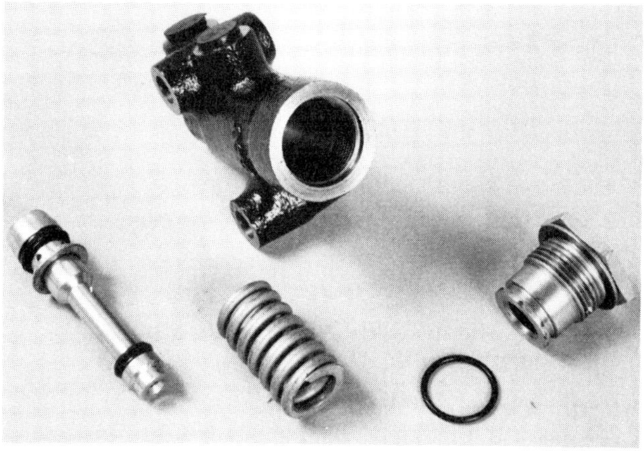

FIGURE 37
The inside of a proportioning valve. Do not attempt service, because parts are not available. Replace the entire unit if it does not work.

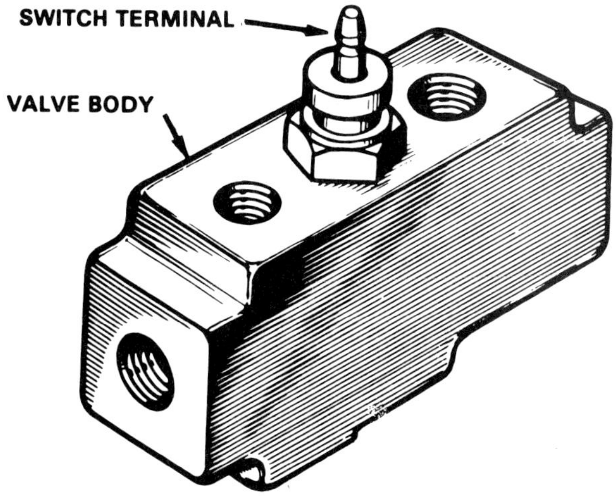

FIGURE 38
Separate pressure differential valve and brake warning light switch. (Courtesy of Allied Aftermarket Division, Allied-Signal Corporation)

Pressure Differential Valve

Some vehicles use a separate pressure differential valve (Figure 38). Often, it is incorporated into a combination valve (Figure 39). On certain front wheel drive cars, the warning light switch assembly is located in the master cylinder.

When pressure is lost in one side of a dual hydraulic system, the other side of the system causes the differential valve to move. This closes a warning light switch, which causes the brake warning light to come on (Figure 40).

On some vehicles, the pressure differential valve automatically recenters as the brake pedal is released. In other words, the warning light turns on only when the brakes are applied.

On other vehicle makes, the pressure differential valve remains off-center after a hydraulic failure. This keeps the warning light on until repairs are made. Then the switch must be "re-armed" by resetting (centering) the valve. This is usually accomplished by depressing the brake pedal after the hydraulic failure has been repaired.

If the pressure differential valve does not have centering springs, it will be necessary to open a bleeder screw to recenter the valve. Turn the ignition key to *on*. Open a bleeder screw in the system opposite the side that failed. Have a helper slowly press on the brake pedal until the warning light goes *off*.

Warning Light Switch Checks

Disconnect the wire lead end from the switch terminal and ground it. Turn the ignition key to *on*. The brake warning light should come on.

To test the warning light switch, open a bleeder screw while a helper depresses the brake pedal and observes the warning light. If the light does not come on, inspect for a burned out bulb, a defective socket, an open in the wiring, or a defective switch. If the bulb is not burned out and the wiring continuity checks out, replace the warning light switch.

Proportioning Valve

This valve is also known as a "P" valve, pressure-ratio valve, pressure control valve, pressure regulating valve, or balance valve.

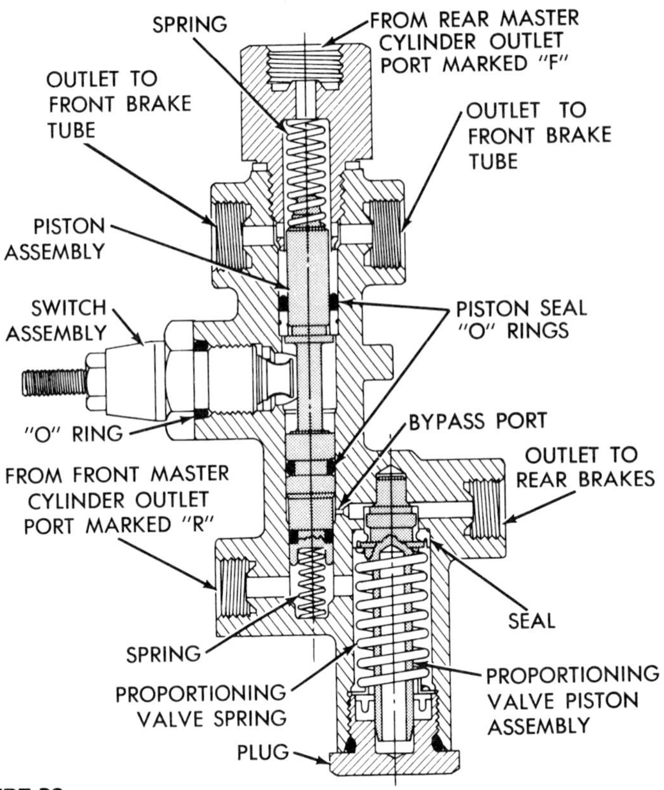

FIGURE 39
Combined brake warning light switch assembly/proportioning valve. (Courtesy of Chrysler Corporation)

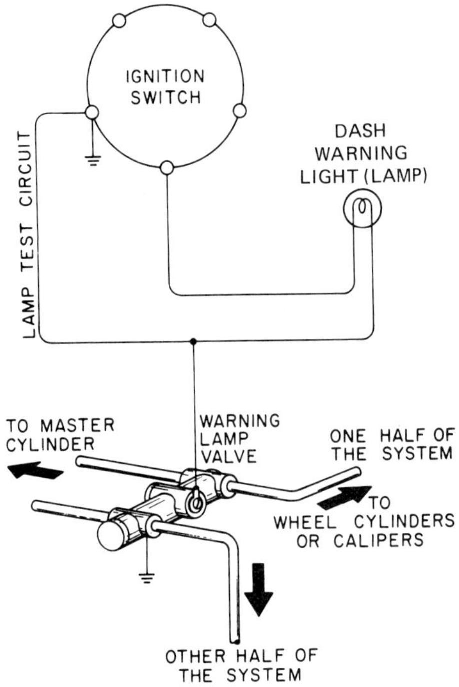

FIGURE 40
Brake warning light circuitry. (Courtesy of EIS Division, Parker Hannifin Corporation)

A proportioning valve is used on most cars with front disc brakes and rear drum brakes. This valve is installed in the line to the rear brakes. Some different proportioning valves are shown in Figure 41.

On diagonally split brake systems, a dual body proportioning valve may be used. The valve is generally located in the engine compartment (Figure 42). The proportioning valves for GM front wheel drive cars are screwed into the master cylinder (Figure 43).

The function of a proportioning valve is to reduce the hydraulic pressure at the rear drum brakes during heavy brake application. Thus, the valve helps to prevent rear wheel skid and provides better braking balance.

FIGURE 41
Different styles of proportioning valves.

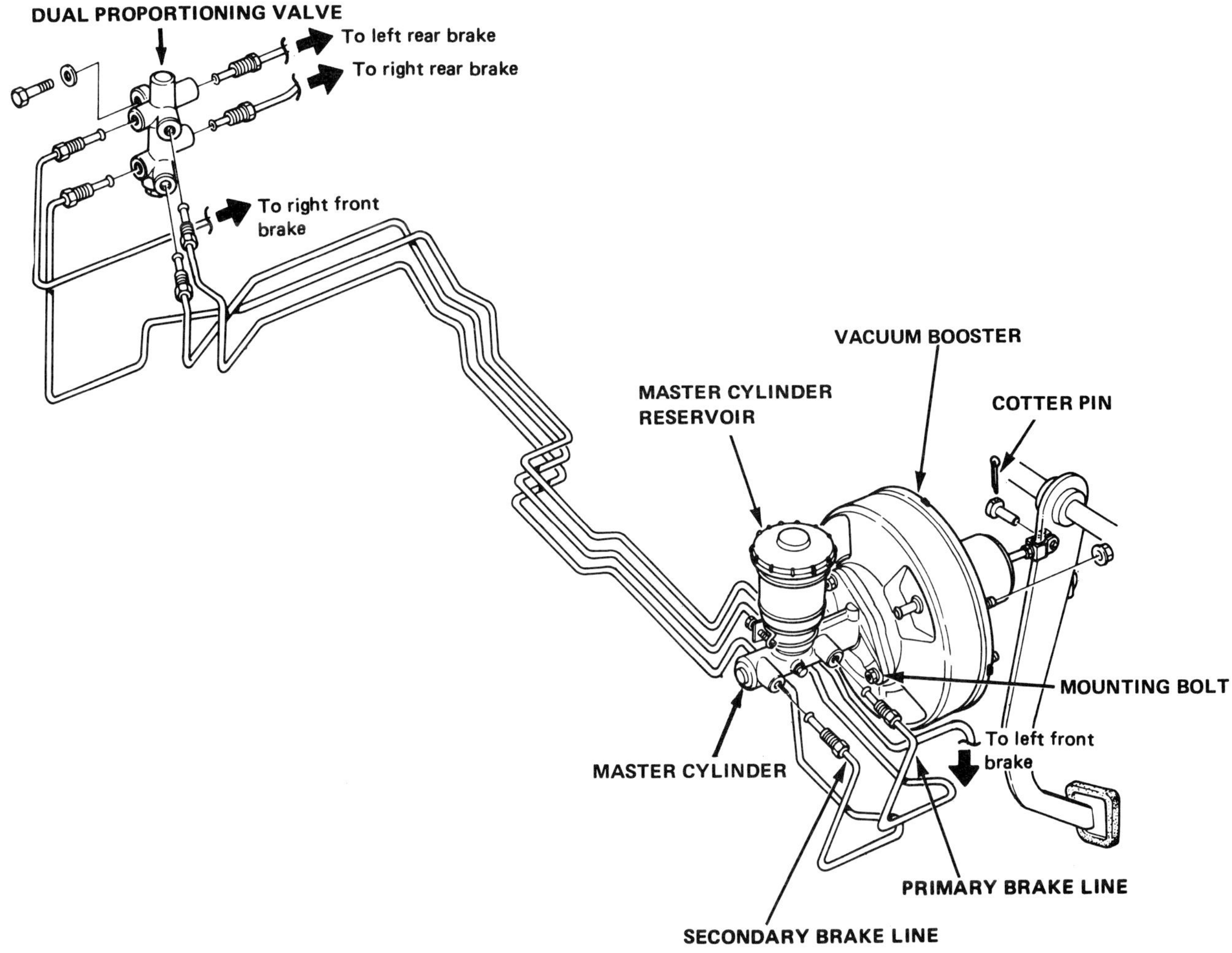

FIGURE 42
Diagonally split brake system with a dual proportioning valve. (Courtesy of American Honda Motor Company, Inc.)

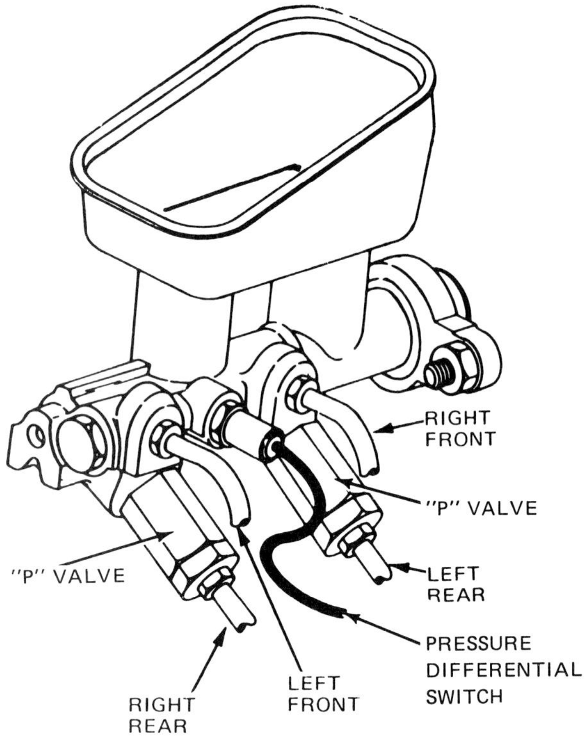

FIGURE 43
GM "X" body master cylinder with attached proportioning valves and pressure differential switch. Early "P" valves were silver-colored steel. Occasionally a clunking noise is encountered under moderate to heavy braking. The noise comes from the "P" valves, but does not affect braking efficiency. Initially, new steel proportioning valves were specified to correct the noise. These were then replaced with aluminum valves that are gold in color. (Courtesy of Wagner Division, McGraw-Edison Company)

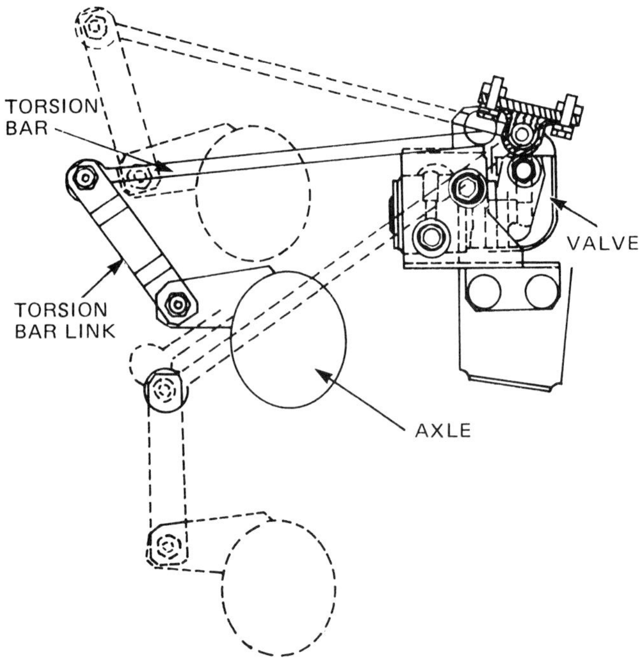

FIGURE 44
Height sensing proportioning valve. (Courtesy of Wagner Division, Cooper Industries, Inc.)

Height Sensing Proportioning Valve

This type of valve is found on some passenger cars and light trucks (Figure 44). If there is a light load on the rear wheels, the linkage is moved by the body angle. This restricts the passage of pressure through the valve and reduces the chance of rear wheel lock-up.

CAUTION: The use of lift kits is not recommended on vehicles that are equipped with a height sensing proportioning valve.

SHOP TIP

Use caution when bleeding brakes if a height-sensing "P" valve is present. If the vehicle is raised on a lift and the axle is allowed to hang down, the valve will bottom. This can hamper the bleeding process and may damage the valve.

Proportioning Valve Servicing

The proportioning valve(s) should be inspected whenever the brakes are serviced. If the valve is leaking, it is defective and should be replaced.

To analyze a proportioning valve, measure the hydraulic system pressures at the rear and front brakes at the same time (Figure 45). A typical pressure curve for a vehicle equipped with a proportioning valve is shown in Figure 46.

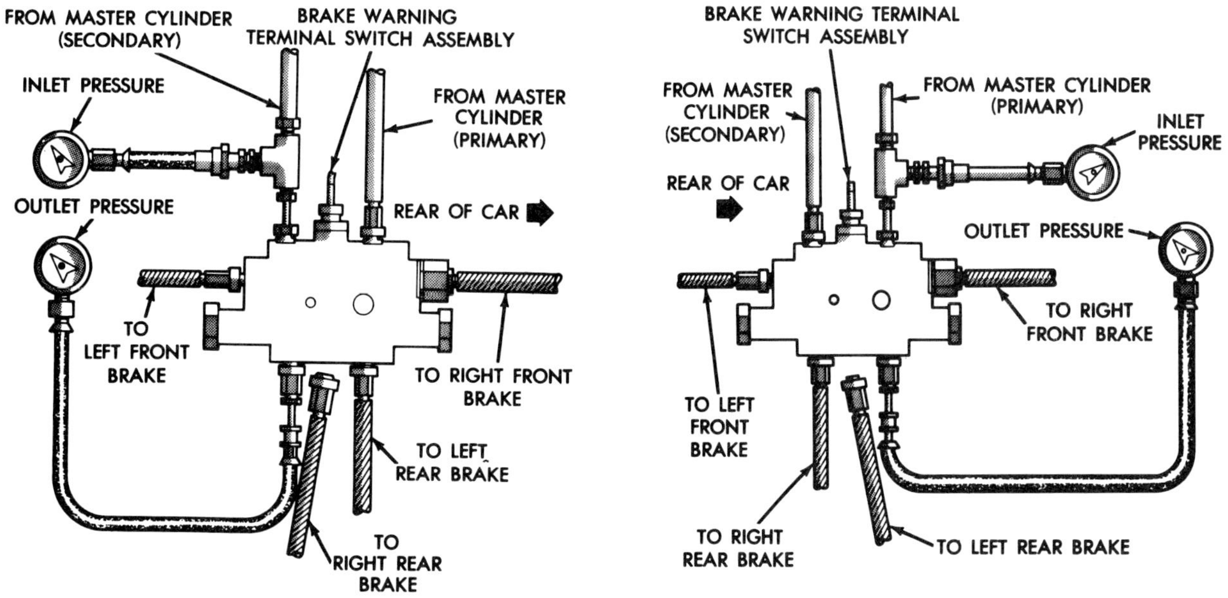

FIGURE 45
Gauges can be used to measure "P" valve pressures. However, the hook-up can be involved and time-consuming. This "P" valve is designed with two separate systems. One half controls the right rear brake, and the other half controls the left rear brake. (Courtesy of Chrysler Corporation)

Metering Valve

Many vehicles use a metering (hold-off) valve to delay pressure application to the front axle disc brakes (Figure 47). The valve is installed in the line to the disc brakes (Figure 48).

The basic purpose of this valve is to improve braking balance, particularly when the brakes are applied lightly. The valve prevents the front pads from coming into use until 75 to 135 pounds per square inch (psi) pressure has built up in the hydraulic system. This

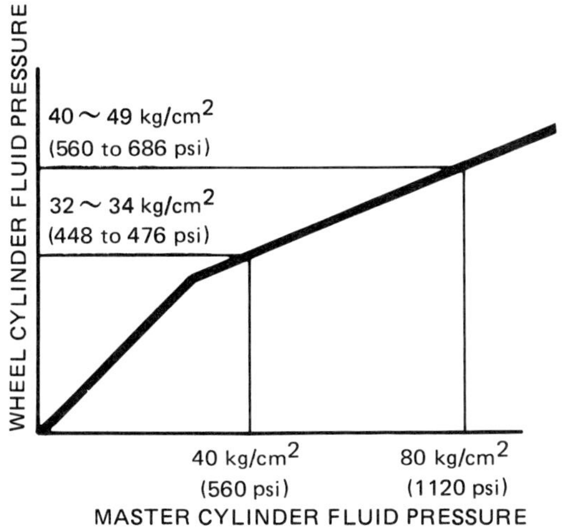

FIGURE 46
Typical proportioning valve pressure curve.

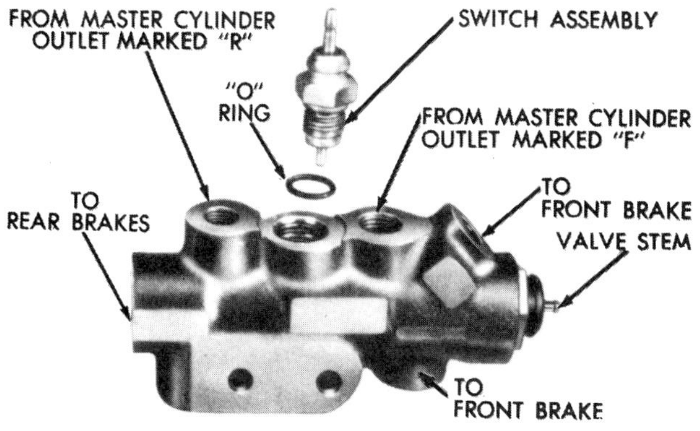

FIGURE 47
Combined brake warning light switch assembly/metering valve. (Courtesy of Chrysler Corporation)

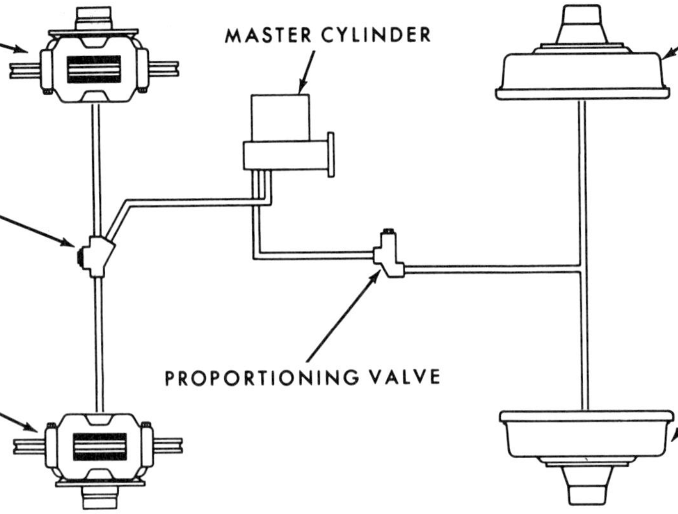

FIGURE 48
Front disc/rear drum brake system with a separate metering valve. (Courtesy of Wagner Division, Cooper Industries, Inc.)

allows the line pressure to the rear drum brakes to overcome the force exerted by the shoe retraction springs. The idea is to have both front and rear brakes take hold at the same time.

COMMENT: The metering valve has no effect on front brake pressure during hard stops.

Metering Valve Servicing

The metering valve should be inspected whenever the brakes are serviced. If the valve is leaking, it is defective and should be replaced.

NOTE: A slight amount of moisture inside the boot is normal and does not indicate a defective valve.

Most metering valves can be checked as follows:

1. The valve stem should move slightly as the brakes are applied and released.

2. A slight "bump" can usually be felt by the foot as the brake pedal is depressed. This bump will occur after the pedal has been depressed about 1 inch.

NOTE: If a "bump" is not felt, it is not a sign that the valve is defective.

The metering valve must be held in an open position if the front calipers are being bled manually by gravity or by using a bleeder tank. Depending upon the design, it can usually be opened by pushing or pulling a pin at the end of the valve (Figure 49).

If the brakes are bled manually using the pedal, the pressures developed are generally great enough to open the metering valve. However, if the system has enough air in it, it may not be possible to develop enough pedal pressure to open the metering valve. In such cases, the valve must be held open.

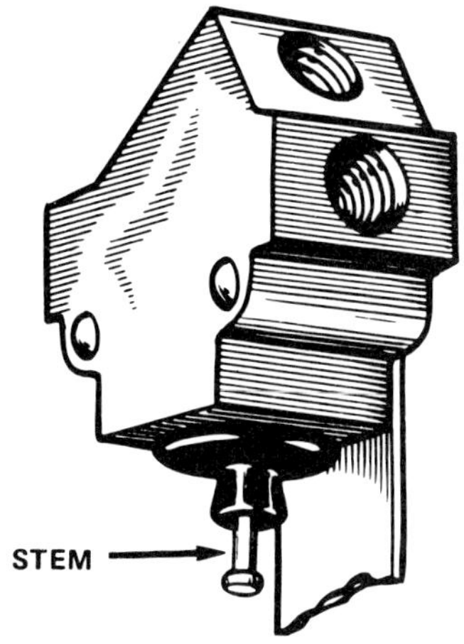

FIGURE 49
Metering valve with pull type stem. Special clips are available for holding the stem open.

Combination Valve

Today, most cars equipped with front disc brakes and rear drum brakes use a combination valve. Combination valves can be three-function valves or two-function valves.

Three-Function Combination Valve

A typical three-function combination valve is shown in Figure 50. The valve contains a pressure differential valve/brake warning light switch, a proportioning valve, and a metering valve.

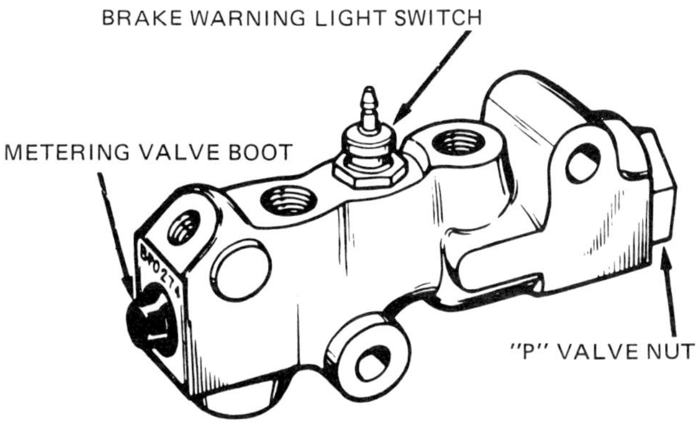

FIGURE 50
Three-function combination valve.

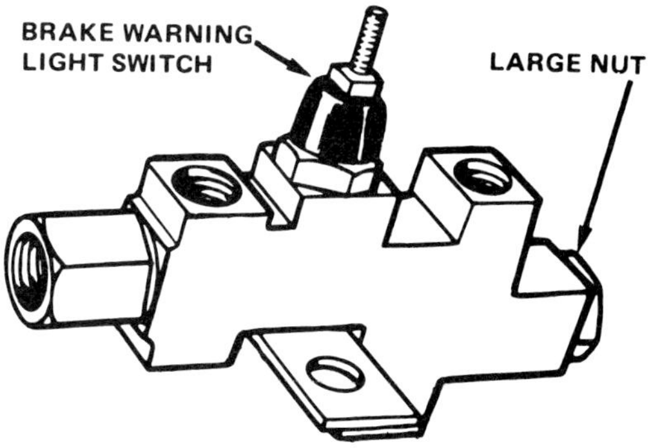

FIGURE 51
Two-function combination valve (proportioning and brake warning light).

Two-Function Combination Valve

There are two types of two-function combination valves. One type performs the proportioning and brake warning light switch functions (Figure 51). The other performs the metering valve and brake warning light switch functions (Figure 52).

Combination Valve Servicing

Combination valves are nonadjustable and nonrepairable. If a valve is defective, the whole valve assembly must be replaced.

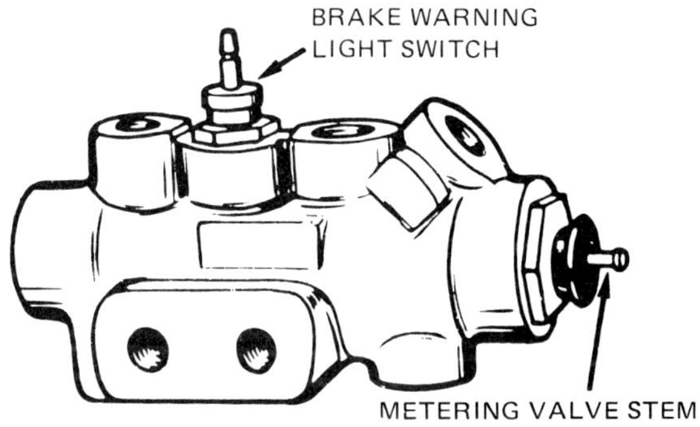

FIGURE 52
Two-function combination valve (metering and brake warning light).

HYDRAULIC SYSTEM VALVE TROUBLESHOOTING

Refer to Chart 2 for a list of possible hydraulic system valve problems.

BLEEDING

When brakes are serviced, there is a possibility of introducing air into the hydraulic system. Bleeding is a service procedure that expels this air along with the fluid.

Bleeder Screw Locations

Bleeder screws are located at high points in the hydraulic system. The air and fluid can be pushed out through them (Figure 53). Bleeder screws are located as follows:

1. In each drum brake wheel cylinder.

2. In each disc brake caliper. Some calipers include two bleeder screws.

3. Adjacent to the outlet port of some master cylinders.

4. On some combination valves used with disc brakes.

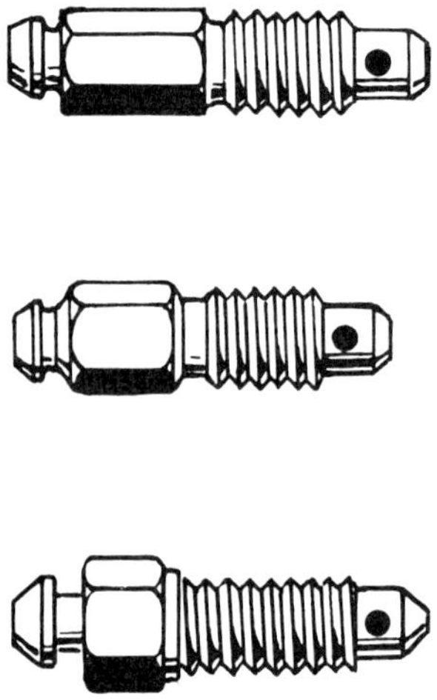

FIGURE 53
Bleeder screws.

Bleeding Sequence

Bleed the hydraulic system in the following order:

1. Master cylinder, if it has bleeder screws.

2. Combination valve, if there is one and it has bleeder screws.

CHART 2 TROUBLESHOOTING CHART FOR HYDRAULIC SYSTEM VALVES

Premature Front Brake Lock-Up
1. Defective metering valve

Rear Brake Lock-Up
1. Defective proportioning valve

Brake Warning Light Does Not Come On
1. Defective pressure differential valve
2. Burned out bulb
3. Faulty bulb socket
4. Defective warning light switch
5. Open wiring

Clunking Noise Under Moderate to Heavy Braking (GM Front Wheel Drive Cars)
1. Wrong proportioning valve material. Replace steel valves with aluminum ones (gold in color)

Front Pad "Burn-Out" on Right or Left Side Only
1. Defective metering valve

Dragging Brakes (All Wheels)
1. Defective metering valve

Dive (Front End Dips Excessively)
1. Defective metering valve

3. Wheel cylinders or calipers. On many newer cars, start with the wheel closest to the master cylinder and work your way out (Figure 54). If the system is diagonally split, a different sequence is usually followed (Figure 55).

NOTE: GM specifies that their quick take-up diagonally split system be bled in a different order (RR, LF, LR, RF).

> **SHOP TIP**
> If a firm brake pedal cannot be obtained, the wrong bleeding sequence is often the cause. Always refer to the car maker's recommendations for the correct order.

Manual Bleeding

This method uses the brake pedal and master cylinder as a hydraulic pump to expel air and fluid when a bleeder screw is "cracked" open. The following steps apply to one bleeder screw and should be repeated at all other bleeder points:

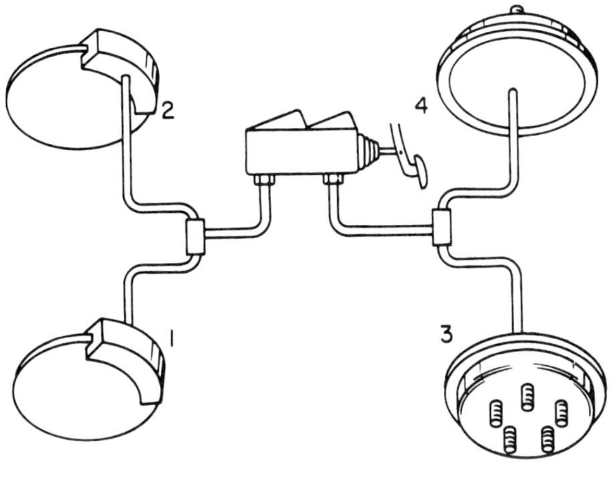

FIGURE 54
The bleeding sequence for many newer cars requires you to start with the wheel closest to the master cylinder and work out. For many years, the rule was to start with the wheel cylinder farthest from the master cylinder and work in. (Courtesy of Wagner Division, Cooper Industries, Inc.)

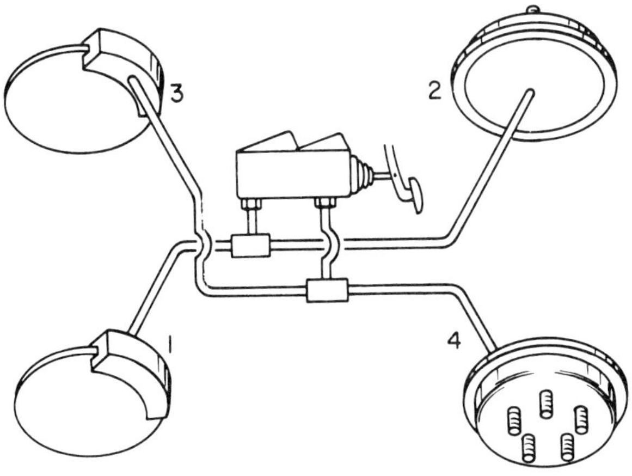

FIGURE 55
The usual bleeding sequence for a diagonally split braking system. (Courtesy of Wagner Division, Cooper Industries, Inc.)

1. Attach a bleeder hose to the bleeder screw. Submerge the other end of the hose in a glass jar partially filled with brake fluid (Figure 56).

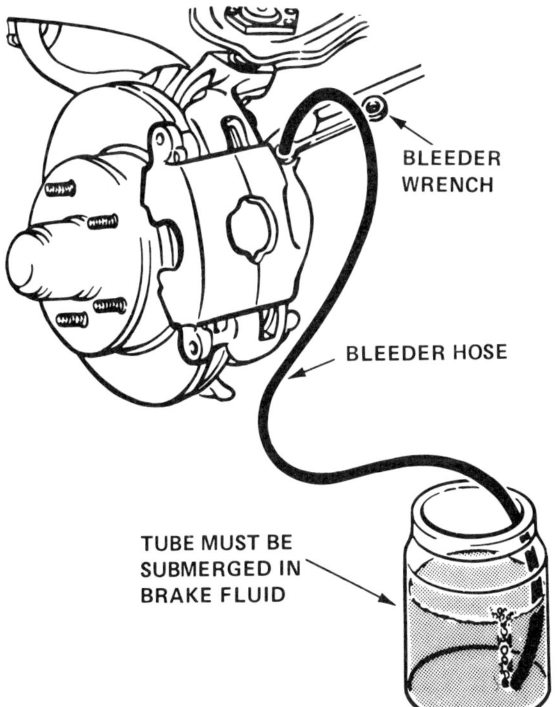

FIGURE 56
When bleeding, slip a hose over the end of the bleeder screw. Then put the free end of the hose into a jar partially filled with brake fluid. Keep the hose submerged when bleeding the wheel. (Courtesy of Pontiac Motor Division, General Motors Corporation)

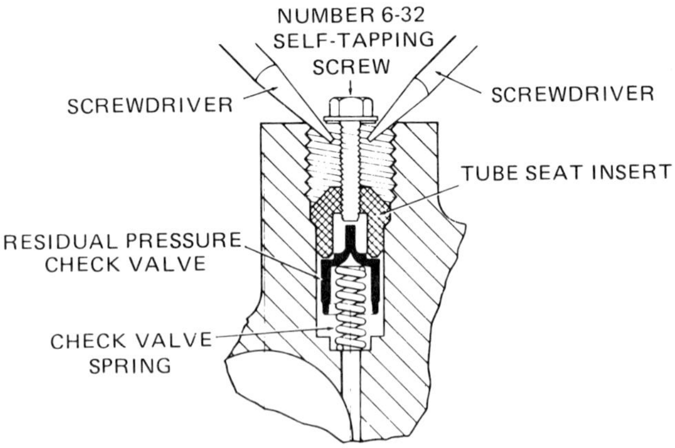

FIGURE 57
Master cylinder residual check valves can be removed once the tube seat inserts are pried out. (Courtesy of Raybestos-Manhattan, Inc.)

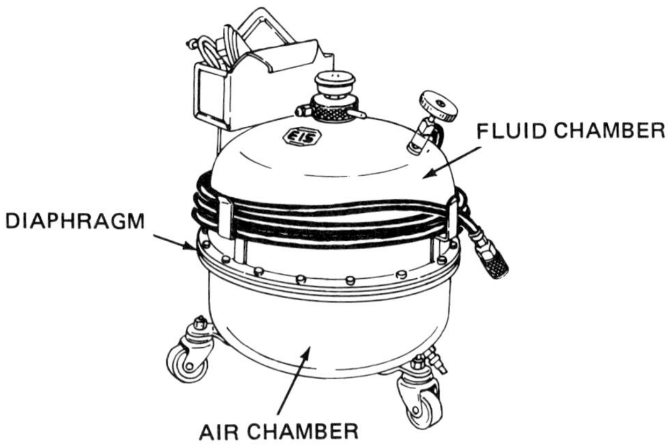

FIGURE 58
If using a pressure bleeder, make sure it has a built-in diaphragm that separates the air chamber from the fluid chamber. (Courtesy of EIS Division, Parker Hannifin Corporation)

2. Apply moderate, steady pressure (40 to 50 pounds) on the brake pedal. Then open the bleeder screw.

3. When the fluid coming from the submerged end of the hose is clear and free of air bubbles, close the bleeder screw and release the brake pedal.

4. If the brake pedal goes to the floor without removing all of the air bubbles, close the screw and release the brake pedal slowly. Then repeat steps 2 and 3.

SHOP TIP

Use of residual pressure check valves removed from used master cylinders will permit you to bleed a system manually by yourself (Figure 57). Push the round end of a valve over the end of a "cracked" open bleeder screw. Then stroke the pedal, and air will be pushed out through the valve. These valves are one-way, and air cannot be drawn in through them.

Pressure Bleeding

A pressure bleeder must be used for pressure bleeding (Figure 58). The fluid supply hose is connected to the master cylinder by means of an adapter. These adapters have different configurations (Figures 59 and 60).

The basic procedure for using a pressure bleeder is as follows:

1. Using shop air, inflate the pressure bleeder to 15 to 20 psi. Link the pressure hose to the adapter.

2. Open the metering valve if there is one.

3. Loosen the bleeder screws, one at a time, until the trapped air is exhausted.

SHOP TIP

If the master cylinder is not equipped with bleeder screws, you can remove trapped air by "cracking" open the tube nuts at the outlet ports, one at a time.

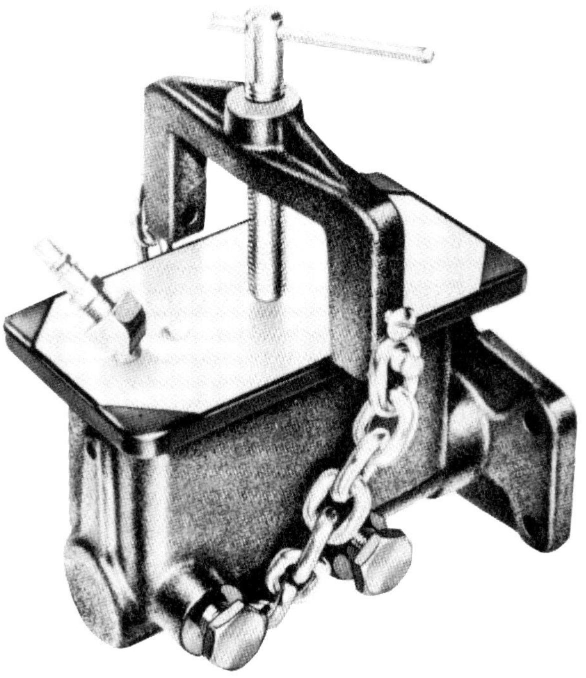

FIGURE 59
Installation of a pressure bleeder adapter plate on a dual master cylinder.

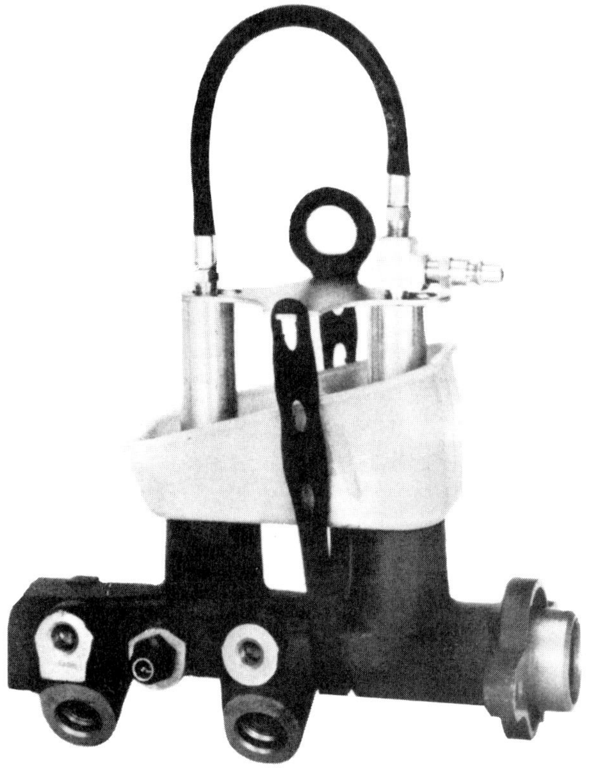

FIGURE 60
Pressure bleeder adapter for a GM "X" body master cylinder. (Courtesy of Raymark Industries, Inc.)

Surge Bleeding

This method can be used with either manual or pressure bleeding, when the recommended method does not expel all the air bubbles.

Surge bleeding stirs up the fluid in the system, rapidly moving it and removing any last bits of air. The procedure is as follows:

1. Bleed the system in the usual manner.

2. Open a bleeder screw. Keep the free end of the bleeder hose submerged in brake fluid.

3. Pump the pedal vigorously several times, then close the bleeder screw. Any remaining air should be expelled in the form of fluid with air bubbles in it.

4. As you finish with each bleeder screw, go back and bleed it once more by pressure or manual bleeding to make sure all air bubbles are removed.

Vacuum Bleeding

Sometimes low and "spongy" pedal problems can be corrected only by vacuum bleeding the hydraulic system. The procedure is as follows:

1. Slip a piece of tubing between the vacuum pump and the lid of the reservoir jar at the port marked "TO PUMP" (Figure 61).

2. Attach a piece of tubing to the other reservoir jar port.

3. Select the appropriate snap-over adapter and fit it over the bleeder screw. Attach the adapter to the reservoir tubing.

4. Place a wrench on the brake bleeder screw and operate the vacuum pump handle ten to fifteen times.

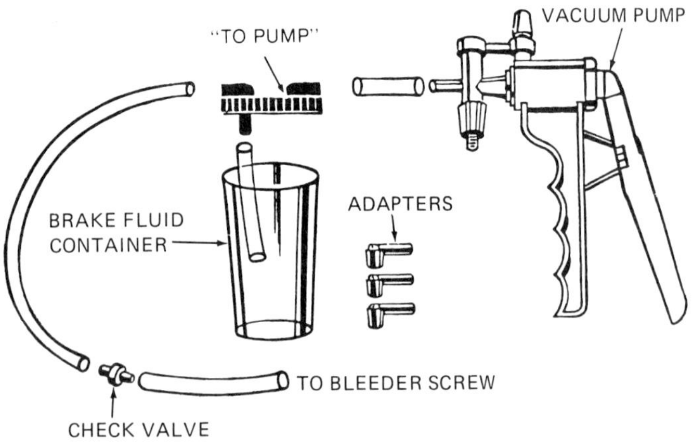

FIGURE 61
Kit for vacuum bleeding a brake system. (Courtesy of Neward Enterprises Inc.)

5. Open the bleeder screw just enough to cause fluid to enter the reservoir jar.

6. After draining about 1″ of fluid into the jar, tighten the bleeder screw.

7. Proceed to the remaining wheels.

Some Bleeding Hints

1. A special procedure is required to manually bleed the quick take-up system used on many GM cars. You must depress and release the brake pedal slowly. Also, you must wait 15 seconds after releasing the pedal before opening a bleeder screw.

NOTE: Failure to bench bleed this cylinder thoroughly is a common cause for a low and spongy pedal.

2. When pressure bleeding a quick take-up master cylinder, it may be necessary to loosen it and make level. This master cylinder mounts at a 17 degree angle, and air entrapment can be a problem.

3. Bleed power brake systems with the engine off (no vacuum present).

4. You may find it difficult to bleed some rear calipers. Lightly tapping the calipers with a hammer will usually regroup the air bubbles so they can be expelled.

5. If a pressure bleeder cannot be attached to the master cylinder reservoir, disconnect the brake line(s) at the master cylinder and attach the pressure hose directly. After bleeding the system, fill the line(s) to the top with brake fluid and connect them. Then manually bleed the master cylinder by loosening the tube nuts.

6. Many systems can be manually bled by gravity. Open a bleeder screw, allowing fluid to drain from the master cylinder until all air is displaced in that line. Continue with the remaining bleeder screws.

FLUSHING

Flushing the system is done in much the same way as bleeding. Basically, brake fluid is let out from each bleeder screw until all of the old or contaminated fluid is removed. Only new brake fluid, or denatured alcohol followed by new brake fluid, is recommended for flushing.

CAUTION: Any alcohol left in the system will dilute the new brake fluid and lower its boiling point.

SHOP TIP

If a vehicle is equipped with a brake fluid level warning switch, make sure the spark plug wires are kept at least 2 inches away from the master cylinder. Electrical radiation from the plug wires can generate a signal that is picked up by the warning switch, triggering the brake warning light. (Figure 62).

MASTER CYLINDER FLUID LEVELS

Always check and adjust the master cylinder fluid levels after bleeding and/or flushing. The reservoir(s) should be filled to approximately 1/4″ from the top (Figure 63). *Overfilling can cause the front pads to drag when a cover and diaphragm assembly is reinstalled. Underfilling can introduce air into the system.*

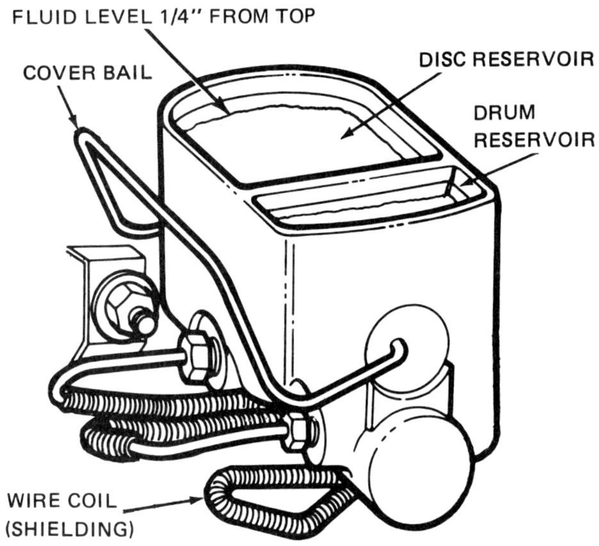

FIGURE 63
Low brake fluid level in the disc reservoir can be caused by worn pads. (Courtesy of Ford Parts and Service Division)

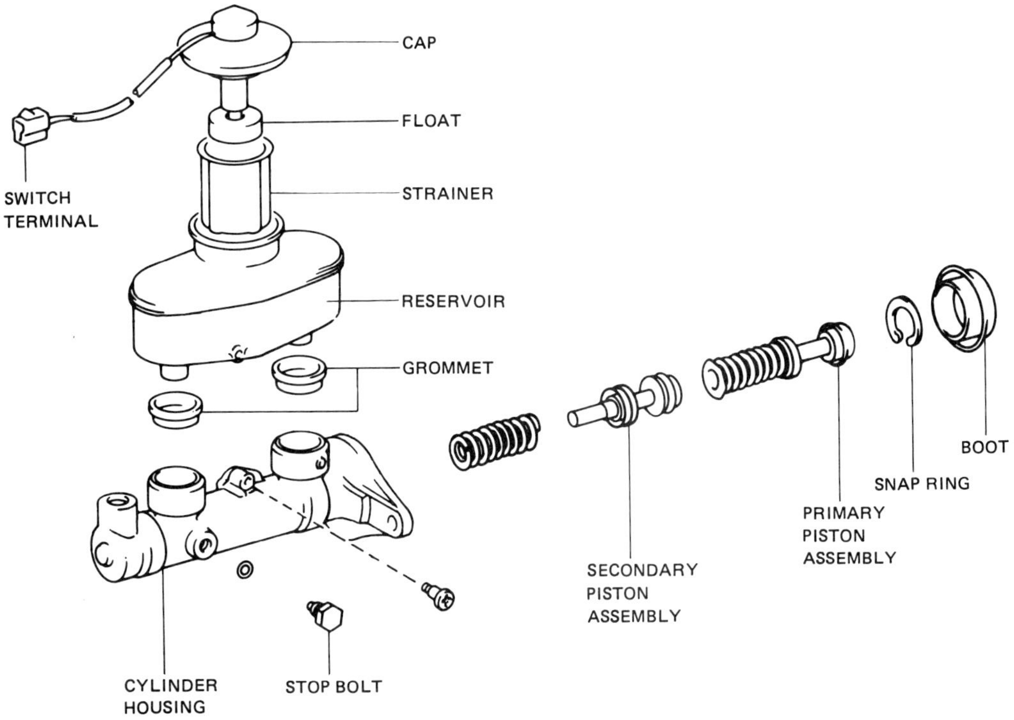

FIGURE 62
A master cylinder with a brake fluid level warning switch. (Courtesy of Toyota Motor Sales, U.S.A., Inc.)

TRUE CASE HISTORY

Vehicle	Honda Civic Station Wagon
Engine	Four-cylinder (CVCC), 1500 cubic centimeters
Mileage	61,000
Problem	1. Dragging brakes 2. Severe pad and rotor wear
Cause	Not enough free pedal travel. This resulted after a new master cylinder was installed.
Cure	Adjusted free pedal travel to specifications; replaced front pads and rotors; replaced front caliper boots and seals.

STUDY QUESTIONS

1. **Technician A** says that brake drag can be caused by using the wrong brake fluid.

   Technician B says that brake drag can be caused by swollen rubber cups.

   Who is right?
   a. A only **c.** Both A and B
   b. B only **d.** Neither A nor B

2. A firm, steady pressure on the brake pedal results in the pedal sinking slowly to the floor. No external fluid leaks can be seen.

   Technician A says that this could be caused by a defective master cylinder residual check valve.

   Technician B says that this could be caused by an internal master cylinder leak.

   Who is right?
   a. A only **c.** Both A and B
   b. B only **d.** Neither A nor B

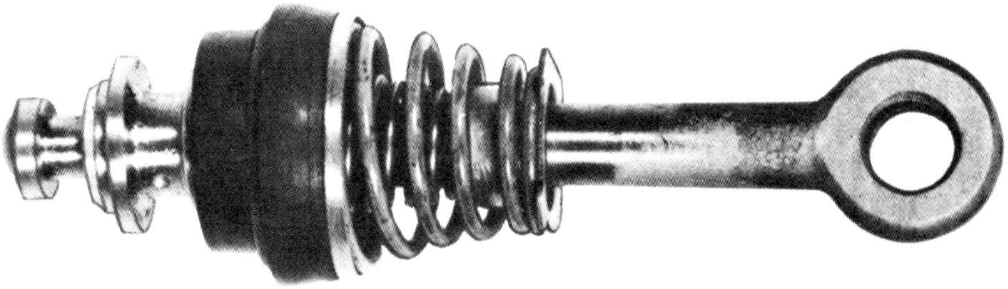

3. **Technician A** says that the above part is adjusted with shims.

   Technician B says that the above part is adjusted with the engine running.

   Who is right?
   a. A only **c.** Both A and B
   b. B only **d.** Neither A nor B

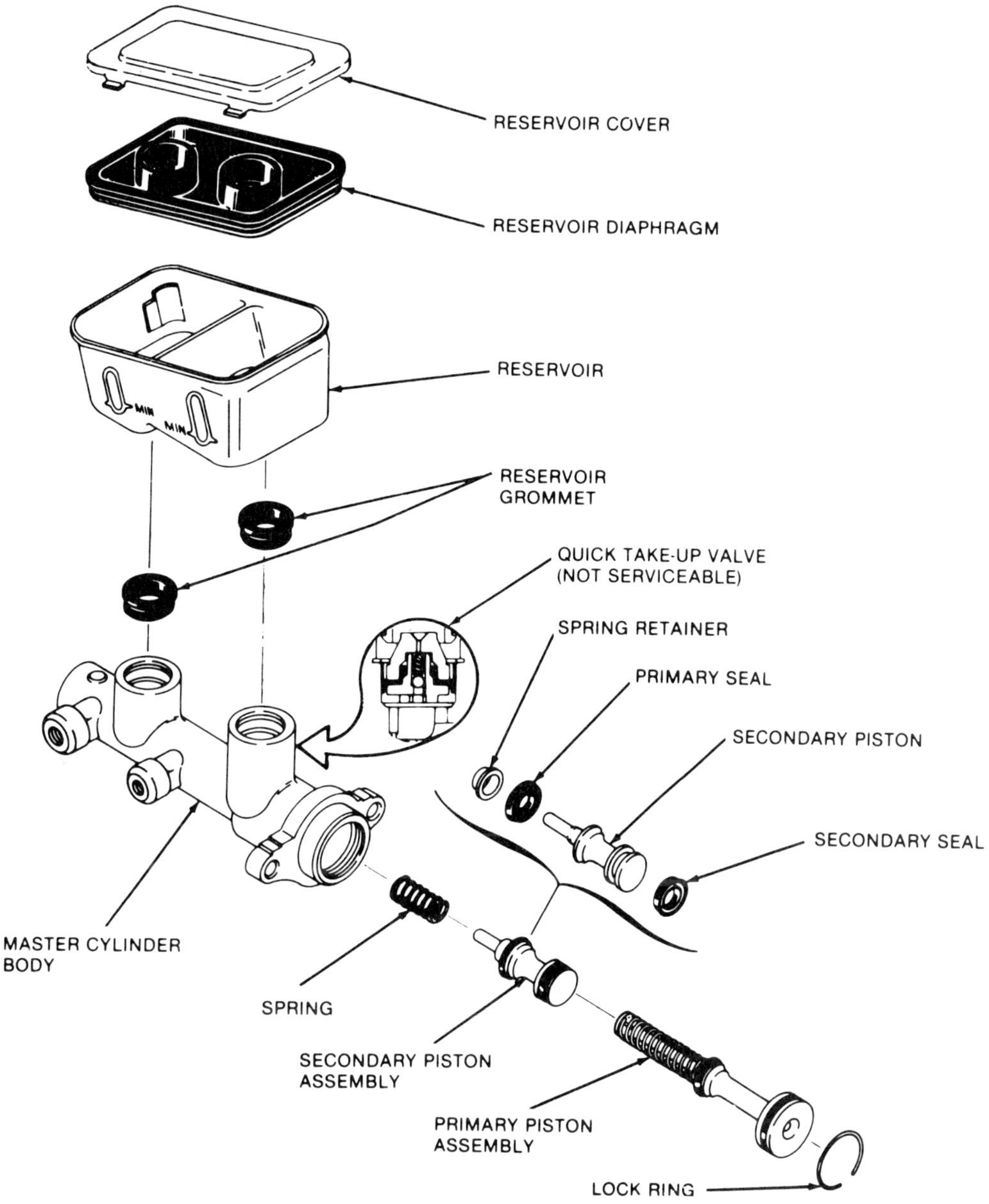

4. The brake pedal is low and spongy on a vehicle equipped with the above master cylinder.

Technician A says that this could be caused by improper bleeding procedure.

Technician B says that this could be caused by a faulty quick take-up valve.

Who is right?

a. A only
b. B only
c. Both A and B
d. Neither A nor B

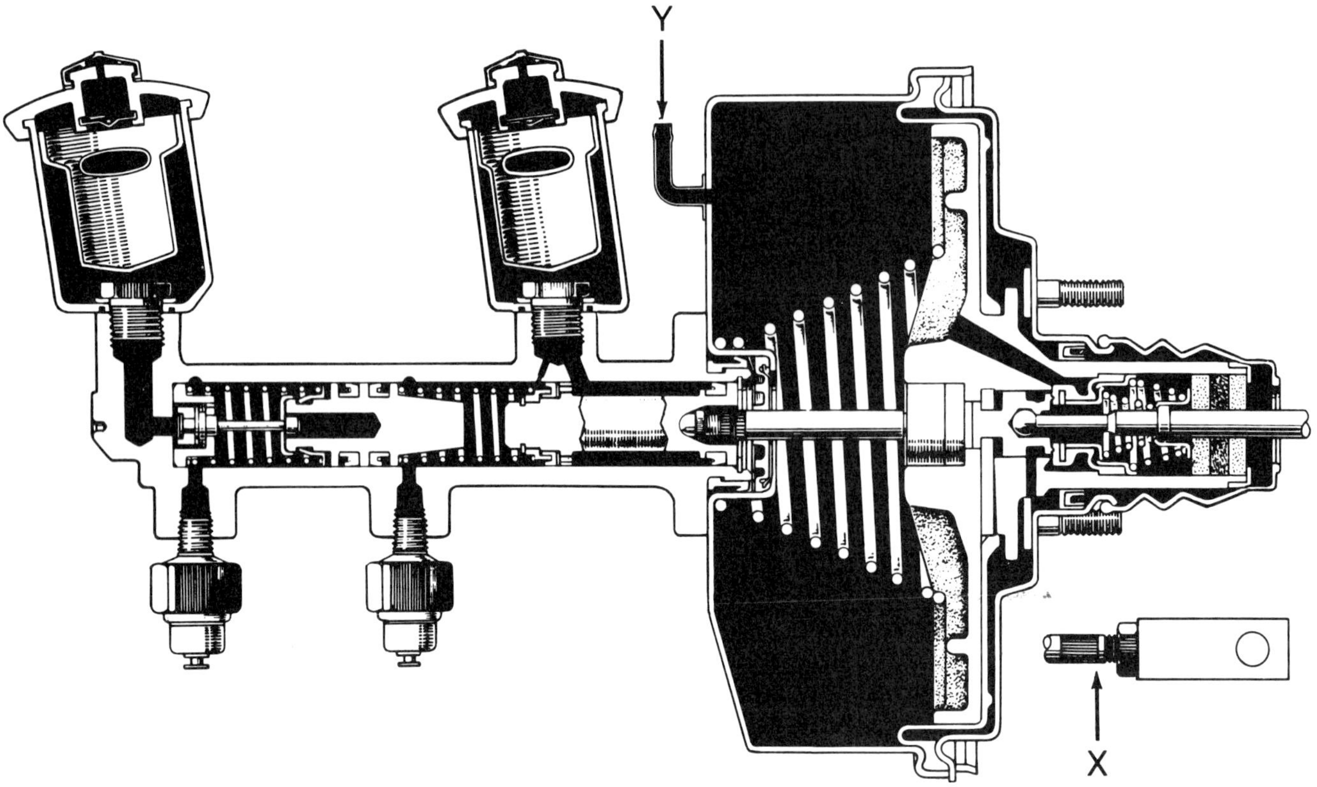

5. The above master cylinder must be removed and replaced.

   Technician A says to check the adjustment at point X.

   Technician B says to bleed the system at point Y.

   Who is right?

   a. A only **c.** Both A and B
   b. B only **d.** Neither A nor B

6. A car is losing fluid from the master cylinder. There is brake fluid on the bulkhead (firewall).

   Technician A says that this could be caused by a defective primary cup on the secondary piston.

   Technician B says that this could be caused by a defective secondary cup on the primary piston.

   Who is right?

   a. A only **c.** Both A and B
   b. B only **d.** Neither A nor B

7. A new tandem master cylinder has just been installed on a vehicle.

   Technician A says to watch for a spurt of brake fluid in the front reservoir when the brake pedal is depressed.

   Technician B says to lubricate the piston assembly with oil.

   Who is right?

   a. A only **c.** Both A and B
   b. B only **d.** Neither A nor B

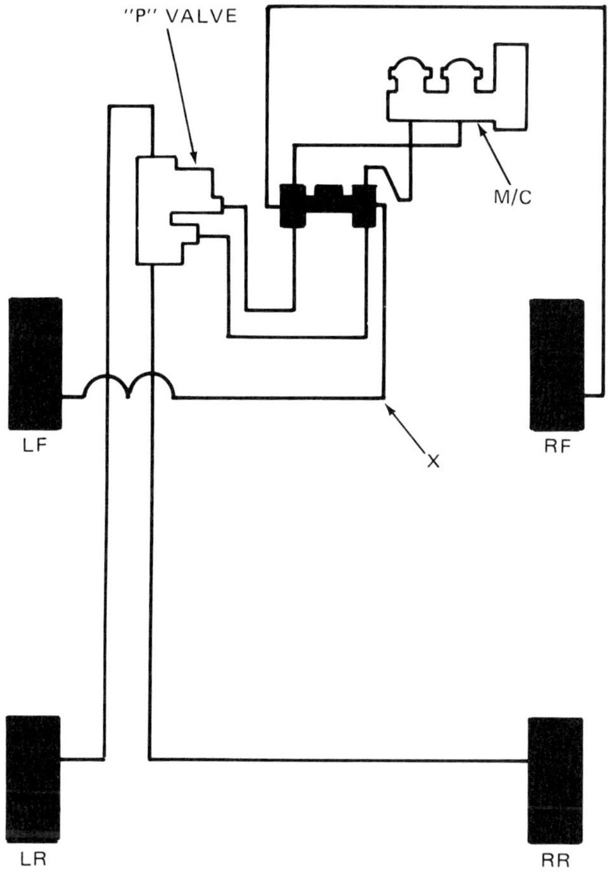

8. A vehicle is equipped with the hydraulic brake system pictured above. There is a hole in the line at point X.

   Technician A says that this will cause a pressure loss to the RF brake.

   Technician B says that this will cause a pressure loss to the RR brake.

   Who is right?
   a. A only **c.** Both A and B
   b. B only **d.** Neither A nor B

9. A master cylinder is being overhauled.

   Technician A says to use solvent to clean the parts.

   Technician B says to use brake fluid to clean the parts.

   Who is right?
   a. A only **c.** Both A and B
   b. B only **d.** Neither A nor B

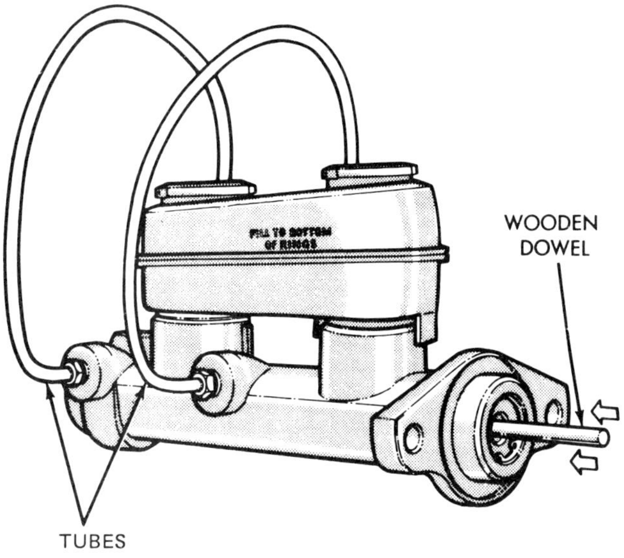

10. Refer to the above operation.

    Technician A says that both reservoirs must be filled.

    Technician B says that depressing the wooden dowel will remove air.

    Who is right?
    a. A only **c.** Both A and B
    b. B only **d.** Neither A nor B

11. A car pulls to one side when the brakes are applied.

    Technician A says that this could be caused by kinked brake lines.

    Technician B says that this could be caused by air in the brake lines.

    Who is right?
    a. A only **c.** Both A and B
    b. B only **d.** Neither A nor B

12. **Technician A** says that silicone brake fluid is hygroscopic.

    Technician B says that silicone brake fluid will attack and remove paint.

    Who is right?
    a. A only **c.** Both A and B
    b. B only **d.** Neither A nor B

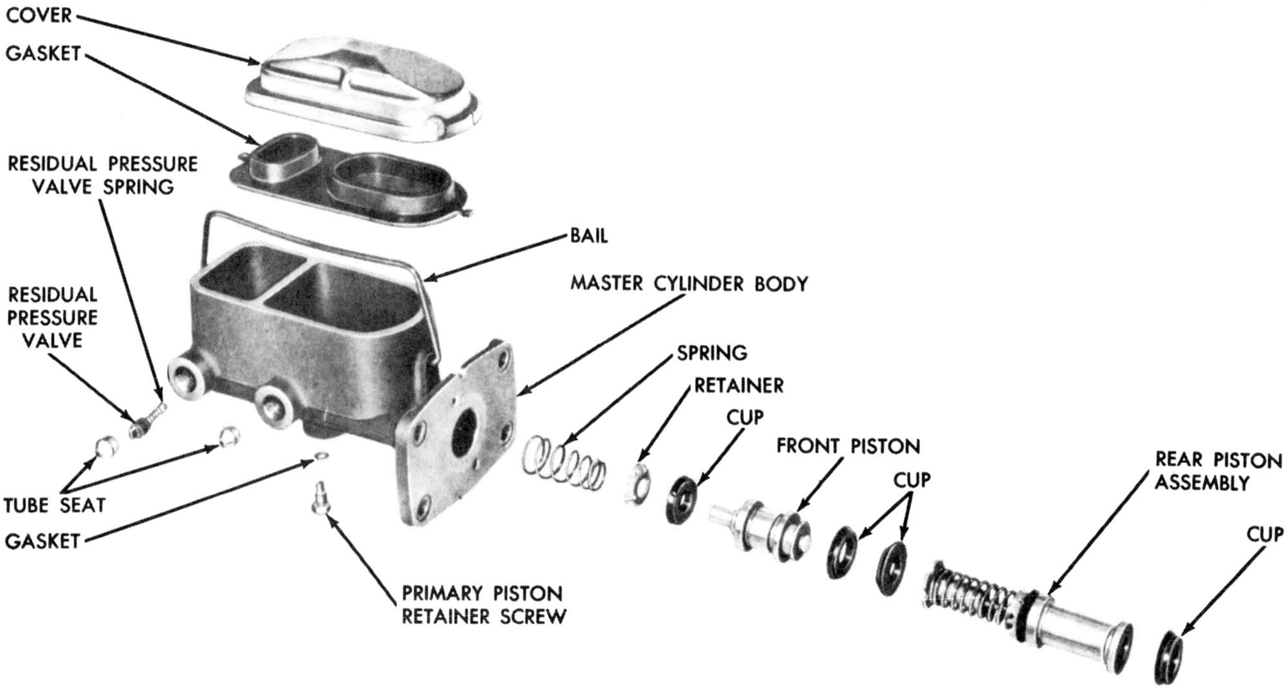

13. **Technician A** says that the front piston in the above illustration is positioned backwards.

    Technician B says that the residual pressure valve in the above illustration is positioned backwards.

    Who is right?
    a. A only
    b. B only
    c. Both A and B
    d. Neither A nor B

14. The tool pictured above is being used to
    a. Test the proportioning valve
    b. Reset the pressure differential valve
    c. Adjust the bypass valve
    d. Hold the metering valve stem open

15. A hydraulic brake line on a car needs to be replaced.

    Technician A says to use copper tubing with single flared ends.

    Technician B says to use a flare wrench.

    Who is right?
    a. A only
    b. B only
    c. Both A and B
    d. Neither A nor B

16. A car has a disc/drum system. The rear wheels lock up too quickly under hard brake application. Which of these is the MOST LIKELY cause?
    a. Trapped air in the hydraulic system
    b. Booster unit not working properly
    c. Plugged compensating port
    d. Defective proportioning valve

17. **Technician A** says that it is normal for a flexible brake hose to bulge a slight amount when the brakes are applied.

    Technician B says that whenever a flexible brake hose is replaced the next size longer should be installed.

    Who is right?
    a. A only **c.** Both A and B
    b. B only **d.** Neither A nor B

18. The brakes on a vehicle that has a height sensing-type "P" valve must be bled.

    Technician A says to disconnect the rear shock absorbers.

    Technician B says to raise the vehicle and allow the axle to hang down.

    Who is right?
    a. A only **c.** Both A and B
    b. B only **d.** Neither A nor B

19. Which of the following can cause the brake warning light to come on?
    a. Leak in brake line
    b. Worn master cylinder
    c. Shorted light circuit
    d. All of the above

20. A power brake system is going to be bled.

    Technician A says to reuse the old fluid.

    Technician B says to maintain vacuum in the power booster.

    Who is right?
    a. A only **c.** Both A and B
    b. B only **d.** Neither A nor B

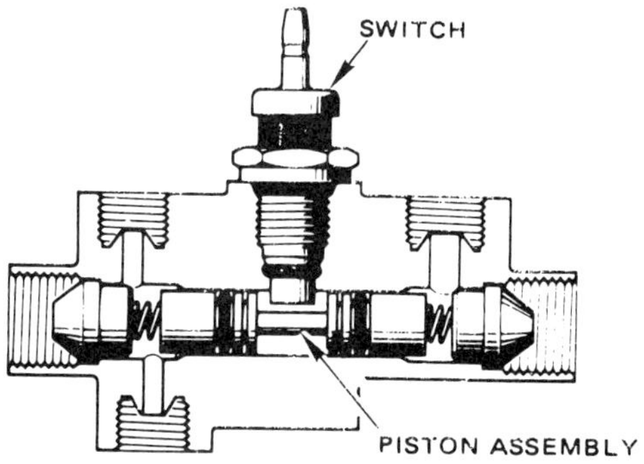

21. **Technician A** says that the piston shown above will move during bleeding, but will recenter itself after the brake pedal is released.

    Technician B says that the piston shown above causes a ground circuit to be completed.

    Who is right?
    a. A only **c.** Both A and B
    b. B only **d.** Neither A nor B

22. A car owner says that fluid drips on the carpet when the brake pedal is depressed. Which of the following is the MOST LIKELY cause?
    a. Torn push rod boot
    b. Leaking piston stop bolt
    c. Overfilled master cylinder
    d. Defective cup

23. After a master cylinder cover is removed, it is observed that the primary reservoir is almost empty of fluid and the secondary reservoir contains too much fluid.

    Technician A says that this could be caused by a defective seal in the middle of the cylinder bore.

    Technician B says that this could be caused by a defective seal on the primary piston.

    Who is right?
    a. A only **c.** Both A and B
    b. B only **d.** Neither A nor B

Section B

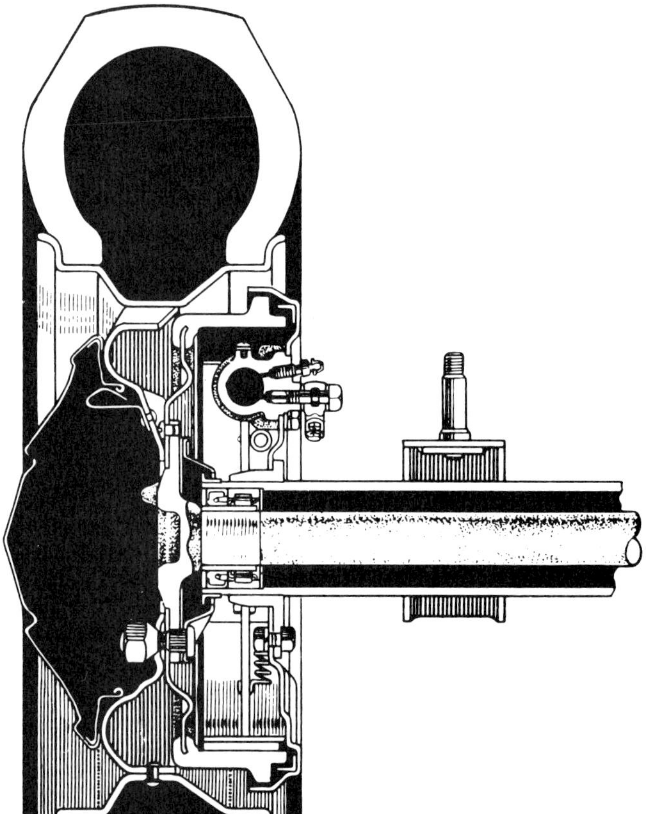

DRUM BRAKE DIAGNOSIS AND REPAIR

After studying this section, and with hands-on practice on a "live" vehicle, you should be able to perform the following tasks:

- Diagnose poor stopping, pulling, or dragging caused by problems in the drum brake hydraulic assembly; determine needed repairs.
- Diagnose poor stopping, noise, pulling, grabbing, dragging, or pedal pulsation caused by problems in the drum brake mechanical assembly; determine needed repairs.
- Remove, clean, inspect, and measure brake drums; follow manufacturers' recommendations in determining need to machine or replace.
- Mount brake drum on lathe and machine according to manufacturers' procedures and specifications.
- Remove, clean, and inspect brake shoes/linings, springs, pins, clips, levers, adjusters/self-adjusters, and other related brake hardware; determine needed repairs.
- Clean and remove loose dirt, rust, or scale on brake backing (support) plates using proper safety procedures; inspect; R&R if necessary.
- Disassemble and clean wheel cylinder assembly; inspect parts for wear, rust, scoring, and damage; hone cylinder (if necessary and recommended by manufacturer); replace all cups, boots, and any damaged or worn parts.
- Lubricate brake shoe support pads on backing (support) plate, adjuster/self-adjuster mechanisms, and other brake hardware.
- Install brake shoes and related hardware.
- Adjust brake shoes and reinstall brake drums or drum/hub assemblies and wheel bearings.
- Reinstall wheel, torque lug nuts, and make final checks and adjustments.

DRUM BRAKE CONSTRUCTION

A typical drum brake is made up of a backing plate with two brake shoes, a wheel cylinder, return springs, hold-down parts, an adjustment mechanism, and a drum. Figures 1 and 2 show different examples of these parts with the drum removed.

REMOVING BRAKE DRUMS

1. Safely support the vehicle.

2. Release the parking brake and put the transmission in neutral.

3. Use a piece of chalk to index each wheel to its hub. This will ensure that wheel balance is not disturbed in case the wheels have been balanced "on the car."

4. Remove the lug nuts, wheel, and drum.

***NOTE:** Sometimes, a worn drum is difficult to remove because the shoes wear and expand into the low spots. If this has happened, the shoes will have to be adjusted inward to permit removal of the drum (Figure 3).*

Cleaning

Dust can be cleaned from the drums, backing plates, and brake assemblies using an approved industrial vacuum cleaner equipped with a high-efficiency filter system. After vacuum cleaning, any remaining dust should be removed by using a rag soaked in water.

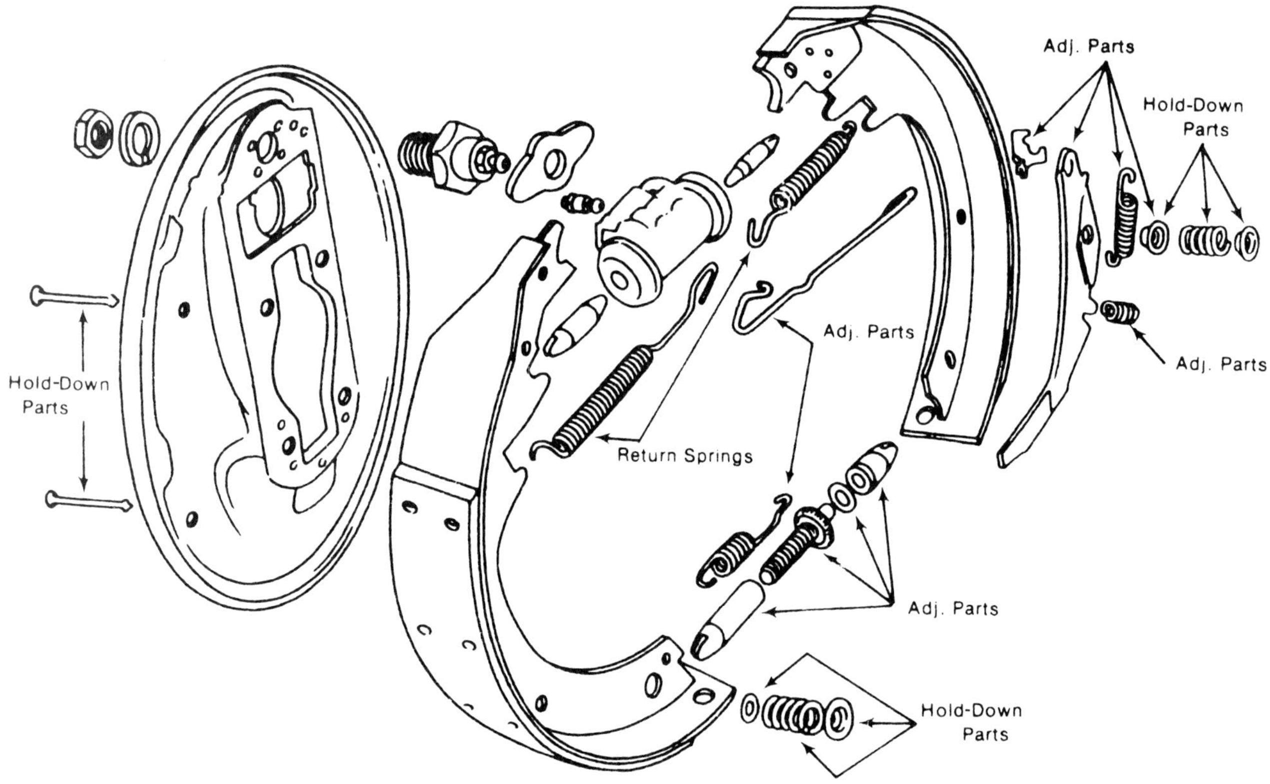

FIGURE 1
Composite illustration of a Bendix self-adjusting brake. The adjuster mechanism operates only when the brake is applied with the car movingin reverse.

Compressed air or dry brushing should never be used for cleaning. Breathing asbestos brake dust can be hazardous to your health.

Inspecting

Look at the friction surface of the drum (Figure 4). If it is scored or otherwise damaged it will need to be reconditioned or replaced (Figure 5 [a], [b], [c], [d], and [e]).

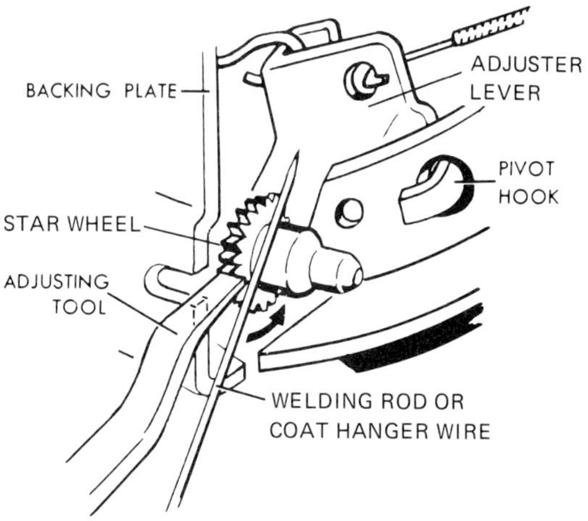

FIGURE 3
Loosening brake shoe adjustment. The adjusting tool (brake spoon) can turn the star wheel once the adjuster lever is pushed away.

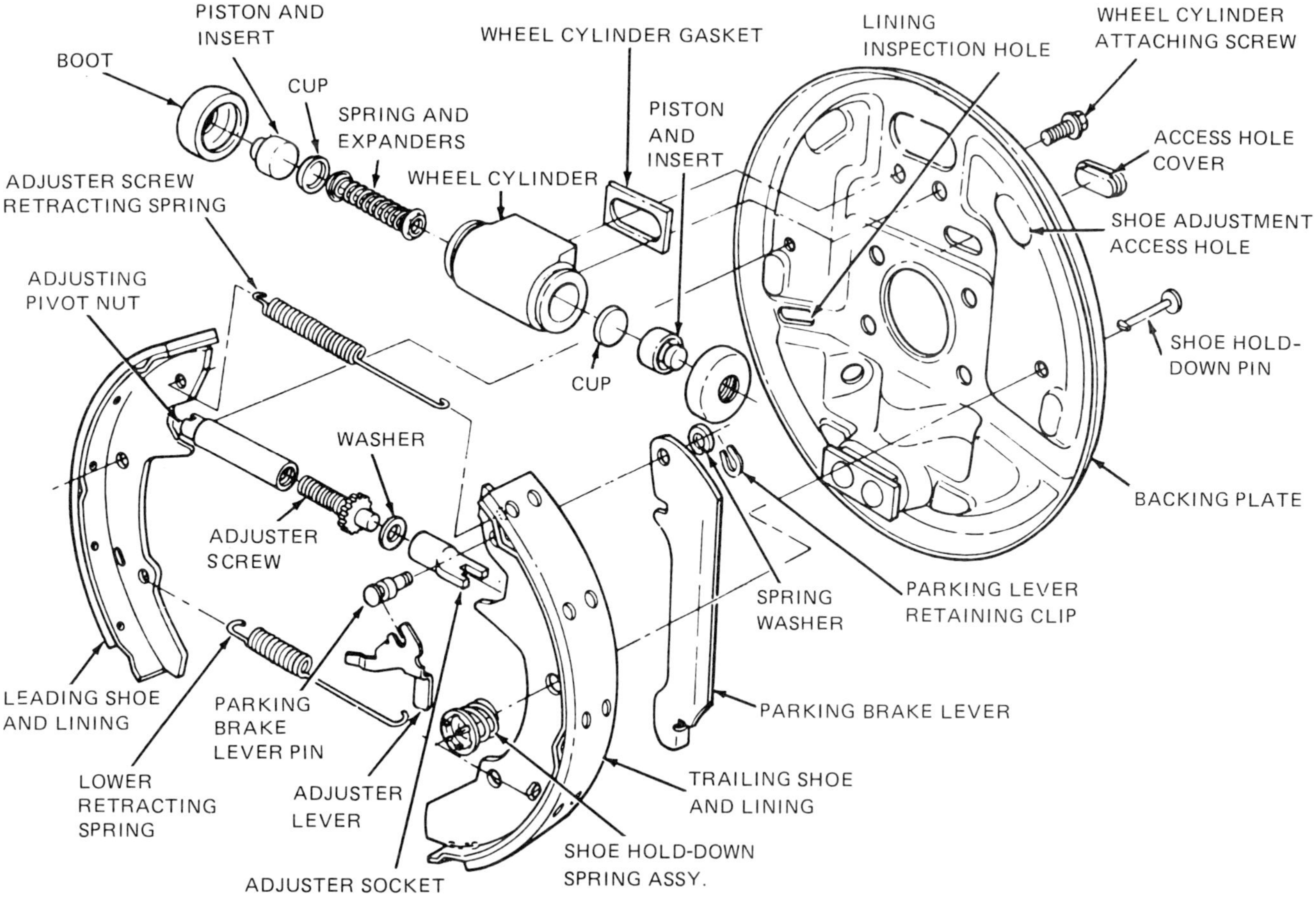

FIGURE 2
Example of a drum brake assembly used at the rear wheels of certain Ford front wheel drive cars. Shoe adjustment takes place, if needed, whenever the parking brake is applied. (Courtesy of Ford Parts and Service Division)

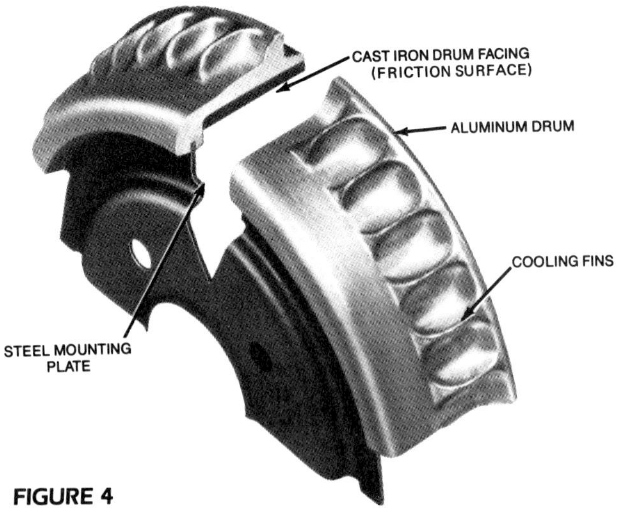

FIGURE 4
The brake drum must provide a good friction surface that is able to dissipate heat as rapidly as possible. (Courtesy of Chrysler Corporation)

Measuring

Measure the drum with a brake drum micrometer (Figure 6). Out-of-roundness, bell-mouthing, scoring, and taper must be corrected by cutting material from the drum friction surface (Figure 7).

> **SHOP TIP**
> Out-of-round drums can be caused by uneven tightening or overtightening of lug nuts. Another cause, which is often overlooked, is applying the parking brake with too much force when the drums are hot.

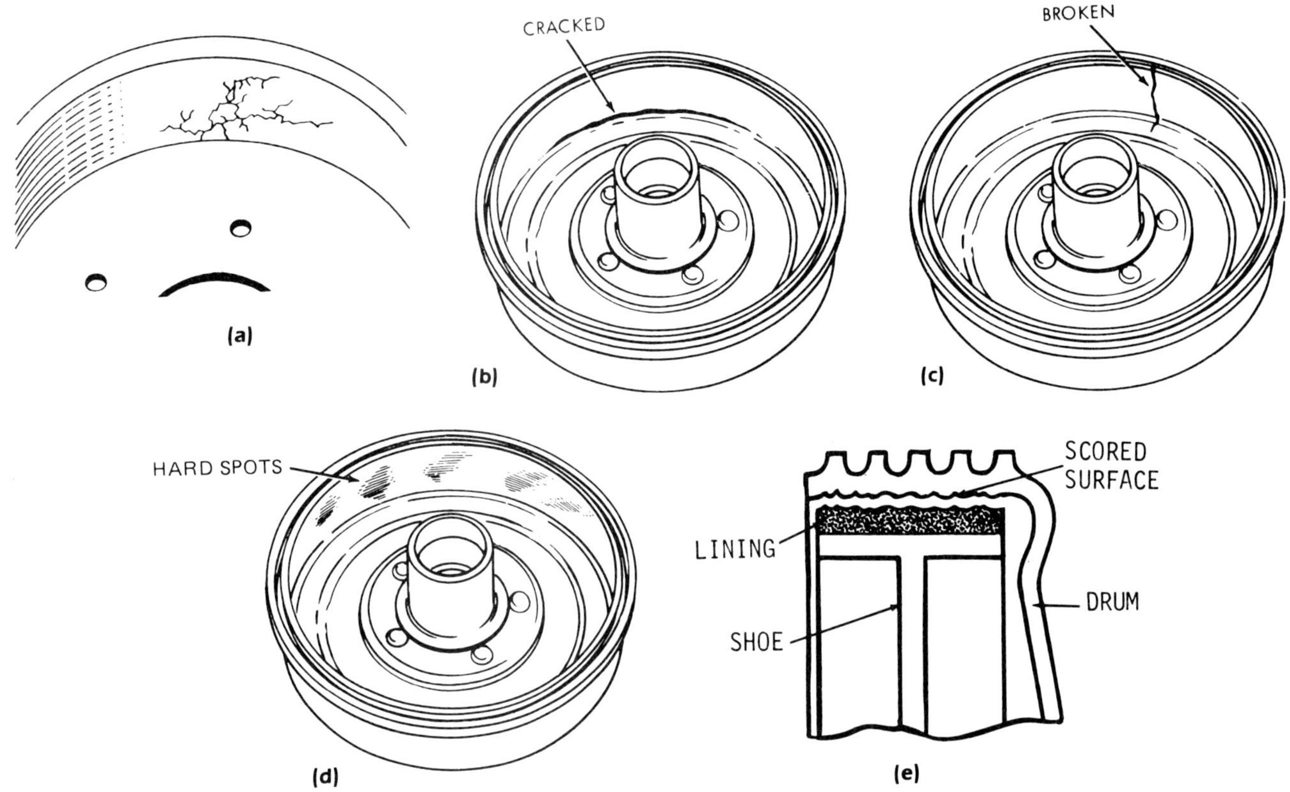

FIGURE 5
(a) Excessive heat checks on the friction surface are a hazardous condition. The drum must be replaced. (b) Cracked drums such as this are often the result of machining the drum beyond the minimum thickness. Do not use a cracked drum. (c) A broken drum is usually caused by some form of accident. This kind of damage has ruined this drum and it must be replaced. (d) Hard spots are "islands" of steel that develop in the cast iron friction surface from braking heat. These spots can be removed by grinding. However, only the raised surface is removed, and they often reappear when heat is applied. The drum should be replaced. (e) Lining wear that is too great, exposing rivet heads or shoe steel, will score the drum friction surface. Scoring can also be caused by the buildup of brake dust between lining and drum. If grooves are not too deep, the drum may be machined.

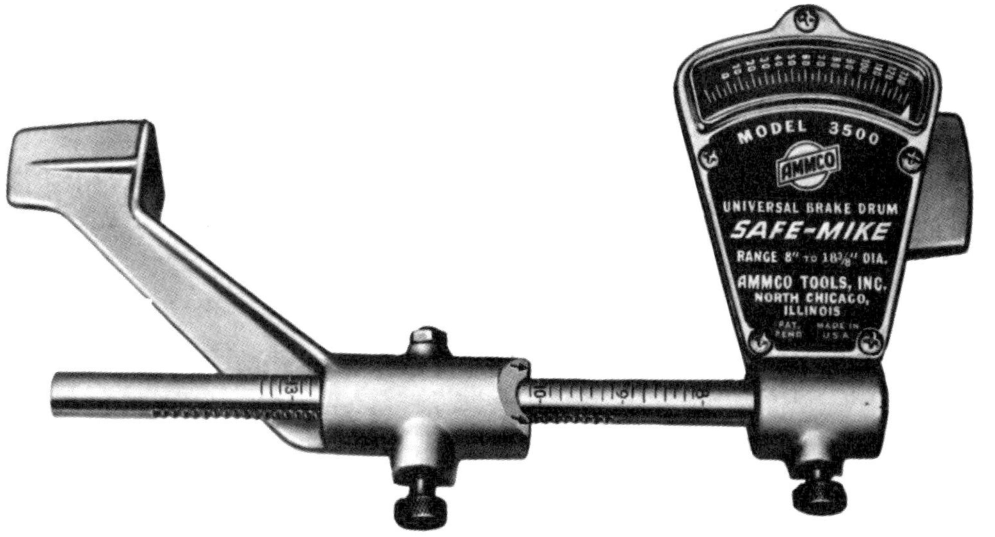

FIGURE 6
Brake drum micrometer. (Courtesy of Ammco Tools, Inc.)

MACHINING BRAKE DRUMS

The friction surface of a brake drum can be refinished by machining (commonly called "turning") on a brake drum lathe (Figure 8). The drum lathe has a rotating arbor (shaft). The drum is mounted on this arbor. As the arbor turns the drum, a tool bit is moved across the inside surface of the drum. The

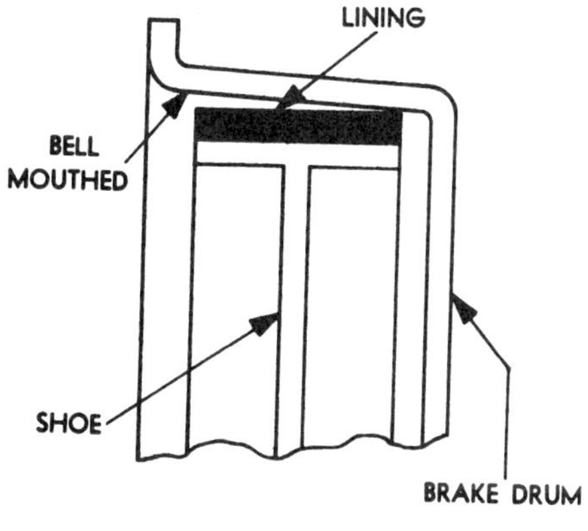

FIGURE 7
If a drum is bell-mouthed, full drum-to-lining contact cannot be achieved and brake fading may be expected. The drum should be resurfaced or replaced if bell-mouthing exceeds .010". (Courtesy of Raybestos-Manhattan, Inc.)

FIGURE 8
Machining a brake drum with the tire attached. The inset at the bottom of the picture shows a brake drum grinder. The brake drum should always be machined before grinding. (Courtesy of Ammco Tools, Inc.)

drum must be perfectly centered and securely held on the arbor. This is accomplished by using the proper adapters and fittings (Figures 9 and 10).

Wrap the drum snugly with a dampening belt (Figure 11). This belt is used to dampen (silence) vibrations that may be produced in the drum as it is being machined. *If the dampening belt is not installed properly, you can hear a vibration "ring" when machining. This vibration will create tool bit chatter marks on the drum wall. These chatter marks usually appear as a herringbone pattern.*

After the drum is secured to the arbor, check the mounting accuracy. This can be done quickly by following these steps:

1. Turn the drum by hand to be sure that everything is clear, and put on eye protection. Turn the lathe on.

2. Advance the tool bit by hand until it just contacts the drum friction surface for a moment and makes a scratch cut (Figure 12).

3. Back the tool bit off, stop the lathe, and loosen the arbor nut.

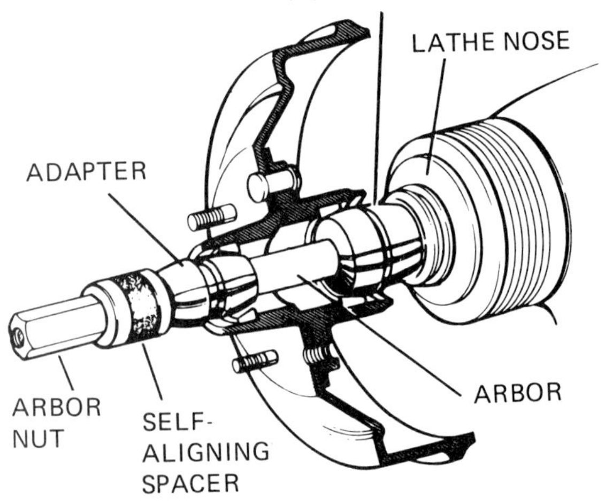

FIGURE 9
The mounting method for a typical **hubbed** drum. Adapters fit in the bearing races, and various spacers fill out the shaft of the arbor so that the arbor nut can be tightened. (Courtesy of Ammco Tools, Inc.)

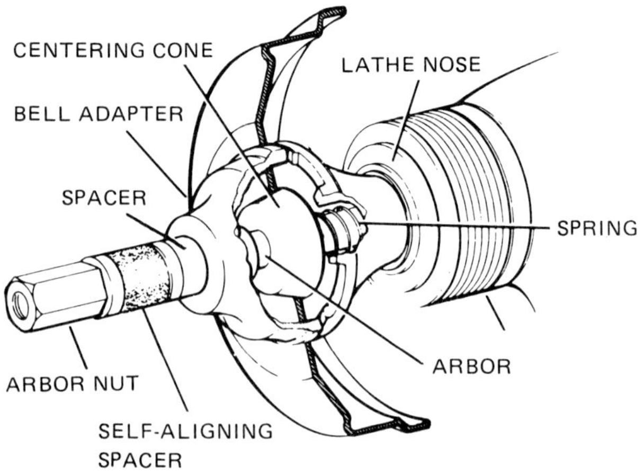

FIGURE 10
The mounting method for a typical **hubless** drum. Note that the centering cone fits into the drum hole from the inside, and the bell adapters clamp the drum from both sides. (Courtesy of Ammco Tools, Inc.)

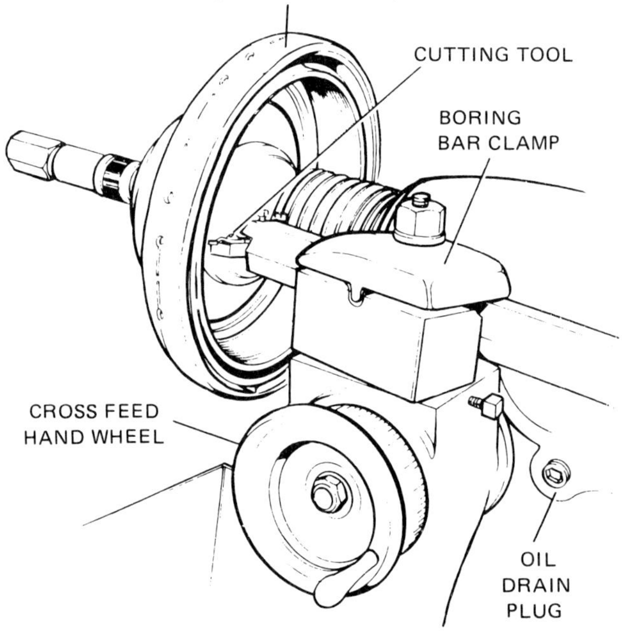

FIGURE 11
A dampening belt is stretched around the outside of the brake drum to reduce chatter during the machining process. The band should be placed as close to the inside edge of the brake drum as possible. (Courtesy of Ammco Tools, Inc.)

FIGURE 12
After a drum is mounted, make a scratch cut. (Courtesy of Ammco Tools, Inc.)

4. Rotate the drum one-half turn on the arbor.

5. Retighten the nut and turn the lathe *on*.

6. Make a second scratch cut and stop the lathe (Figure 13). If the scratches are side by side, go ahead and machine the drum. If the first and second scratch cuts are opposite one another (180 degrees apart), remove the drum from the arbor. Check the mounting adapters, fittings, and arbor for burrs or nicks.

FIGURE 13
Loosen the lathe arbor nut and rotate the drum one-half turn. Make a second scratch cut. If the scratches are side by side, the drum has been accurately mounted. (Courtesy of Ammco Tools, Inc.)

If the drum diameter after machining is larger than the manufacturer's specified rebore limit, *it must be replaced*.

Drums that are too thin are dangerous! They do not dissipate heat well and can come apart (Figure 14). Studies show that, if drum thickness is reduced to 80 percent of original, the ability to withstand pressure is reduced to 64 percent of original strength.

FIGURE 14
This drum was too thin and came apart while braking.

FIGURE 15
Sand the drum wall with sandpaper or emery cloth after machining in order to break up the threaded surface. (Courtesy of Chrysler Corporation)

COMMENT: Most manufacturers warn against machining a drum to more than .060″ oversize. Brake drums manufactured after December 31, 1970, are stamped or cast with a discard dimension. This is the allowable wear dimension and not the allowable machining dimension. There must be .030″ left for wear after turning the drum.

Here are some drum machining guidelines:

1. Use only tool bits in good, sharp condition. Chipped or dull tool bits produce a poor finish and require heavy cutting pressure, which can result in untrue drums or rotors.

2. An adequate rough cut *feed* is usually somewhere between .010″ and .015″ for most drum lathes. The tool bit *depth* for a rough cut should be .005″ to .010″.

3. A finish cut is usually made with a *feed* of no more than .006″. The *depth* for a finish cut should be no more than .005″.

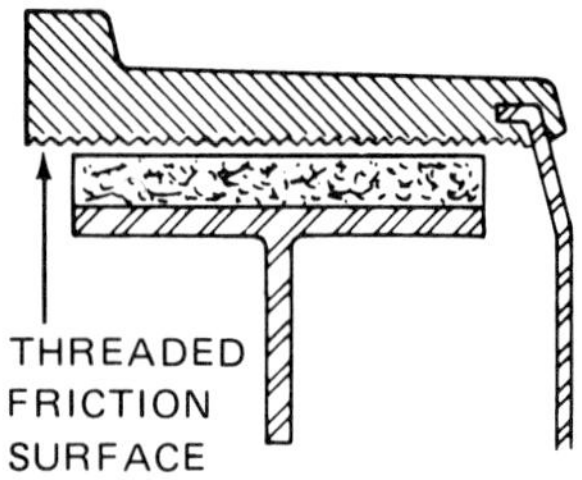

FIGURE 16
Shoe "knock" will result if the lining is pulled outward by drum threading, then "snapped" back by the hold-down springs. The noise will occur just as the brakes are applied.

4. After the drum has been machined, sand the surface produced by the tool bit (Figure 15). This prevents shoe "knock" when the brakes are applied. See Figure 16. Be sure to wipe from the drum any bits of material produced from machining and sanding.

5. Front drums should be machined within .010″ of each other to avoid the possibility of vehicle pull when braking.

BRAKE SHOE REMOVAL

The brake assembly pictured in Figure 17 is a typical Bendix self-adjusting brake that is widely used. There are brakes that have other designs, but similar parts (Figures 18 and 19).

Disassembly (Bendix Self-Adjusting Brake)

1. Remove the brake shoe retraction (return) springs. Refer to Figure 20. Observe which spring is installed first and the position of the self-adjusting cable eye (Figure 21).

2. Slide the eye of the self-adjusting cable off the anchor (Figure 22). Then un-

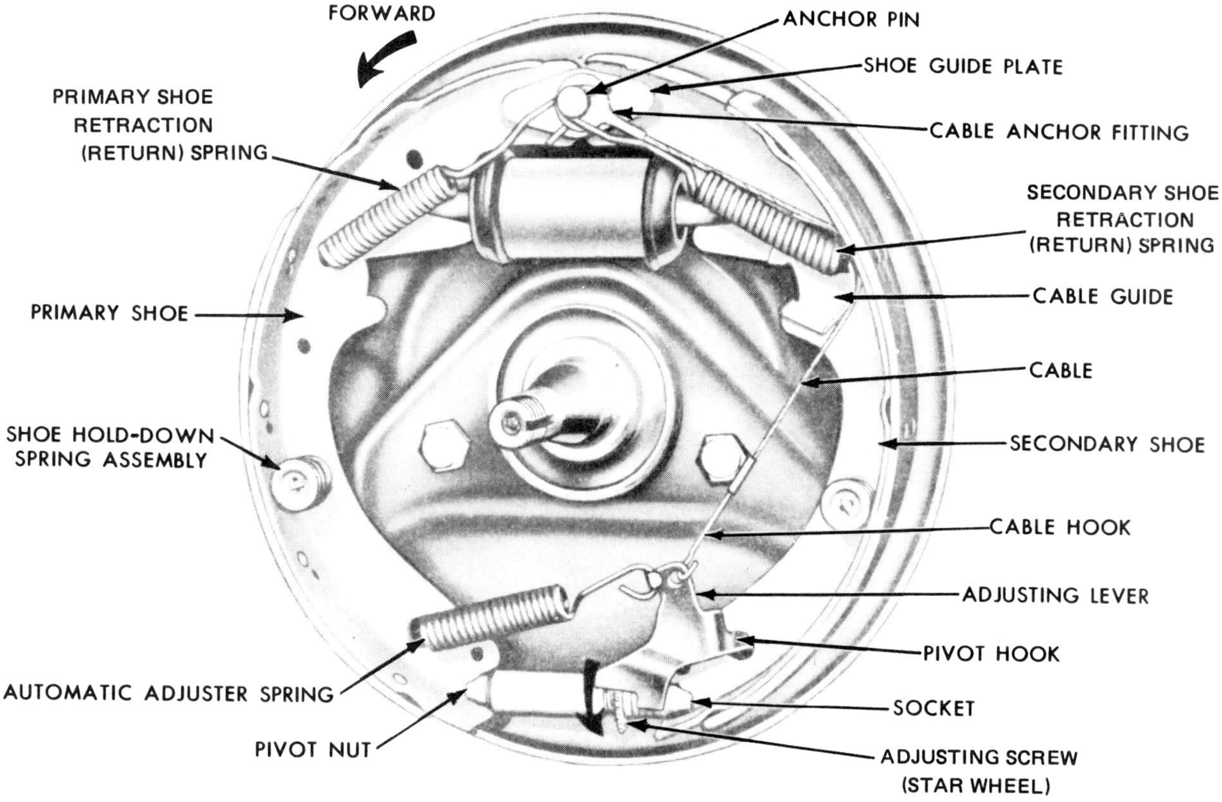

FIGURE 17

The Bendix self-adjusting brake. The adjusting mechanism is operated by movement of the secondary shoe during brake application while the car is moving in reverse. Adjustment takes place only when enough lining has been worn away to permit the adjusting lever to engage the next tooth of the star wheel. This adjusts the shoes closer to the brake drum. (Courtesy of Ford Parts and Service Division)

SHOP TIP

When disassembling an unfamiliar brake assembly, work on one wheel at a time and use the other wheel for reference.

hook the cable from the adjusting lever. Remove the cable, cable guide, and shoe guide plate.

3. Disengage the adjusting lever from the automatic adjuster spring by sliding it forward to clear the pivot hook. Remove the adjusting lever.

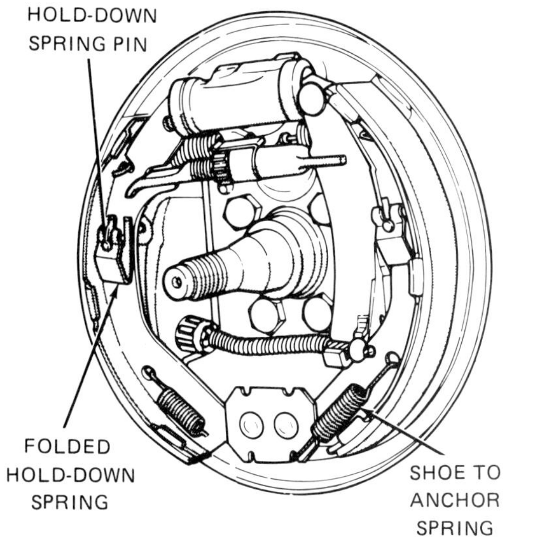

FIGURE 18

A folded hold-down spring assembly is used in some brake designs to hold the shoes against the backing plate. (Courtesy of Chrysler Corporation)

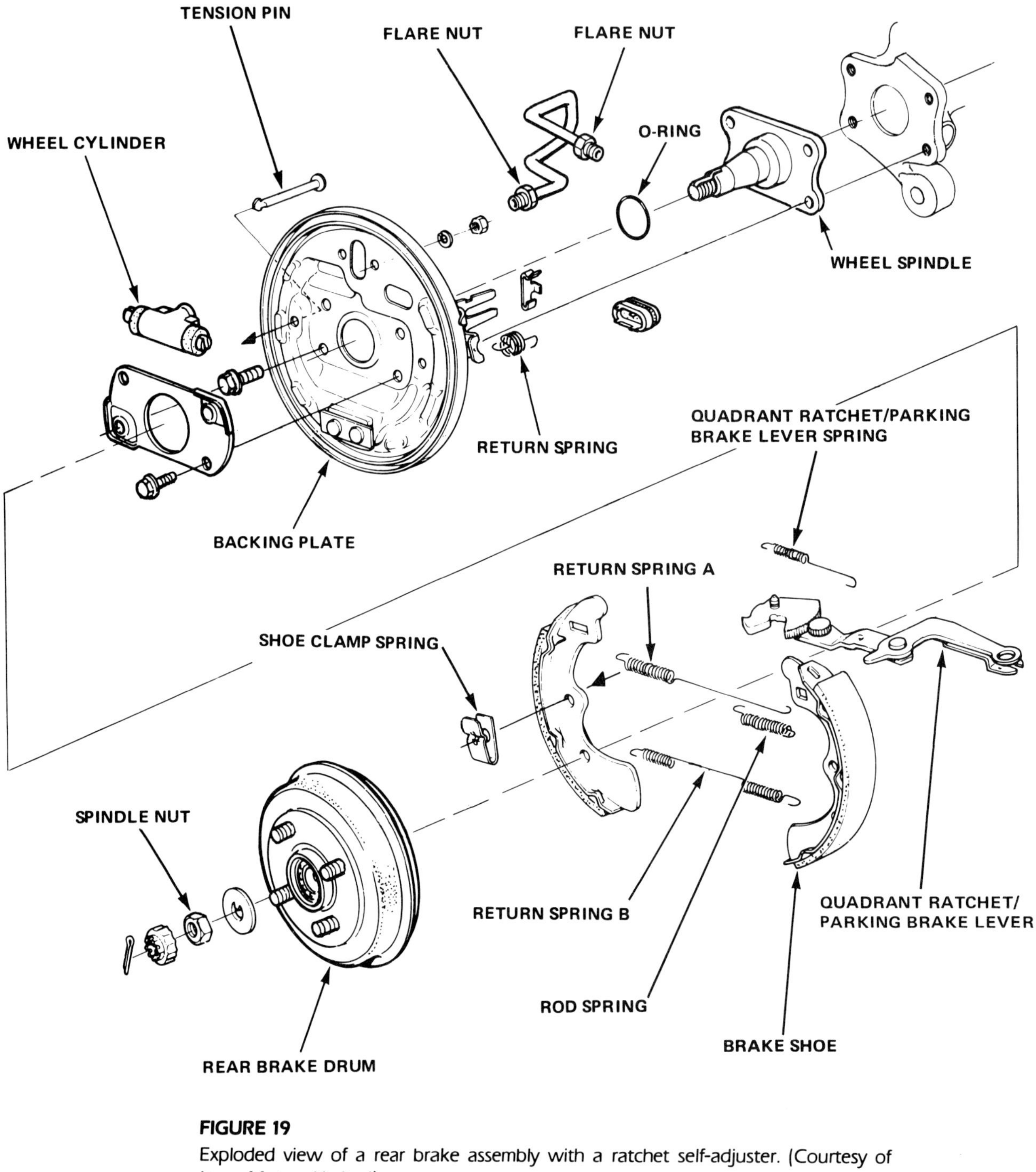

FIGURE 19
Exploded view of a rear brake assembly with a ratchet self-adjuster. (Courtesy of Isuzu Motors Limited)

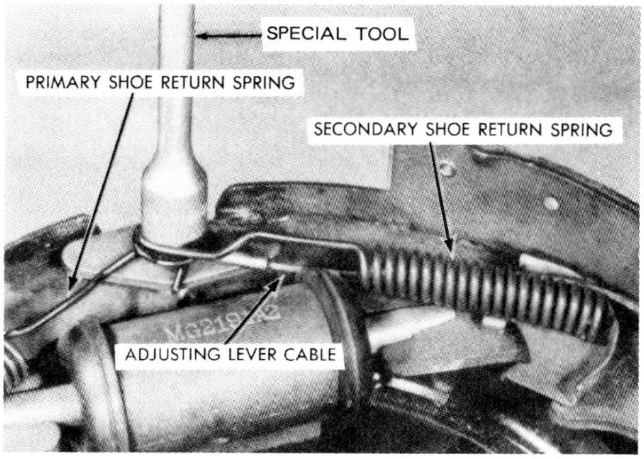

FIGURE 20
Using a special tool to remove the secondary shoe return spring.

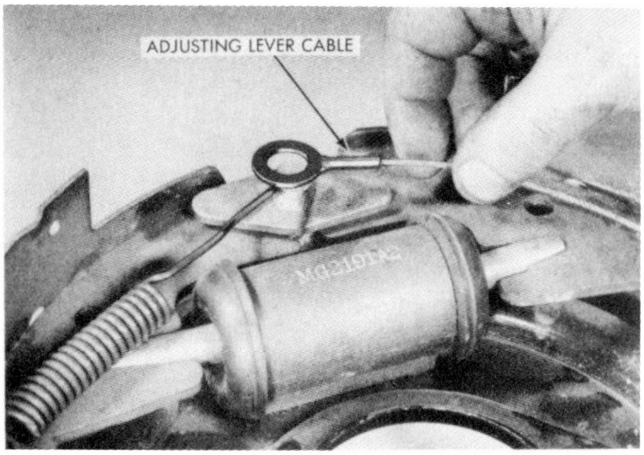

FIGURE 22
Removing self-adjusting cable eye from anchor pin.

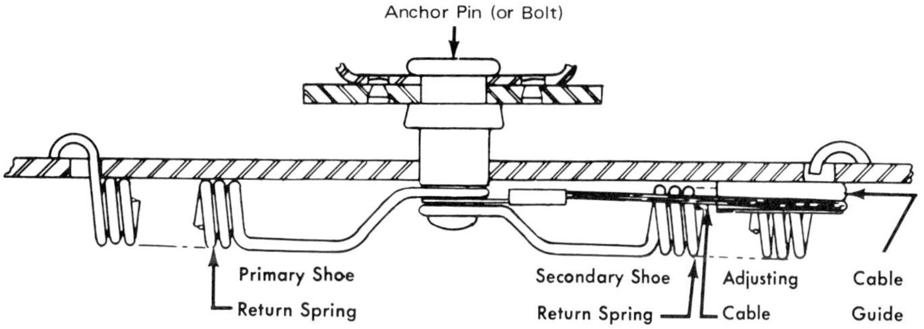

FIGURE 21
Top view of brake shoe return springs. Notice that the primary shoe return spring has been installed first on the anchor pin. This is the usual rule of thumb.

4. Remove the automatic adjuster spring by working it out from the primary shoe.

5. Remove the brake shoe hold-down spring assembly from the backing plate (Figure 23 [a] and [b]).

6. Disengage the primary and secondary shoes from the wheel cylinder and remove the adjusting screw assembly from the shoes.

7. Spread the anchor ends of the primary and secondary shoes and remove the parking brake lever strut and antirattle spring.

8. Disengage the parking brake cable from the parking brake lever.

Cleaning and Inspection

Wipe the metal portion of the brake shoes and check for broken welds, cracks, and wear (Figure 24). Examine the lining contact pattern to check whether or not the shoes are distorted or twisted. The lining should show contact across its entire width. Lining showing contact only on one side should be replaced.

Clean the backing plate, using a suitable solvent. Inspect for grooves at the shoe contact areas, cracks, damage from road salt, and bent metal (Figure 25). The backing plate should be replaced if these defects are found. Check the anchor bolt to be sure that it is tight.

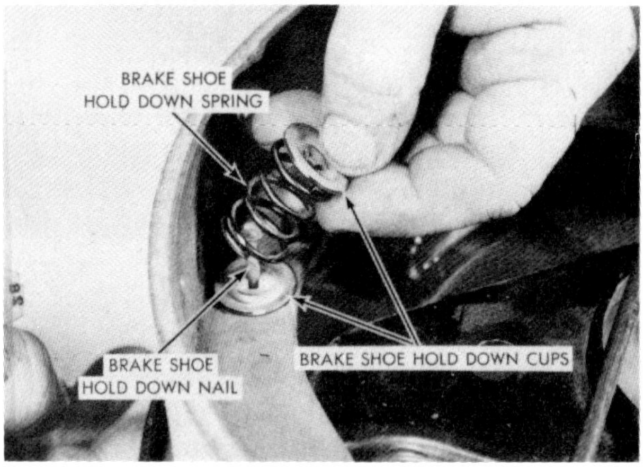

(a)

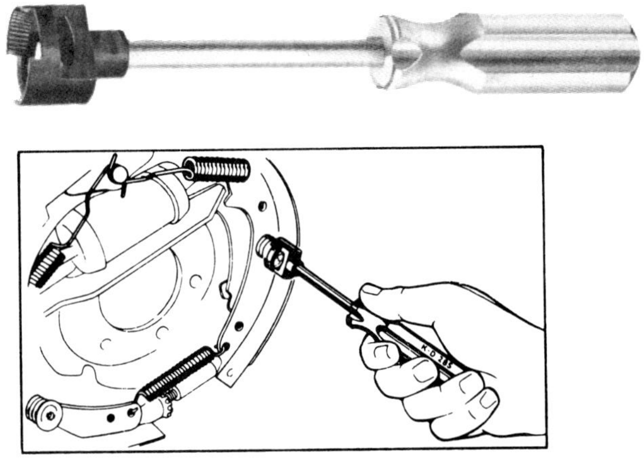

(b)

FIGURE 23
(a) Removing brake shoe hold-down spring assembly. (b) Special hold-down spring removal tool.

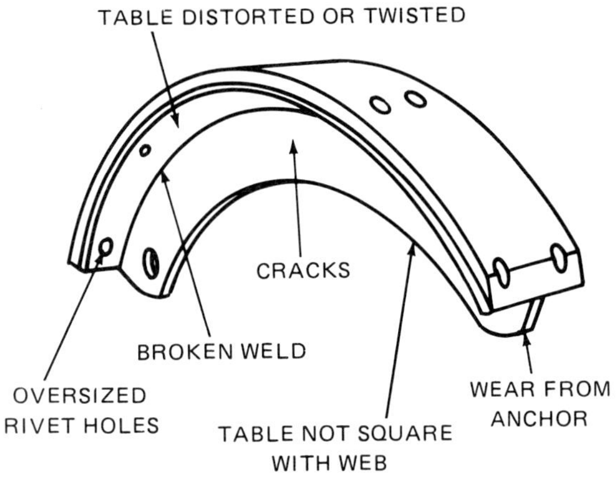

FIGURE 24
Brake shoe inspection areas.

SHOP TIP

Hidden cracks in cast aluminum shoes can be extremely difficult to detect, unless the hammer method is used. Tie a string into a hole in the shoe and suspend the shoe while tapping it with a hammer (Figure 26). A cracked shoe will have a dull sound, while a good shoe will have a clear, ringing sound.

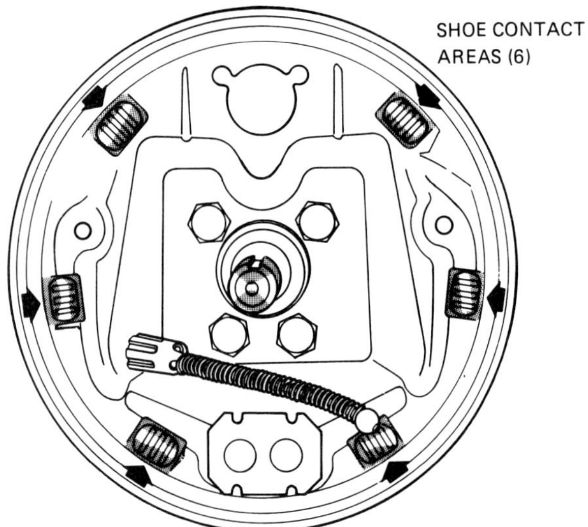

FIGURE 25
Shoe contact areas on backing plate. (Courtesy of Chrysler Corporation)

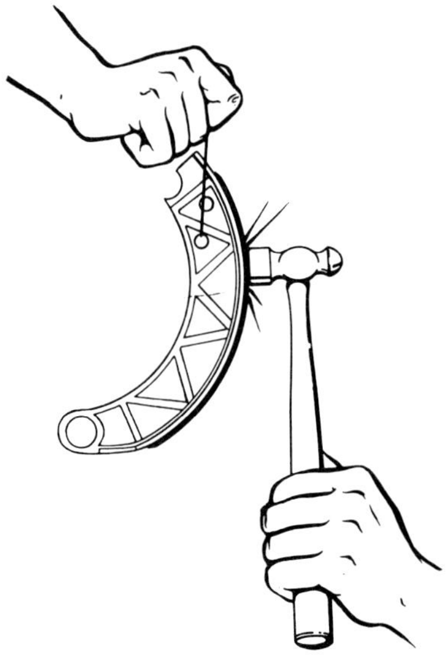

FIGURE 26
Checking cast or malleable aluminum shoes.

FIGURE 27
A washer being used to clean the brake assembly. (Courtesy of Ammco Tools, Inc.)

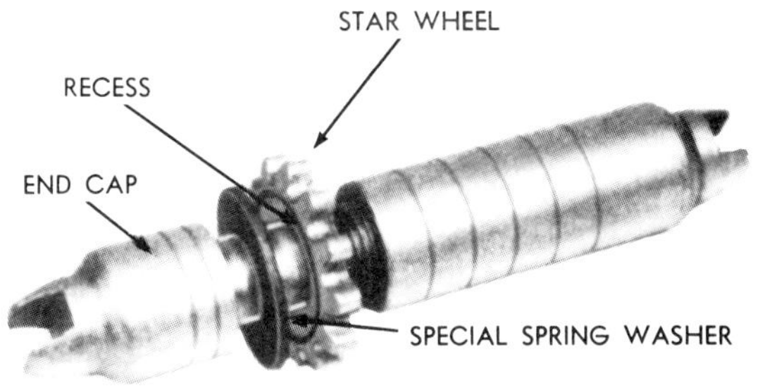

FIGURE 28
Adjusting screw assembly. The screw is usually marked with an "L" or an "R" to indicate a left- or right-hand thread. (Courtesy of Raybestos-Manhattan, Inc.)

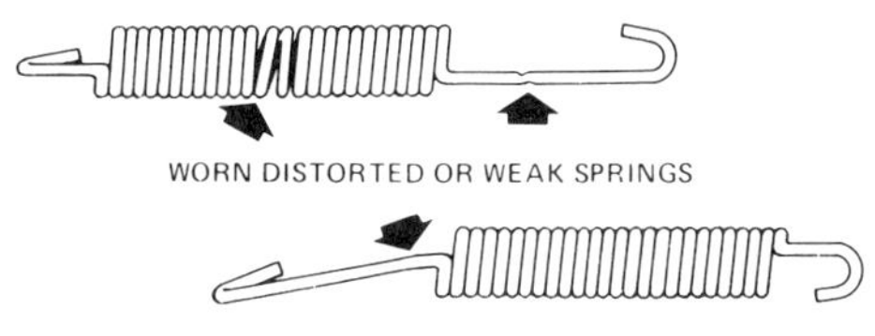

FIGURE 29
Inspect return springs closely. Drop spring on a cement floor to determine whether or not tension is weak. A weak spring will bounce and "ring" when it contacts the floor. A spring with good tension will hit the floor with a "thud." (Courtesy of Wagner Division, Cooper Industries, Inc.)

***NOTE:** Many shops prefer to use a brake washer (Figure 27). The liquid concentrate will remove dirt, oil, grease, and asbestos fibers before they can get into the air and become a health hazard.*

Clean and inspect the adjusting screw for pulled or stripped threads (Figure 28). If the threads are in good condition, the parts should be lubricated with a high-temperature lubricant and reassembled.

New brake shoe return springs and hold-down springs should be installed when the old springs have been overheated or distorted (Figure 29). Discolored paint on a spring indicates that it has been overheated. Overheated springs lose their tension, and if they are not replaced they can cause new linings to wear prematurely.

Closely check the hold-down clips, pins, wire links, adjusting levers, cables, cable guides, supports, and other related hardware for damage (Figures 30 and 31). Replace them if necessary.

DRUM BRAKE TROUBLESHOOTING

Refer to Chart 1 for problems in the drum brake wheel and mechanical assembly.

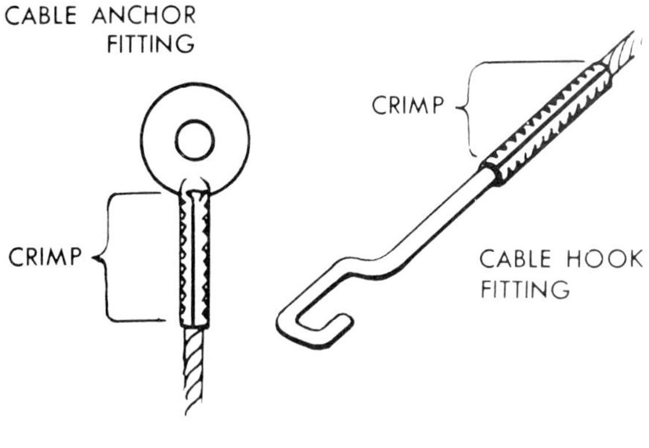

FIGURE 30
Check crimped ends of self-adjusting cables. The cable must fill the crimped section of the fitting. If not, the cable has stretched and must be replaced.

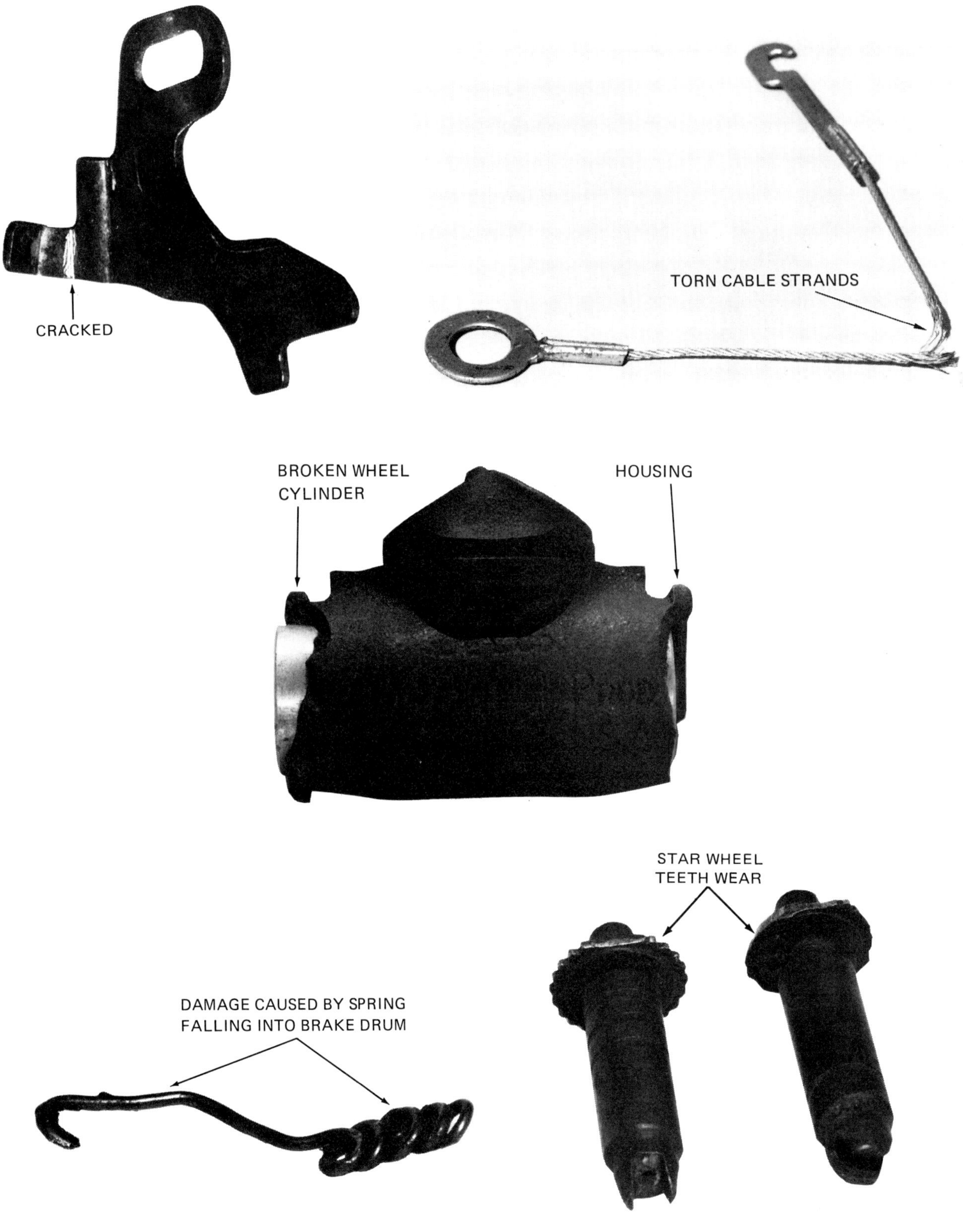

FIGURE 31

Examine all brake hardware closely. Sometimes it is difficult to spot the damage.

FIGURE 31 (Continued)

WHEEL CYLINDERS

When the brake pedal is depressed, fluid pressure is transmitted into the wheel cylinders (Figure 32). This forces the pistons outward, which in turn causes the brake shoes to move against the brake drum.

Wear, corrosion, or dirt may cause a wheel cylinder to leak. If the rubber boot is wet on the outside, or if brake fluid is found on the inside of the boot, the wheel cylinder should be rebuilt or replaced (Figure 33).

Whenever the brake shoes are replaced, the wheel cylinders should be rebuilt or replaced.

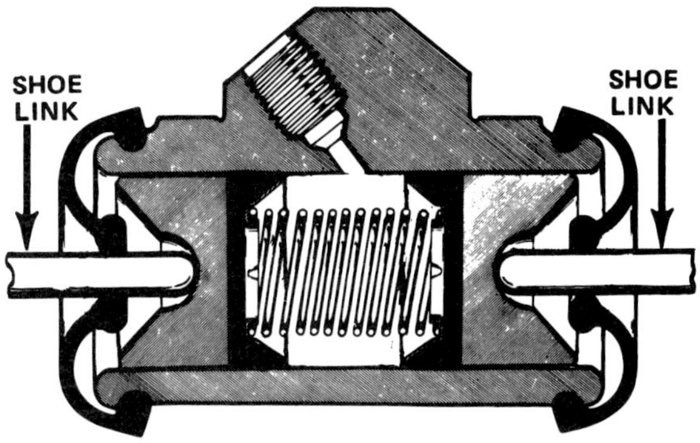

FIGURE 32
Cups and pistons move outward in wheel cylinder bore when the brake pedal is depressed. (Courtesy of Allied Aftermarket Division, Allied-Signal Corporation)

CHART 1 TROUBLESHOOTING CHART FOR DRUM BRAKES

Low Pedal

1. Too much lining-to-drum clearance
2. Excessively thin drum that expands under heat and pressure
3. Wheel cylinder push rod not properly located
4. Self-adjuster not working

Pulling (Out-of-Line Stops)

1. Contaminated lining (oil, grease, or brake fluid)
2. Unmatched lining
3. Bent brake shoe
4. Standard diameter brake drum on one side and oversize drum on the opposite side
5. High-friction lining on one side and lower friction lining on the opposite side
6. Faulty suspension parts
7. Low tire pressure
8. Improper wheel alignment

Dragging

1. Weak or broken shoe return spring
2. Bent or out-of-round brake drum
3. Loose or torn brake lining
4. Binding pedal linkage
5. Brake shoe not releasing properly due to lack of backing plate lubrication at bosses, anchors, or abutments; deep grooves worn in backing plate bosses or worn shoe at boss locations
6. Over-adjusted shoes
7. Contaminated brake fluid causing swelling of rubber parts
8. Loose or worn wheel bearings
9. Master cylinder push rod out of adjustment
10. Improperly adjusted stop light switch (not allowing full return of pedal)

Failure to Self-Adjust

1. Adjuster screw frozen
2. Adjuster screw installed on wrong wheel or installed backward
3. Adjust lever does not engage star wheel
4. Broken adjuster cable

Hard Pedal (Too Much Pedal Pressure Required to Stop)

1. Lining worn or glazed
2. Pedal bind or interference
3. Water film on brake lining surface
4. Cast iron dust from surface of newly turned brake drum imbedded in brake lining
5. Bell-mouthed or barrel-shaped drum
6. Brake fluid on lining surface from leaking wheel cylinder
7. Grease on lining surface from leaking grease seal

Grabbing (Severe Reaction to Pedal Pressure)

1. Contaminated lining (oil, grease, or brake fluid)
2. Primary and secondary shoes and linings installed backward on one side (can also cause pulling)
3. Loose backing plate, anchor, or wheel cylinder
4. Hard spots in drum braking surface (can also cause poor braking and pedal vibration)
5. Shoes not centered in drum
6. Out-of-round drum
7. Worn brake lining

Pulsating Pedal

1. Out-of-round drum
2. Bent hub, drum, or wheel
3. Excessive rear axle flange runout due to bent axle or worn axle bearing
4. Loose wheel bearing adjustment

Squeal Noise

1. Glazed lining
2. Weak or broken shoe return spring
3. Dust or foreign matter in drum
4. Scored drum
5. Bent backing plate
6. Incorrect lining
7. Cheap, low-quality lining
8. Worn lining (can also cause grinding noise)

CHART 1 TROUBLESHOOTING CHART FOR DRUM BRAKES (Continued)

Chatter Noise

1. Bent or out-of-round drum
2. Riveted lining on shoe working loose
3. Hard spots in drum braking surface
4. Loose backing plate
5. Too little clearance at lining toe and heel

Knocking, Clicking, or Snapping Noise

1. Lathe tool marks left in newly turned drum
2. Weak hold-down spring
3. Bent hold-down spring pin
4. Cracked drum
5. Lack of lubrication at backing plate bosses, abutments, blocks, anchors, or any other points that require lubrication
6. Backing plate bosses deeply grooved.

Fade (Falling Away of Pedal Under Steady Foot Pressure)

1. Incorrect lining
2. Overheated brake drum
3. Poor lining-to-drum fit
4. Glazed lining
5. Cheap, low-quality lining

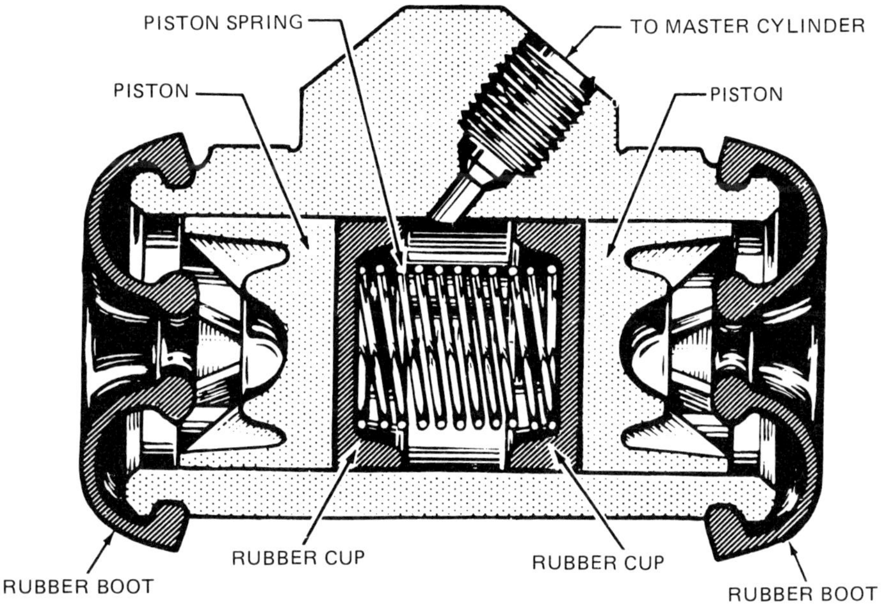

FIGURE 33
Inspect wheel cylinder for leaks by peeling back the rubber boot from the housing. If there is leakage, the wheel cylinder should be rebuilt or replaced. (Courtesy of Allied Aftermarket Division, Allied-Signal Corporation)

Mounting

Wheel cylinders are attached to the backing plate by one of the following methods:

1. Mounting bolts.
2. Spring locking retainer (Figure 34).
3. Lockplates and shims (Figure 35).

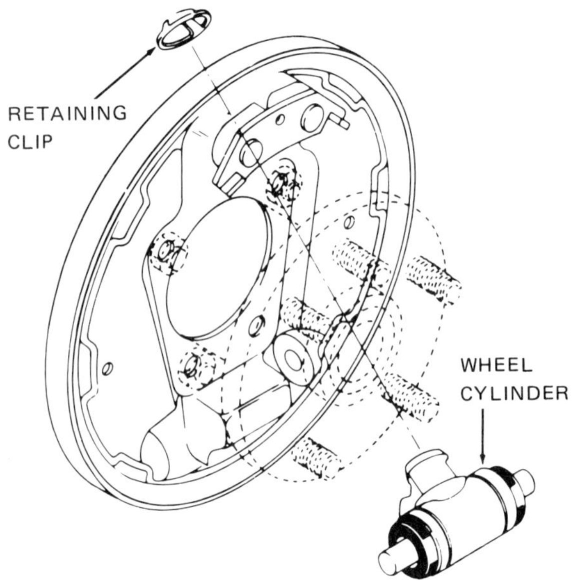

FIGURE 34
Some GM cars use a retaining clip to attach the wheel cylinder to the backing plate. When replacing this wheel cylinder, be sure both clip taps are properly engaged. Cases have been reported of the wheel cylinder falling out while driving. (Courtesy of Oldsmobile Motor Division, of General Motors Corporation)

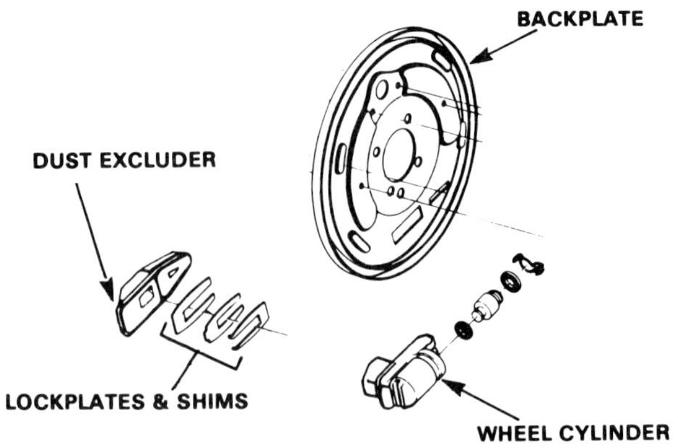

FIGURE 35
Lockplates and shims hold this wheel cylinder to the backing plate. When servicing this brake, remove the wheel cylinder and clean its mounting slot. The slot should be lubricated with an antiseize compound along with the cylinder body. Once installed, the wheel cylinder should move freely by hand. (Courtesy of Allied Aftermarket Division, Allied-Signal Corporation)

The wheel cylinders on many brakes can be taken apart without removing the cylinder from the backing plate.

On some brakes, the wheel cylinder is mounted in a recess in the backing plate, or a piston stop is welded to the backing plate (Figure 36). When these wheel cylinders are being rebuilt, they must be removed from the backing plates before they can be taken apart.

SHOP TIP

"Break" the bleeder screw loose before disconnecting a recessed wheel cylinder in order to hone it. Some bleeder screws are very tight. It is difficult to hold a wheel cylinder in order to loosen the bleeder screw after the cylinder is removed from the backing plate.

Cleaning and Inspection

After the protective rubber boots have been pulled off the cylinder, the internal parts should slide out easily (Figure 37). If necessary, parts can be tapped out with a wooden dowel or blown out by applying compressed air to the fluid inlet port. If the parts cannot be removed easily, severe damage is indicated and the cylinder must be replaced.

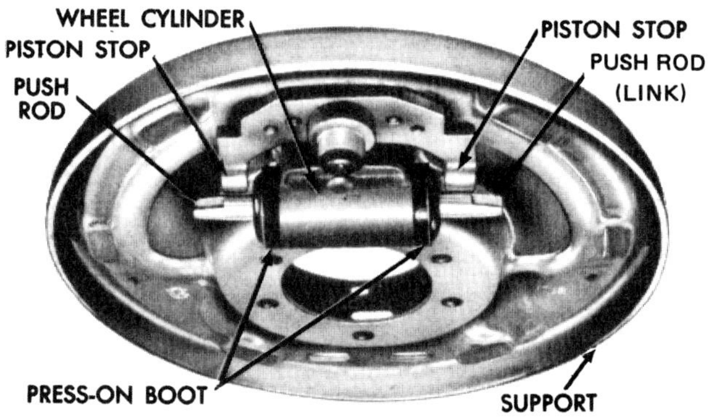

FIGURE 36
Wheel cylinder piston stops. (Courtesy of Chrysler Corporation)

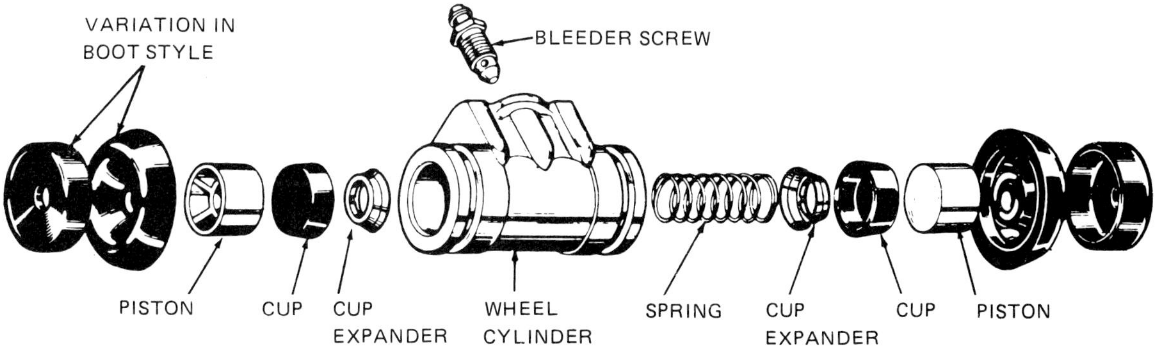

FIGURE 37
Typical wheel cylinder parts. (Courtesy of Ford Parts and Service Division)

CAUTION: Always wear eye protection when using compressed air.

Clean the cylinder and internal parts in denatured alcohol or brake parts cleaner. Inspect the cylinder bore for scoring, pitting, rusting, and corrosion. If any of these conditions are found, a new wheel cylinder should be installed.

Rebuilding

If the wheel cylinder bore has minor imperfections it can usually be honed lightly and rebuilt.

CAUTION: Do not rebuild aluminum cylinders. Honing will remove the wear-resistant anodized coating on the cylinder wall.

Crocus cloth or a cylinder hone may be used to remove light roughness, scratches, deposits, or discoloration from the bore (Figure 38). Hone lightly and use brake fluid as a cutting lubricant. If the bore does not clean up easily, replace the cylinder.

After honing, flush the inside of the cylinder with brake fluid and then wipe dry with a lint-free cloth. Make sure that all grit is removed and the bleeder screw and inlet passages are clean and open.

The clearance between the cylinder bore wall and the pistons must be checked after honing. If a narrow (1/8″ wide) .005″ feeler gauge can be inserted between the bore wall and a new piston, replace the wheel cylinder. See Figures 39 and 40.

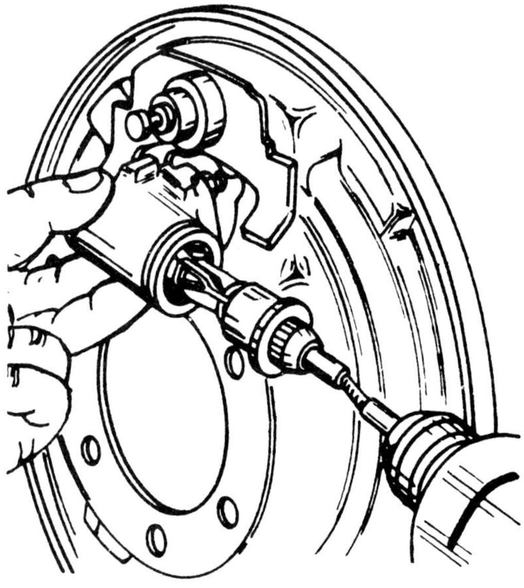

FIGURE 38
Honing a wheel cylinder. If it is honed on the car as shown, bits of abrasive material can collect in the brake line and in the bleeder screw cavity. These particles must be flushed out before assembling the cylinder. (Courtesy of EIS Division, Parker Hannifin Corporation)

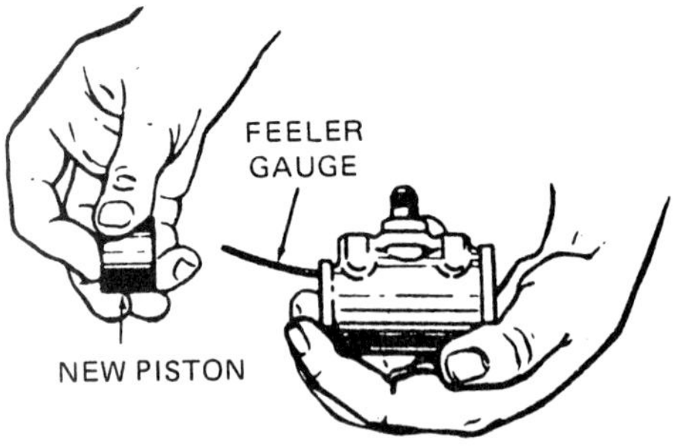

FIGURE 39
Measuring wheel cylinder clearance.

Assemble the cylinder with new internal parts, making sure that the cylinder wall is wet with brake fluid or assembly lubricant (Figure 41). Insert the cups and pistons from each end. *Never slide them through, or damage to the cup lips will occur. Cup lips are always faced inward.* Remount the cylinder (if it was removed) and reassemble the brake shoes.

FIGURE 41
Brake cylinder assembly lubricant.

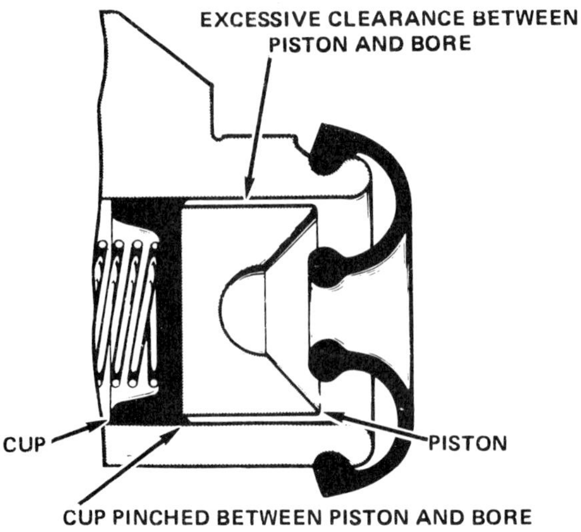

FIGURE 40
Piston drag may occur if piston-to-bore clearance is greater than .005". This can cause rapid cup wear and sluggish piston return when the brakes are released.

BRAKE SHOE APPLICATION

The brake shoe is the backbone of a drum brake assembly (Figure 42). The shoe must support the lining and carry it into the drum when the brakes are applied.

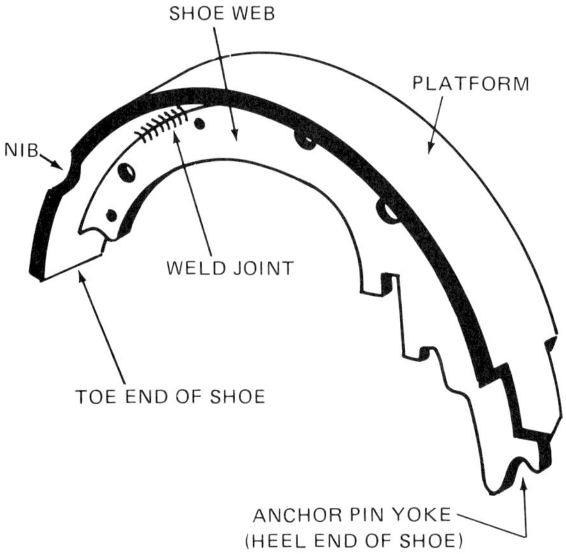

FIGURE 42
Various sections of a secondary brake shoe.

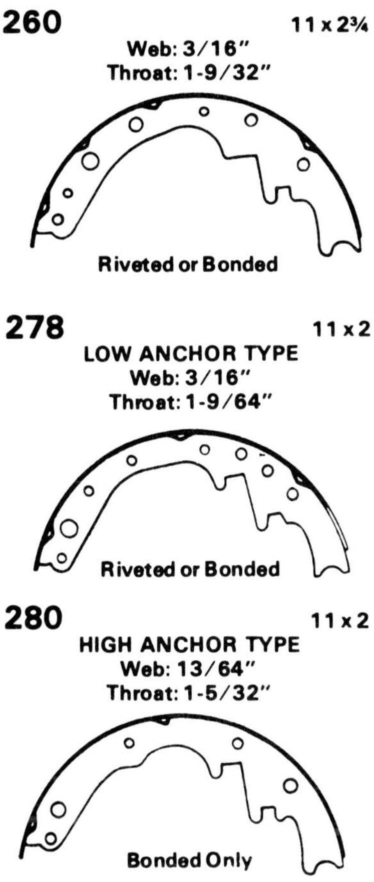

FIGURE 43
Sample brake shoe listings from an FMSI catalog.

Shoes can vary in size and appearance. There can be differences in the anchor, shape of web cut-outs, web thickness, and position of reinforcements. Some vehicles require shoes made of high-strength steel. These shoes are usually coded with a letter symbol stamped on the shoe web.

Correct brake shoe applications can be determined by referring to a Friction Materials Standards Institute (FMSI) catalog. Figure 43 shows some sample listings from this publication.

Lining Edgebranding

The edgebrand on brake lining has three parts. The first identifies the manufacturer and is normally a letter code. The next identifies the material, and the last part is the friction code for the listed materials. Refer to Figure 44.

Lining Material

Semimetallic lining is standard equipment on most new cars.

COMMENT: Some new synthetic fibers are proving to be equal to semimetallic compounds and are starting to be used by car makers. These new fibers have two advantages. They have excellent frictional characteristics, and they make it possible to avoid using asbestos as a frictional material.

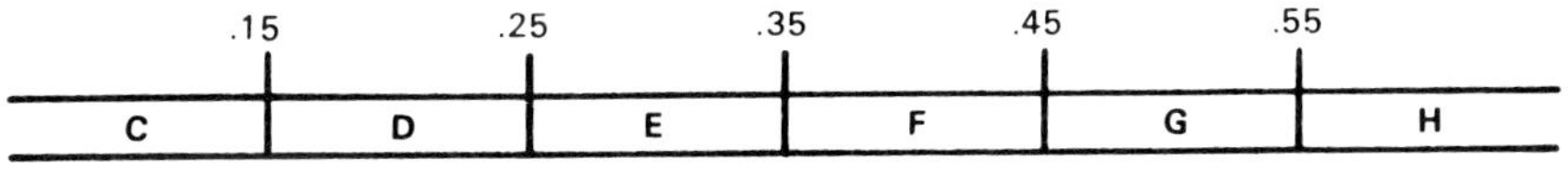

FIGURE 44
Edgebranding provides a letter code that shows the brake lining coefficient of friction (frictional characteristics). If the coefficient of friction is too high, brakes may grab and cause the wheels to slide. If it is too low, excessive pressure on the brake pedal is required to stop the car.

BRAKE SHOE INSTALLATION

Here is some general information that applies to most manufacturers' drum brakes:

1. The shoe toward the front of the vehicle is the primary shoe, and it usually has a shorter lining.

2. The hardware for the self-adjuster is usually attached to the secondary shoe. This shoe is toward the rear of the vehicle, and it usually has a longer lining.

3. The star wheel adjuster is usually toward the rear of the wheel.

4. Make sure all brake hardware is in good condition when installing new linings (Figure 45).

Reassembly (Bendix Self-Adjusting Brake)

1. Coat each backing plate boss and the anchor pin with a thin film of antiseize compound or an approved brake lubricant (Figure 46).

CAUTION: Never allow grease, oil, or brake fluid to get on the shoe linings during reassembly.

2. Install the parking brake lever on the inner side of the secondary shoe web after lubricating the pivot. Secure it with a wave washer and horseshoe clip.

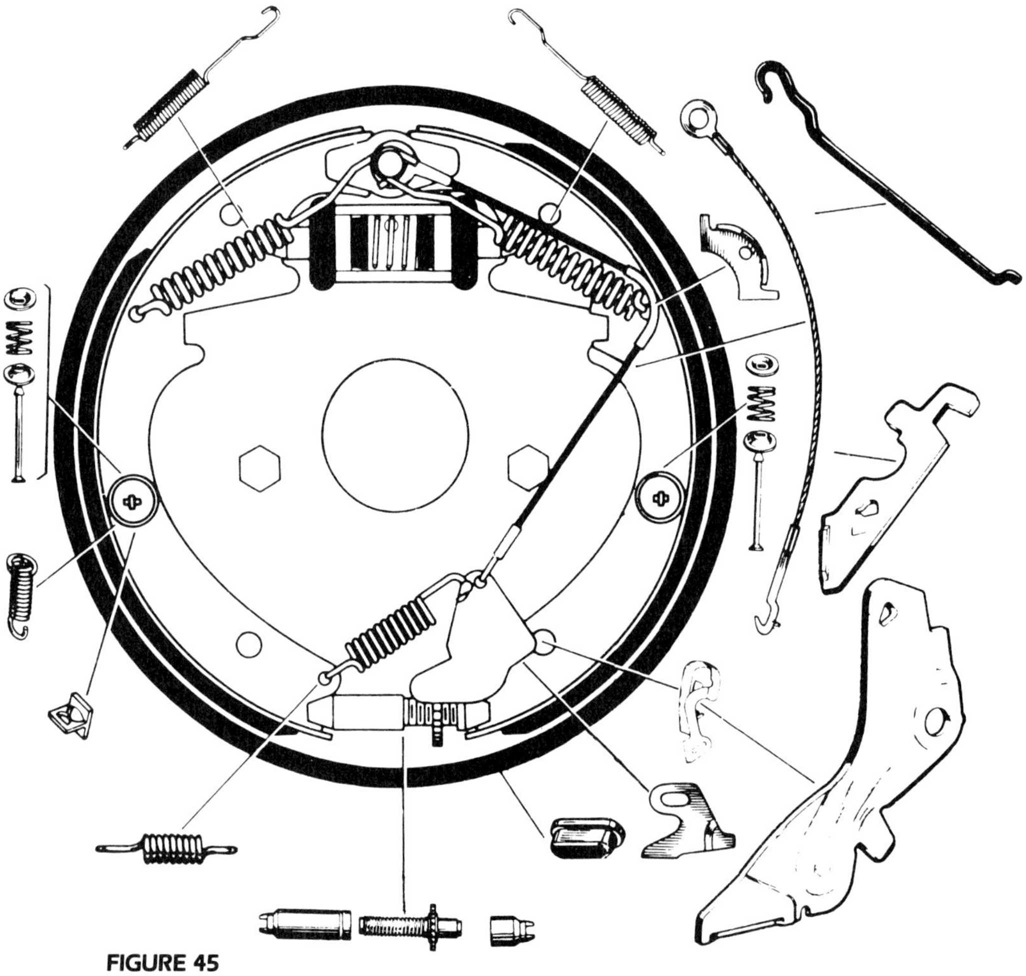

FIGURE 45
Examples of different drum brake hardware. (Courtesy of EIS Division, Parker Hannifin Corporation)

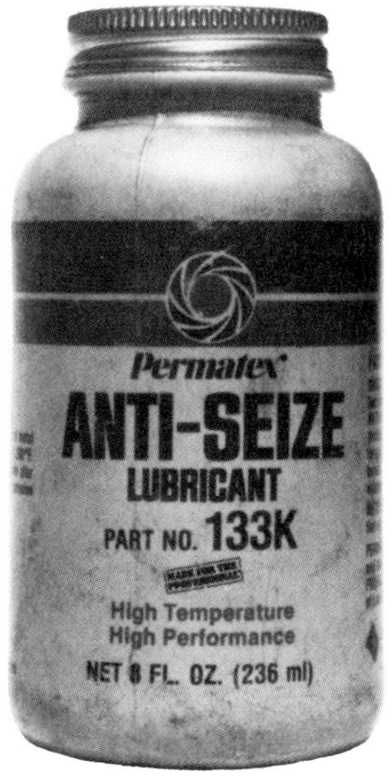

FIGURE 46
Antiseize compound is an excellent high-temperature lubricant.

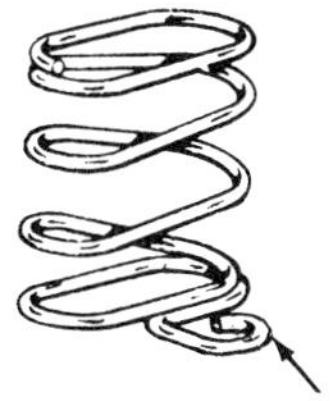

FIGURE 47
Parking brake strut antirattle spring. When installed, the spring loop must rest against the brake shoe web.

3. Engage the parking brake lever with the cable, then slide the secondary shoe against the backing plate. Position the shoe web into the push rod and against the anchor.

4. Slide the parking brake strut behind the axle flange and into the slot in the parking brake lever. Slide the antirattle spring over the free end of the strut (Figure 47).

5. Install the shoe retaining pins, retainers, and springs.

6. Slide the primary shoe into position and position the shoe web into the push rod and the free end of the strut. Install the anchor plate over the anchor, then slip the eye of the self-adjuster cable over the anchor.

7. Install the primary shoe return spring (Figure 48).

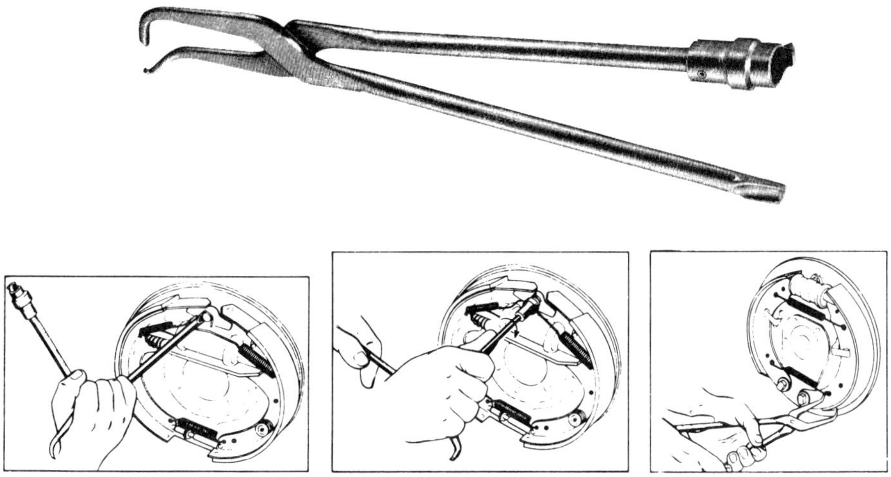

FIGURE 48
A tool for removal and installation of brake shoe return springs.

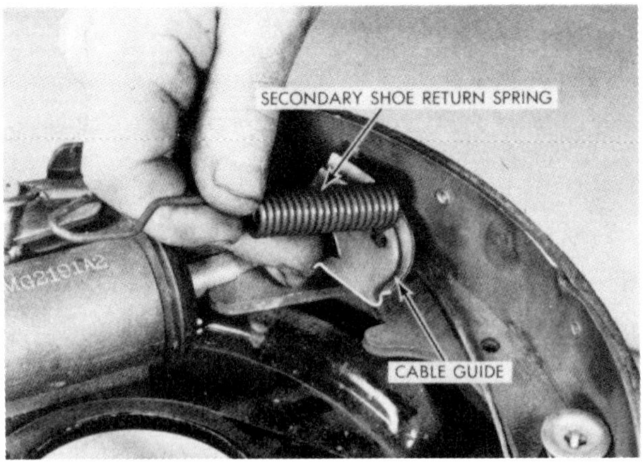

FIGURE 49
Putting secondary return spring into brake shoe web.

8. Put the secondary shoe return spring through the guide and into the web (Figure 49). Install the other end over the anchor pin.

NOTE: Be sure the cable guide remains flat against the shoe web (Figure 50). Using pliers, squeeze the ends of the spring loops (around the anchor pin) until they are parallel.

9. Be sure a thin film of lubricant is applied to the adjusting screw threads, socket, and washer. The adjusting screw must turn freely.

10. Install the adjusting screw assembly between the primary and secondary shoes.

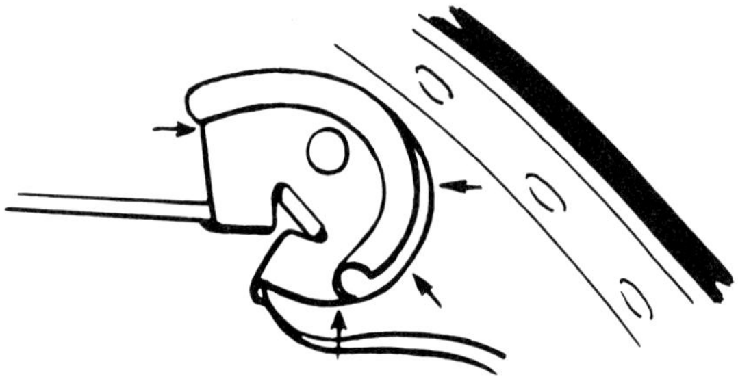

FIGURE 50
The cable guide must lay flat against the shoe web in order for the self-adjuster mechanism to work properly.

NOTE: Self-adjusting screws and related parts are not interchangeable from one side of the car to the other. On most cars, the adjusting screw on one side has right-hand threads, and the adjusting screw on the other side has left-hand threads.

11. Install the adjusting lever and spring.

12. Thread the self-adjuster cable over the guide and hook the end into the adjusting lever.

NOTE: Be sure the eye of the cable is pulled tight against the anchor pin and in a straight line with the guide.

13. Check the operation of the adjusting lever by pulling on the cable. The lever should operate the star wheel (Figure 51).

SHOP TIP

After assembling a Bendix brake with a cable self-adjusting mechanism, always check the position of the adjusting lever. The lever bottom should be parallel to (or slightly above) an imaginary line drawn through the center of the adjusting screw (Figure 52). If not, look for incorrect cable guide seating or a cable problem.

Adjusting Shoes

After the brakes are assembled, a shoe adjustment must be made. This is usually done with a brake shoe gauge. Here is the procedure:

1. Set the gauge to the inside diameter of the drum and tighten the set screw (Figure 53).

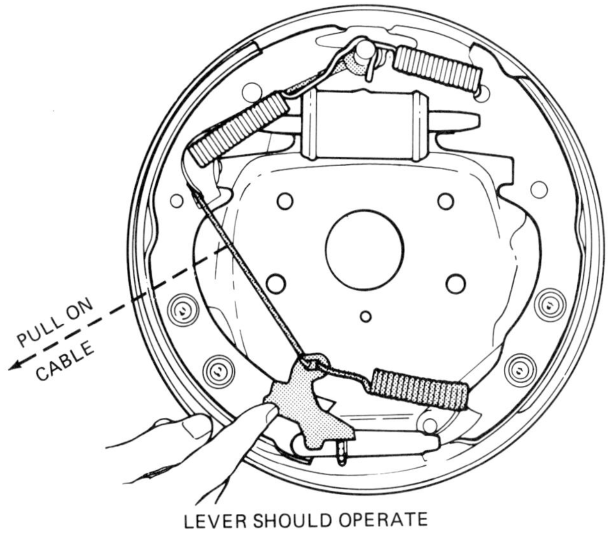

FIGURE 51
After assembling the brake, pull on the cable to see if the adjusting lever turns the star wheel.

FIGURE 53
Set brake shoe gauge to drum diameter. (Courtesy of Raybestos-Manhattan, Inc.)

NOTE: The adjustment is finalized by driving the car in reverse and applying the brakes firmly several times.

2. Position the other end of the gauge over the brake shoe (Figure 54).

3. Expand the shoe by turning the star wheel by hand until the gauge just slides over the lining. Rotate the gauge around the lining to make sure it has the proper clearance.

Some other types of adjusters used are as follows:

1. Strut (Figure 55). This assembly fits between the leading shoe and the parking

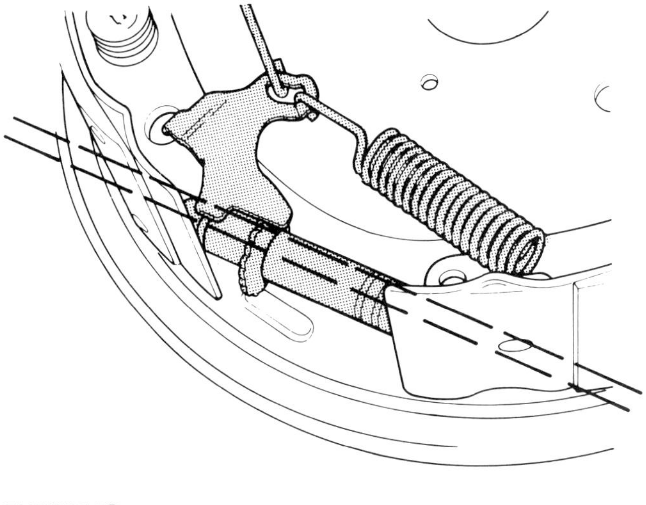

FIGURE 52
The position of the adjusting lever bottom is very important for proper operation.

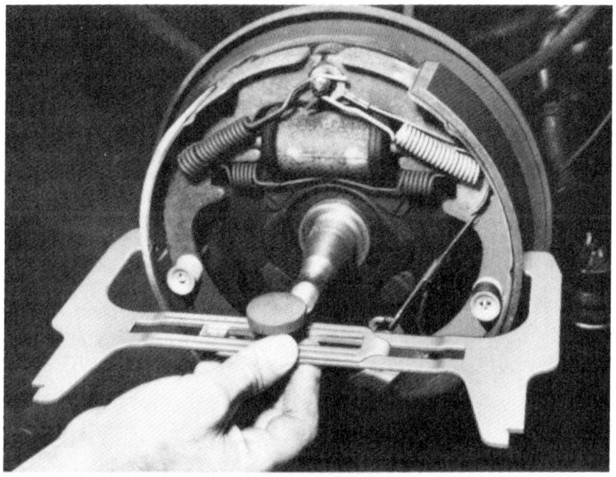

FIGURE 54
Adjust brake shoes to the brake gauge setting. This will set proper drum-to-lining clearance. **Make sure the parking brake cable is fully released**. There should be no drag when you rotate the wheel. (Courtesy of Raybestos-Manhattan, Inc.)

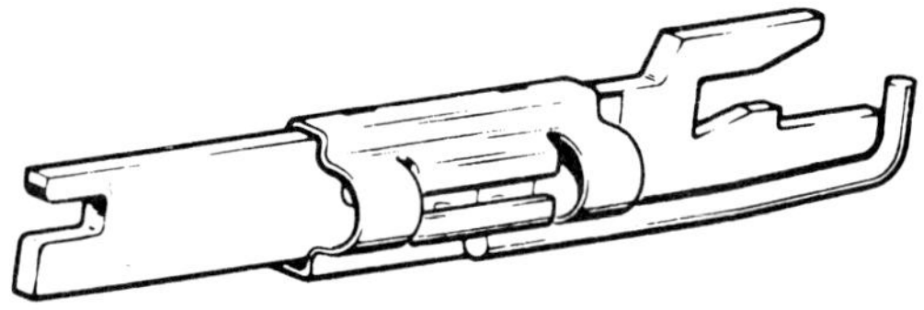

FIGURE 55
Self-adjusting strut assembly. (Courtesy of Lee Manufacturing Company, Division of Rolero, Inc.)

brake lever. Adjustment occurs with parking brake operation.

2. Ratchet (Figure 56). As the lining wears, the large ratchet pivots farther and catches the next tooth on the smaller ratchet and adjusts the shoes.

3. Wedge (Figure 57). This totally manual adjuster is used by many Japanese imports.

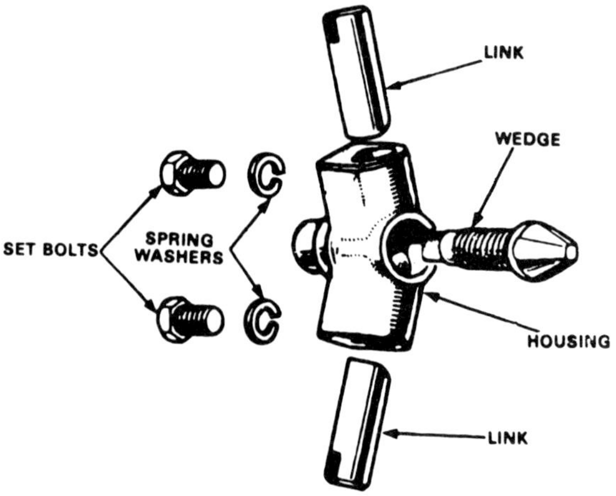

FIGURE 57
Wedge manual adjuster. (Courtesy of Allied Aftermarket Division, Allied-Signal Corporation)

4. Eccentric Cams (Figure 58). These can have either manual or automatic adjustment.

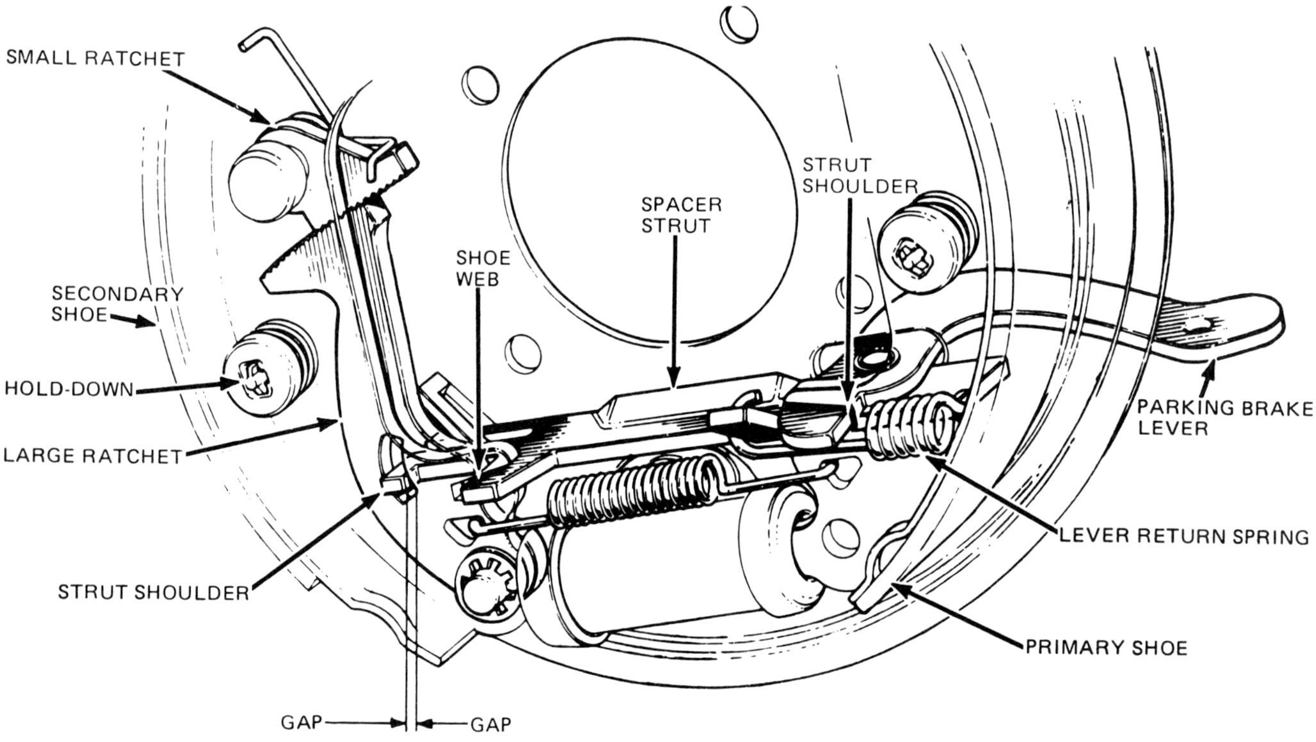

FIGURE 56
Ratchet self-adjusting mechanism. (Courtesy of Ford Parts and Service Division)

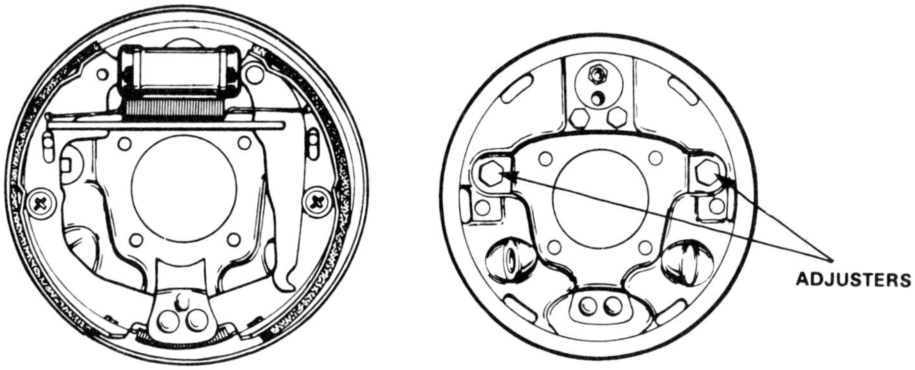

FIGURE 58
Eccentric cam adjusters. (Courtesy of Allied Aftermarket Division, Allied-Signal Corporation)

INSTALLING DRUMS

There are a number of methods of mounting brake drums (Figures 59 and 60). Some important drum installation instructions are as follows:

1. Use solvent to clean rust preventative from new drums. Also, it is good practice to use a drum lathe for a light cut on a new drum to be sure the drum is concentric and to improve upon the friction surface finish.

2. Clean any rust or corrosion from the hub flange.

3. Wheel bearings should be adjusted to specifications (if applicable).

4. Check the condition of any brake drum dampening springs. Replace if there is any damage or sign of being overheated (Figure 61).

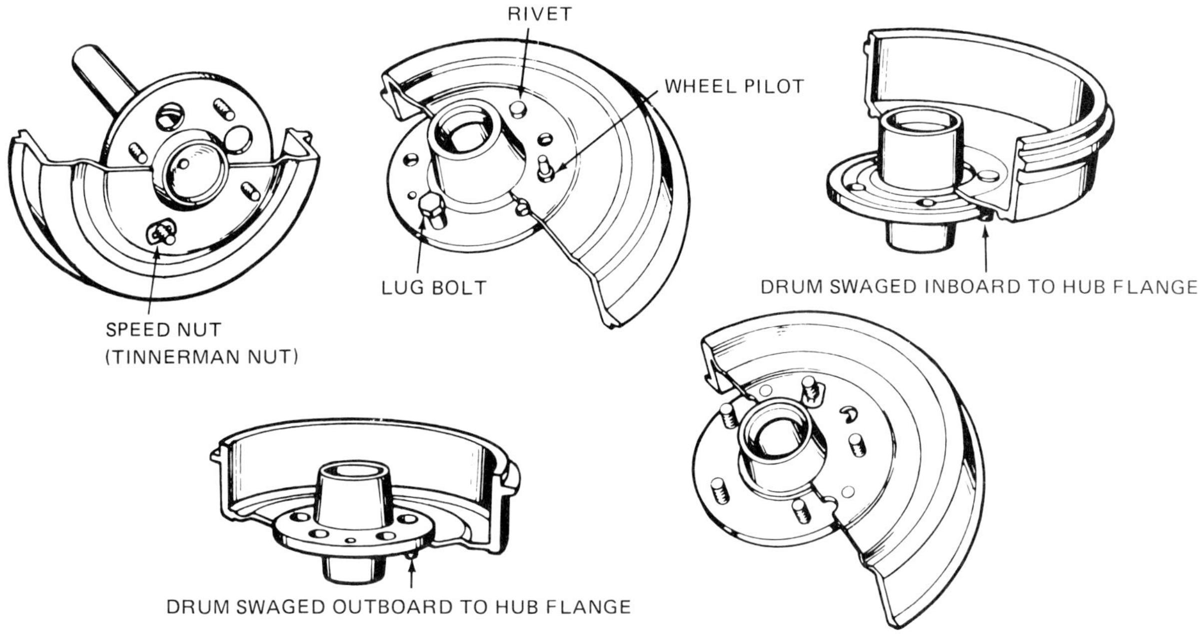

FIGURE 59
Brake drum mounting methods. (Courtesy of Allied Aftermarket Division, Allied-Signal Corporation)

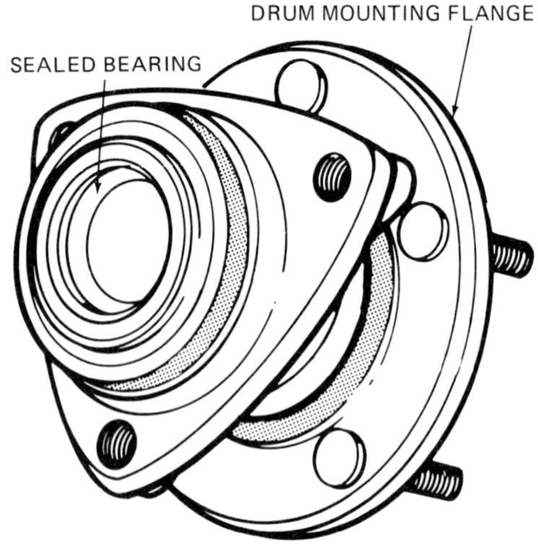

FIGURE 60
Rear drum mounting flange with sealed wheel bearing assembly. Sealed wheel bearings cannot be adjusted or repacked. (Courtesy of Oldsmobile Motor Division, General Motors Corporation)

FIGURE 62
Loose lug nuts caused this damage. This is all that was holding the right rear wheel on the car when it was driven into the shop!

INSTALLING WHEELS

Many problems can be caused by improper wheel installation. For example, the wheel may crack and/or come off if the lug nuts are left too loose (Fig. 62). If the lug nuts are tightened too much, the drum may be distorted and produce brake pedal pulsation.

Most lug nuts have right-hand threads. In some instances, nuts with left-hand threads hold the wheels on the left side. Always look for indentification markings on the studs and nuts before loosening or tightening (Figure 63).

If you use an impact wrench to install the lug nuts, use it only to run the nuts up snug. After the car is lowered to the floor, use a torque wrench to final-tighten the nuts in the proper sequence (Figure 64 [a] and [b]).

FIGURE 61
Some brake drums are equipped with damper springs. If they are weak, braking noise can result.

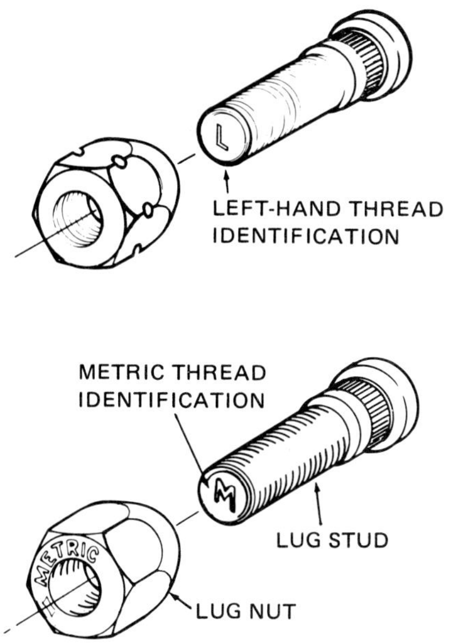

FIGURE 63
Wheel stud and lug nut markings. (Courtesy of General Motors Corporation, Product Service Training Division)

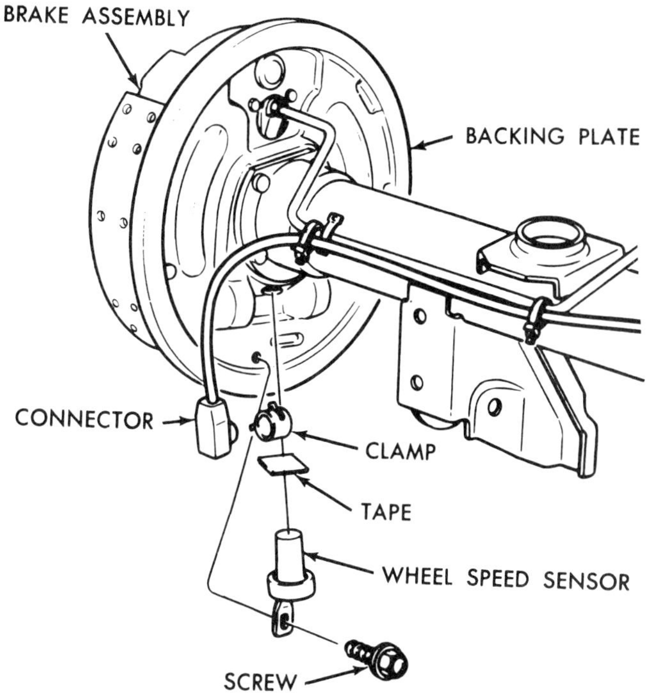

FIGURE 65
Installing wheel with an impact wrench can damage the speed sensor. (Courtesy of Cadillac Motor Division, General Motors Corporation)

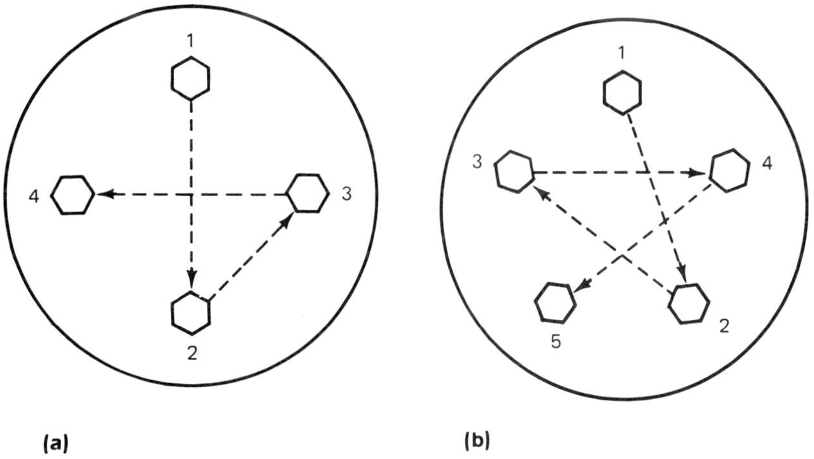

FIGURE 64
(a) Lug nut tightening sequence for a 4-bolt wheel. (b) Lug nut tightening sequence for a 5-bolt wheel.

NOTE: *Use of an impact wrench to remove or install lug nuts on a wheel that has an antiskid sensor (Figure 65) is not recommended. The sensor can be damaged or connections loosened from the vibration.*

As a final procedure, check and adjust the air pressure in all four tires. Many problems are caused by improper tire pressure.

TRUE CASE HISTORY

Vehicle	Ford Escort
Engine	Four-cylinder, 98 cubic inches
Mileage	6,750
Problem	Vibration (roughness) during braking.
Cause	The wheel mounting surface of the brake drum was not flat (Figure 66).
Cure	Replacing the brake drum.

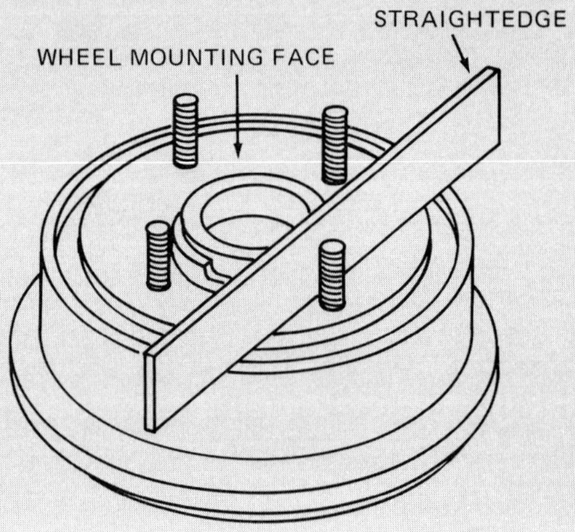

FIGURE 66
Checking the wheel mounting surface to see whether or not it has warped.

STUDY QUESTIONS

1. The rear brakes grab (lock up) on a vehicle.

   Technician A says that this could be caused by brake fluid on the linings.

   Technician B says that this could be caused by out-of-round drums.

   Who is right?
   a. A only **c.** Both A and B
   b. B only **d.** Neither A nor B

2. A heavy drag is felt while rotating a rear wheel that has drum brakes.

   Technician A says that this could be caused by a broken shoe return spring.

   Technician B says that this could be caused by grooves in the backing plate bosses.

   Who is right?
   a. A only **c.** Both A and B
   b. B only **d.** Neither A nor B

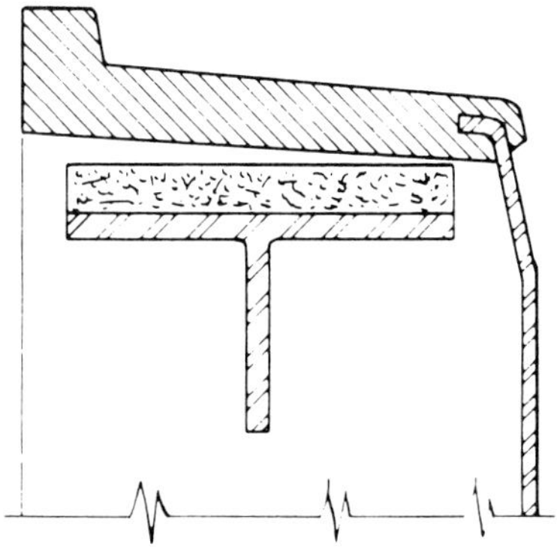

3. The fit between a drum and a shoe is shown above.

   Technician A says that this fit can cause hard pedal.

   Technician B says that this fit shows a bell-mouthed condition.

   Who is right?
   a. A only **c.** Both A and B
   b. B only **d.** Neither A nor B

4. The rubber belt wrapped around a brake drum during machining on the lathe is to
   a. Cool the drum
   b. Deflect flying metal chips
   c. Prevent drum expansion
   d. Silence vibration

5. **Technician A** says that weak brake shoe hold-down springs can cause rapid lining wear.

   Technician B says that weak brake shoe return springs can cause rapid lining wear.

   Who is right?
   a. A only **c.** Both A and B
   b. B only **d.** Neither A nor B

6. **Technician A** says to use compressed air to clean loose dirt from a backing plate.

   Technician B says to use a broom to clean loose dirt from a backing plate.

   Who is right?
   a. A only **c.** Both A and B
   b. B only **d.** Neither A nor B

7. A wheel cylinder needs to be installed on a backing plate that has piston stops.

   Technician A says to remove the piston stops before installing the cylinder.

   Technician B says to put on the dust boots after installing the cylinder.

   Who is right?
   a. A only **c.** Both A and B
   b. B only **d.** Neither A nor B

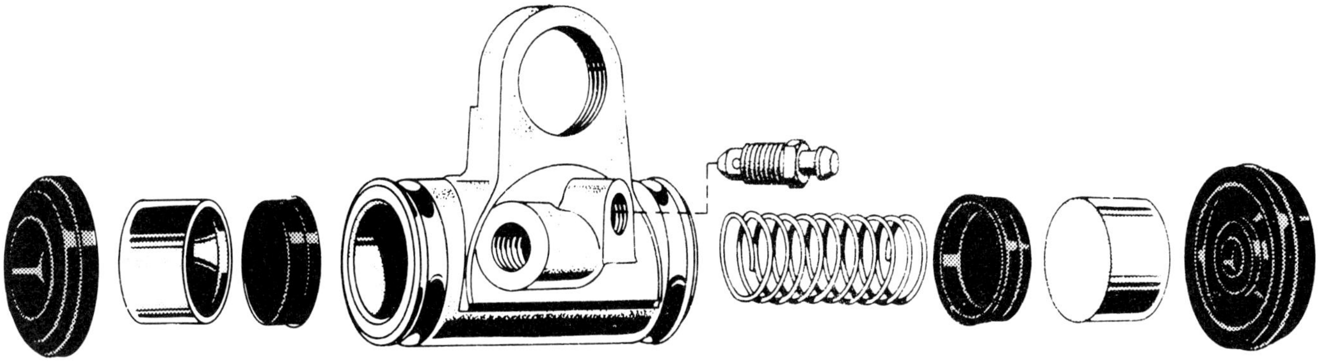

8. The above picture shows the layout of a wheel cylinder before reassembly. Which part is in the wrong position?
   a. Expander
   b. Cup
   c. Boot
   d. Piston

9. A Bendix self-adjusting brake is being assembled.

Technician A says to lubricate the anchor pin.

Technician B says to lubricate the retracting spring coils.

Who is right?
a. A only
b. B only
c. Both A and B
d. Neither A nor B

10. The edgebrand on a brake shoe lining is "R/M 4924-1 FE."

Technician A says that the "R/M" identifies the friction material.

Technician B says that the "FE" is the friction code.

Who is right?
a. A only
b. B only
c. Both A and B
d. Neither A nor B

11. Brake shoes are being replaced on a car.

Technician A says to lubricate the backing plate ledges (bosses).

Technician B says to groove the backing plate ledges (bosses) to match the shoe arc.

Who is right?
a. A only
b. B only
c. Both A and B
d. Neither A nor B

12. A shop does a brake job on a car that has Bendix self-adjusting drum brakes. Several weeks later, the car comes back with a low pedal.

Technician A says that this could be caused by glazed linings.

Technician B says that this could be caused by installing the star wheels on the wrong side.

Who is right?
a. A only
b. B only
c. Both A and B
d. Neither A nor B

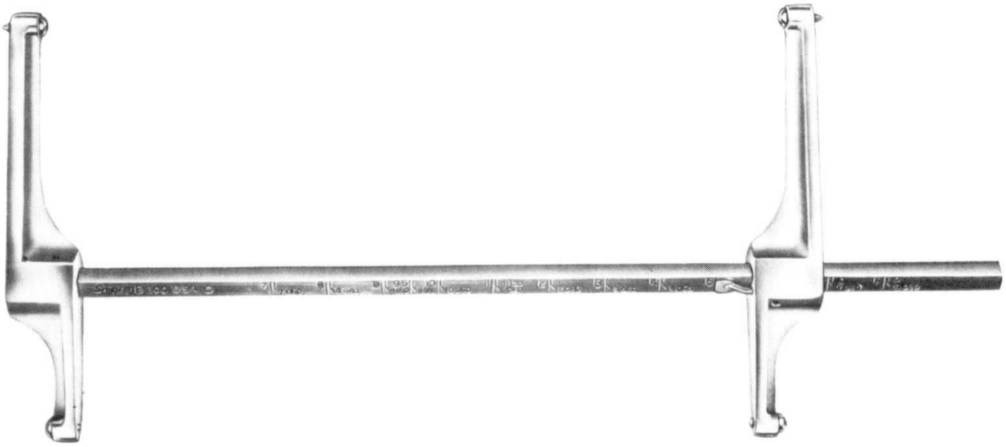

13. **Technician A** says that the above tool sets lining-to-drum clearance.

Technician B says that the above tool checks brake pedal height.

Who is right?
a. A only
b. B only
c. Both A and B
d. Neither A nor B

14. **Technician A** says that lug nuts should be final-tightened with an impact wrench.

    Technician B says that lug nuts should be final-tightened in a certain order.

    Who is right?
    a. A only
    b. B only
    c. Both A and B
    d. Neither A nor B

15. The above wheel has cracks running from stud hole to stud hole.

    Technician A says this could have been caused by loose studs.

    Technician B says this could have been caused by loose lug nuts.

    Who is right?
    a. A only
    b. B only
    c. Both A and B
    d. Neither A nor B

Section C

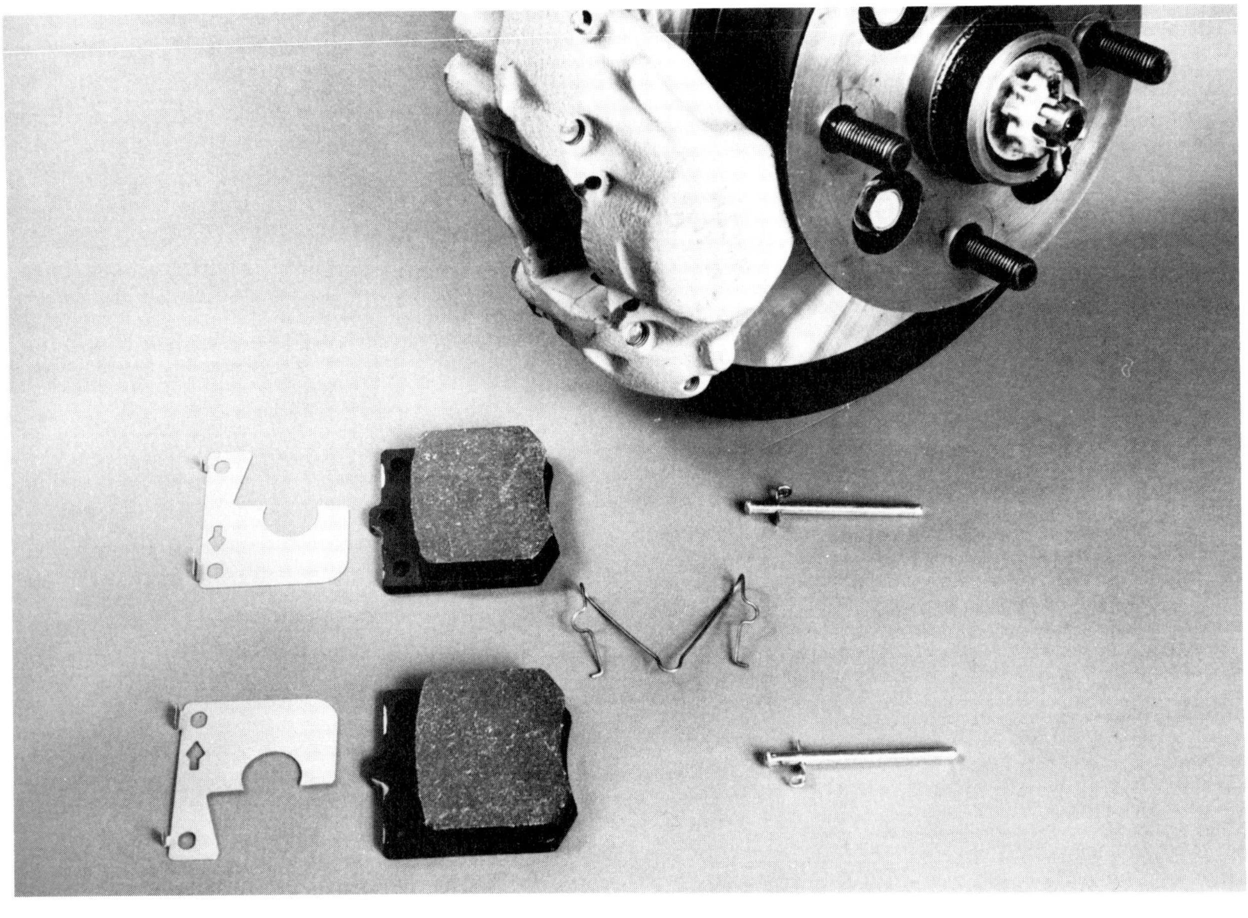

DISC BRAKE DIAGNOSIS AND REPAIR

After studying this section, and with hands-on practice on a "live" vehicle, you should be able to perform the following tasks:

- Diagnose poor stopping, pulling, or dragging caused by problems in the disc brake hydraulic assembly; determine needed repairs.
- Diagnose poor stopping, noise, pulling, grabbing, dragging, or pedal pulsation caused by problems in the disc brake mechanical assembly; determine needed repairs.
- Remove caliper assembly from mountings; clean and inspect for leaks and damage to caliper housing.
- Clean and inspect caliper mountings and slides for wear and damage.
- Remove, clean, and inspect pads and retaining hardware; determine needed repairs, adjustments, and replacements.
- Disassemble and clean caliper assembly; inspect parts for wear, rust, scoring, and damage; replace all seals, boots, and any damaged or worn parts.
- Reassemble caliper.
- Clean, inspect, and measure rotor with a dial indicator and a micrometer; follow manufacturers' recommendations in determining need to machine or replace.
- Remove rotor, mount on lathe, and machine according to manufacturers' procedures and specifications.
- Install pads, calipers, and related attaching hardware in accordance with vehicle manufacturers' specifications.
- Adjust calipers with integrated parking brakes in accordance with vehicle manufacturers' recommendations.
- Fill master cylinder with recommended fluid to proper level and seat pads by depressing pedal; inspect caliper for leaks.
- Reinstall wheel, torque lug nuts, and make final checks and adjustments.

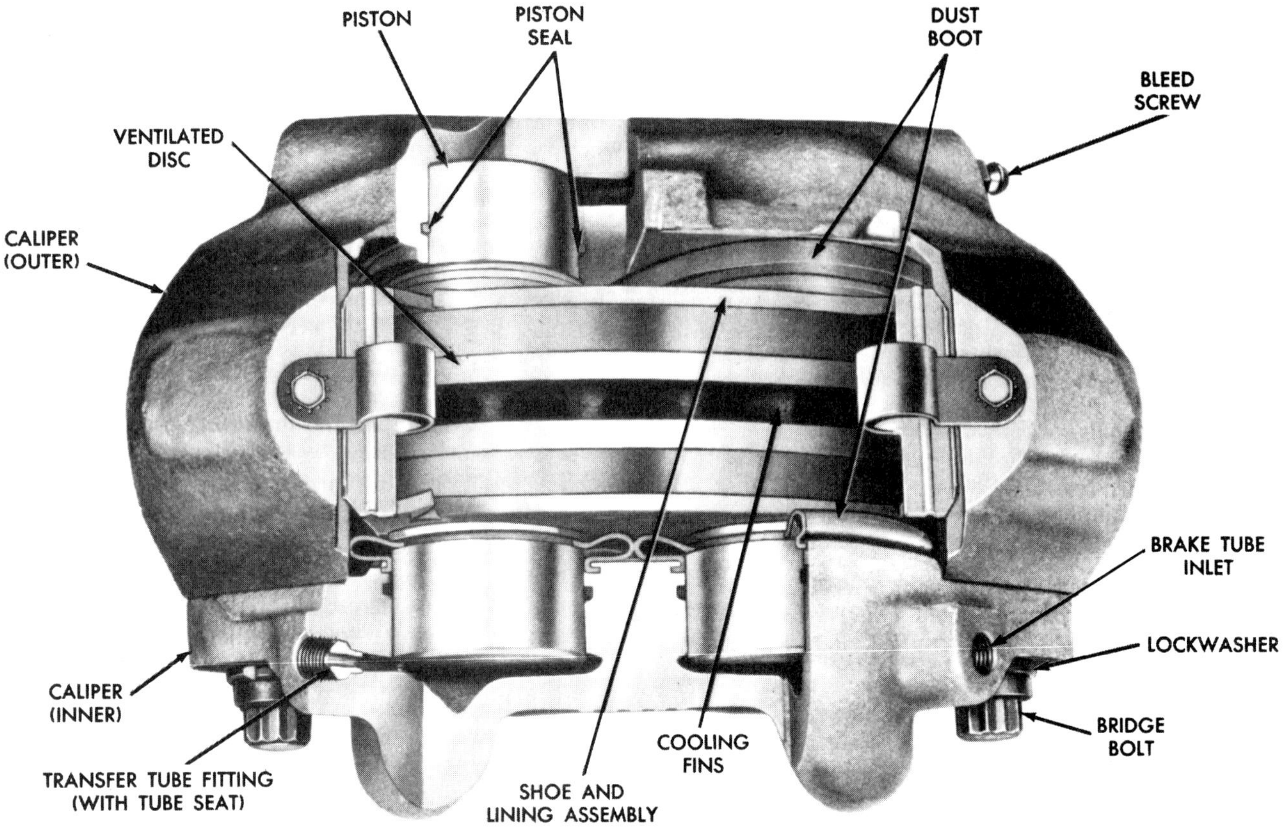

FIGURE 1
Sectional view of a four-piston fixed caliper assembly. (Courtesy of Chrysler Corporation)

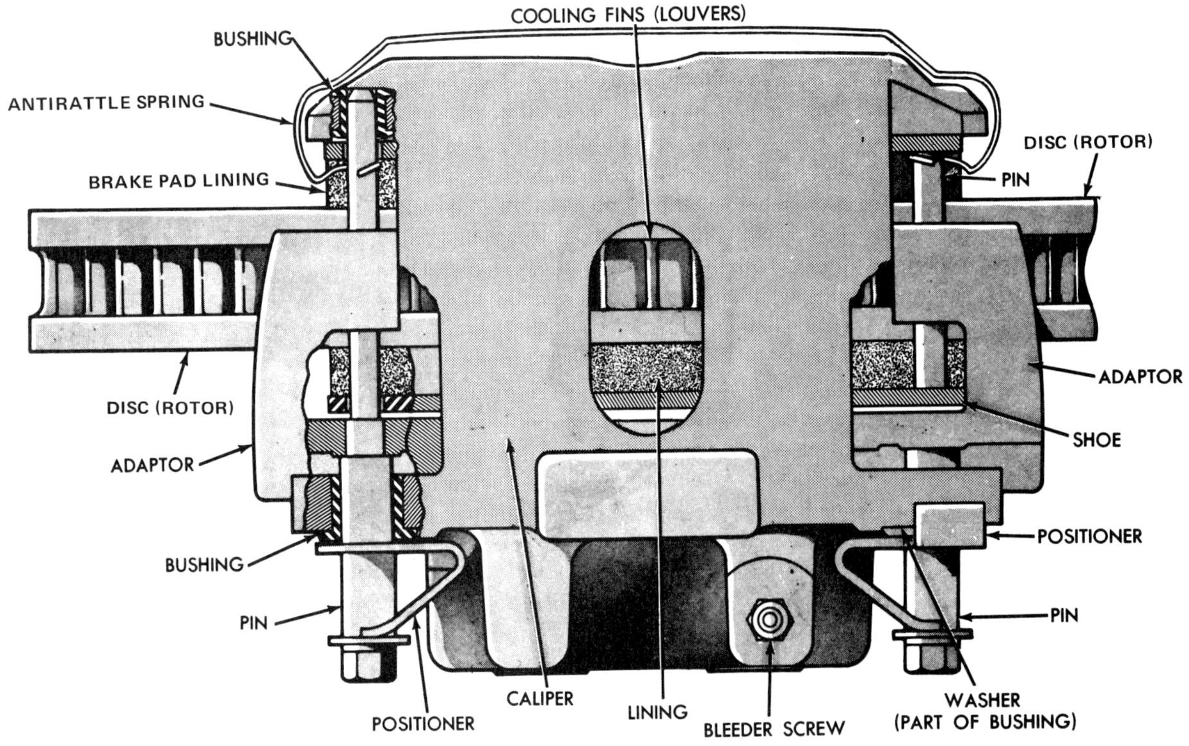

FIGURE 2
Sectional view of a single-piston floating caliper assembly. (Courtesy of Chrysler Corporation)

DISC BRAKE OPERATING PRINCIPLES

All disc brakes operate on the same basic principle. A pair of nonrotating pads pinch a rotating disc (rotor) by means of hydraulic pressure. This produces frictional force to slow down or stop the vehicle.

While there are a number of differences in automobile disc brakes, all of them fall into one of the following three basic designs:

1. Fixed caliper (Figure 1)
2. Floating caliper (Figure 2)
3. Sliding caliper (Figure 3)

Each disc brake, regardless of the design, contains a rotor and a caliper assembly. The caliper assembly includes the pads, mounting hardware, and one or more hydraulic pistons.

When the brakes are applied, the caliper piston seals are deflected a certain amount by hydraulic pressure (Figure 4). When the pressure is released, the seals relax and retract. This pulls the pistons back from the pads a few thousandths of an inch and provides running clearance between the pads and rotor. Built-in rotor runout also helps maintain running clearance.

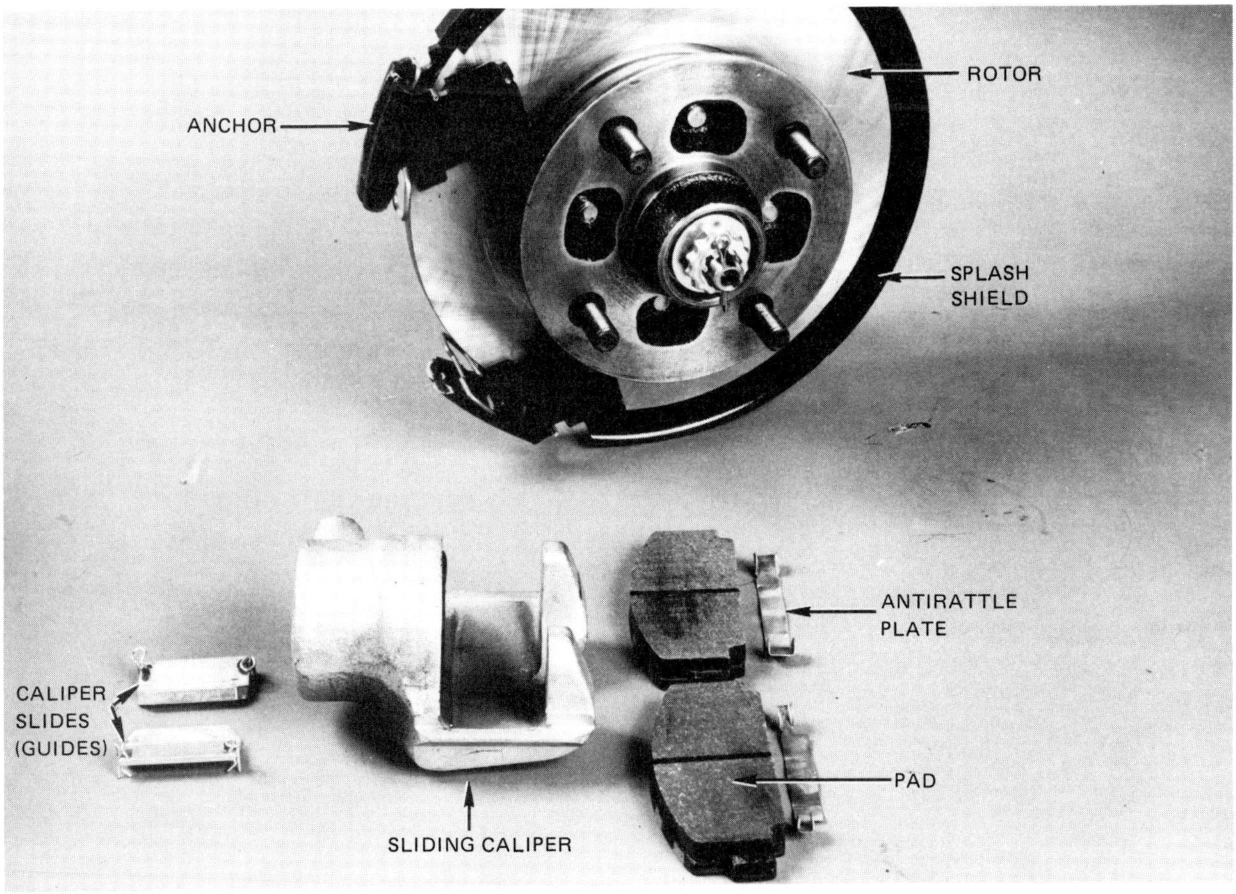

FIGURE 3

A single-piston sliding caliper assembly. The caliper slides should be coated with antiseize lubricant when installed.

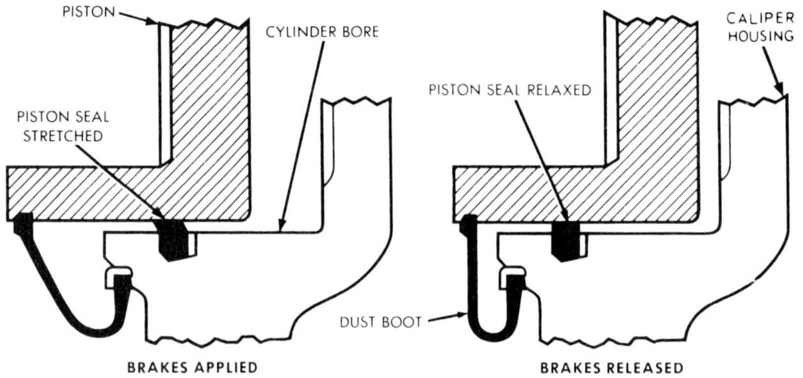

FIGURE 4
Function of piston seal. In a quick take-up caliper, the piston is retracted slightly more during release. (Courtesy of Raybestos-Manhattan, Inc.)

Self-Adjusting Action

The caliper piston seals serve as a self-adjusting mechanism and seal the brake fluid in the caliper (Figure 5). Automatic adjustment is achieved as the pads wear. This is because the pistons will slide farther outward and assume a new position relative to the seals.

On floating or sliding caliper disc brakes, the piston seals control the inboard shoes only.

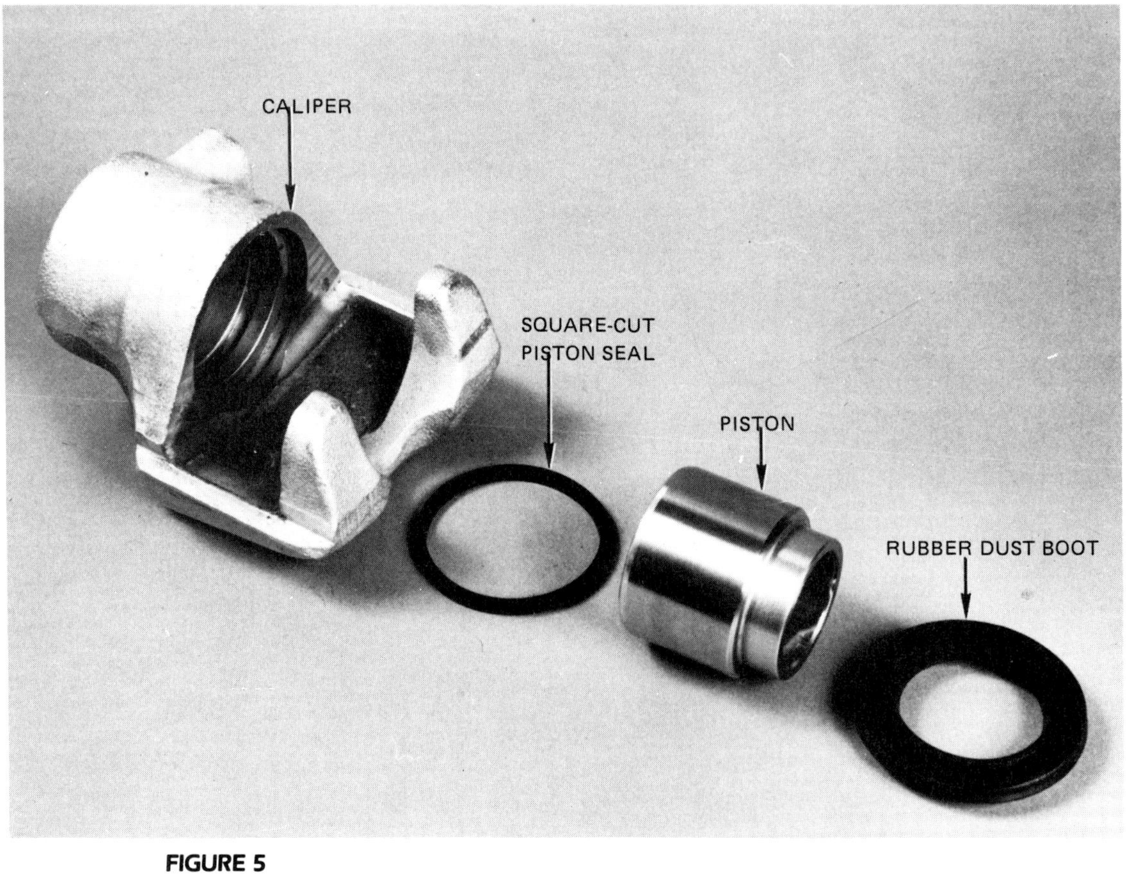

FIGURE 5
Exploded view of a single-piston sliding caliper. The square-cut seal serves as the self-adjusting mechanism.

SHOP TIP

Premature wear of a single piston caliper inboard shoe is often encountered. This can be caused by the piston seal losing its ability to return the piston properly, or the caliper binding on the caliper support. Replace the seal and boot in order to restore the vital action of the piston and inboard shoe, and make sure the caliper is not "frozen."

Hardware

The caliper and shoes are kept lined up properly by hardware parts. These parts can include bushings, sleeves, stabilizers, guide bolts, and springs (Figure 6). Any wear or deterioration of these parts can allow the caliper to go out of alignment. This can cause noise and shorten pad life.

DISC BRAKE ASSEMBLY TROUBLESHOOTING

Refer to Chart 1.

CALIPER REMOVAL

The following steps are typical for removing and servicing most floating or sliding calipers:

1. Remove the cover from the master cylinder and use a syringe to siphon approximately half of the fluid from the disc brake reservoir (Figure 7). This will prevent fluid overflow if it is necessary to push a caliper piston back into its bore. Sometimes this is done to make caliper removal easier (see step 3).

2. Remove caliper guide bolts, support keys, or mounting bolts (Figures 8 and 9).

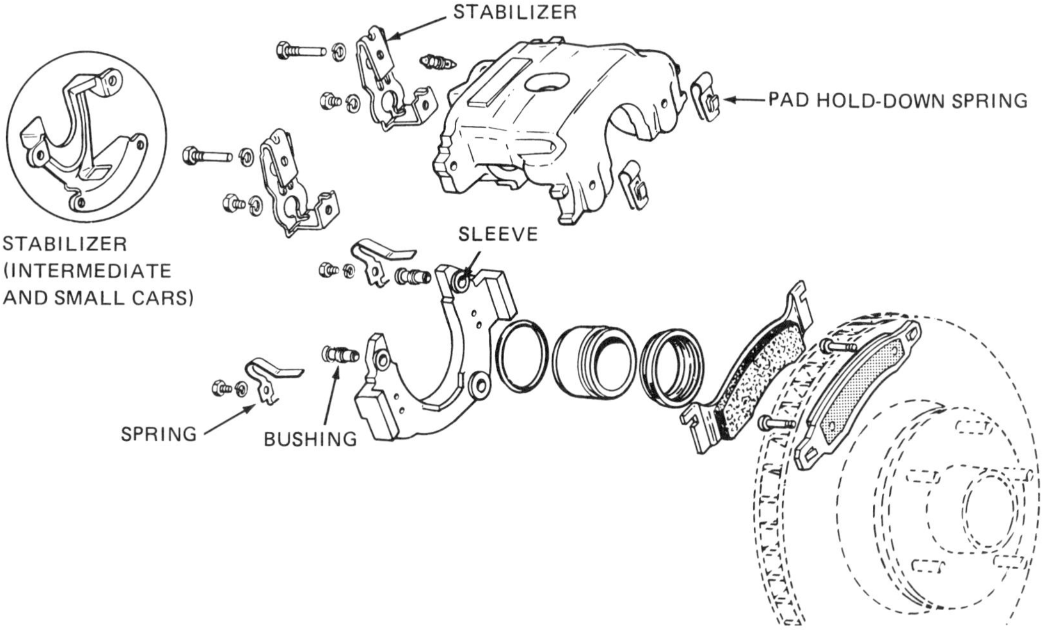

FIGURE 6

Note the caliper hardware for this floating caliper single-piston disc brake. (Courtesy of Wagner Electric Sales Corporation)

CHART 1 TROUBLESHOOTING CHART FOR DISC BRAKE ASSEMBLY PROBLEMS

Excessive Pedal Travel

1. Air or not enough fluid in the system
2. Damaged caliper piston seal
3. Left and right side calipers switched
4. Severe rotor runout
5. Wrong brake fluid
6. Rear brakes out of adjustment
7. Loose wheel bearing adjustment
8. Master cylinder or power brake unit not working properly

Excessive Pedal Effort

1. Frozen or seized pistons
2. Brake fluid, oil, or grease on pads
3. Wrong pad material
4. Glazed pads and/or rotor
5. Pads worn below specified minimum
6. Power brake unit not working properly

Pull

1. Loose caliper mounting bolts or guide pins
2. Collapsing or worn strut rod bushings
3. Caliper not in proper alignment to rotor
4. Restricted hose or line
5. Pad material not matched the same on each wheel
6. Incorrect tire pressure
7. Improper wheel alignment

Leaky Caliper

1. Scores or corrosion on surface of piston
2. Damaged or worn caliper piston seal

Dragging Pads

1. "Frozen" or sticking pistons
2. Caliper piston seals losing their "flex"
3. Power brake unit not working properly
4. Binding (sliding or floating caliper)

Pedal Pulsation

1. Too much lateral runout of rotor (can also cause braking vibration)
2. Too much parallelism of rotor (can also cause braking vibration)

Noise

1. Worn or glazed pads
2. Missing, out-of-position, or damaged hardware
3. Rust or mud build-up on edges of rotor
4. Lack of lubricant (antiseize) on metal-to-metal contact points
5. Bent splash shield permitting the rotor to scrape
6. Improper torque on wheel lugs
7. Improper torque on caliper mounting bolts
8. Low lining indicators touching rotor
9. Outer pad ears not fit snugly to caliper

Grabbing (Uneven Braking Action)

1. Calipers loose on mounting
2. Power brake unit not working properly
3. Contaminated pads (brake fluid, oil, or grease)
4. Faulty metering or proportioning valve
5. Soft or swollen caliper piston seals

3. Position a large "C" clamp over the caliper as shown in Figure 10. Screw in the handle and compress the piston into its bore.

4. Lift the caliper from the rotor. *Do not let the brake hose support the weight of the caliper.* Instead, support the caliper with a strong piece of wire.

Cleaning and Inspection

Check for piston seal leaks. These would be shown by brake fluid in and around the boot area and inboard pad. Look for any cracks or tears of the piston boots(s). *If a boot is damaged or fluid can be seen, both pieces of rubber (boot and seal) should be replaced.*

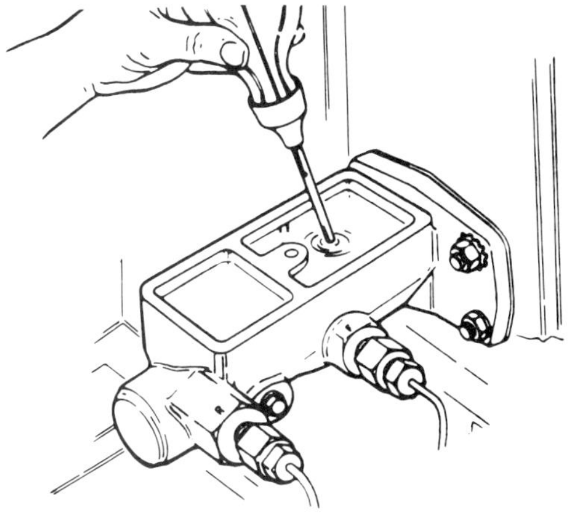

FIGURE 7
Removing brake fluid from the master cylinder before removing the caliper. This will prevent fluid from overflowing if the caliper piston has to be pushed back into its bore. (Courtesy of Ford Parts and Service Division)

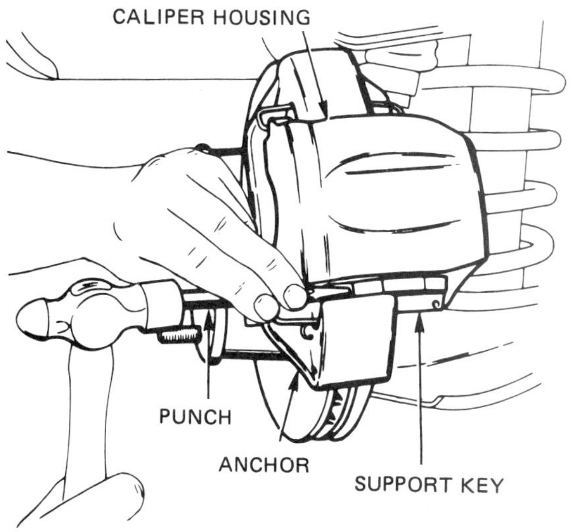

FIGURE 9
Support keys must be tapped out in order to remove this caliper. (Courtesy of Allied Aftermarket Division, Allied-Signal Corporation)

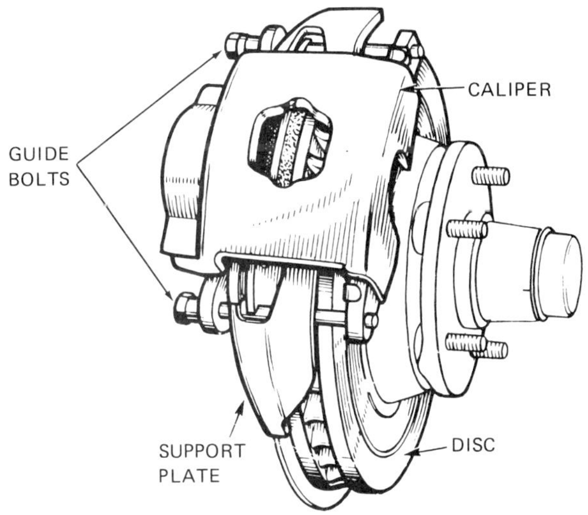

FIGURE 8
This sliding caliper can be lifted off the rotor after screwing out the guide bolts. (Courtesy of Allied Aftermarket Division, Allied-Signal Corporation)

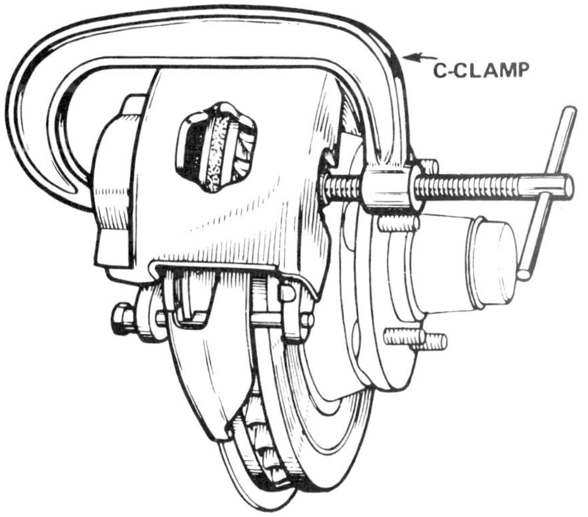

FIGURE 10
Bottoming the caliper piston in its bore to provide rotor clearance. Sometimes this is necessary before the caliper can be lifted from the rotor. (Courtesy of Allied Aftermarket Division, Allied-Signal Corporation)

Inspect the machined mating surfaces of the abutments on the caliper and anchor plate (Figure 11). If they are corroded, burred or rusty, smooth the surfaces with emery cloth and apply a thin film of antiseize lubricant when putting them back together. Replace the caliper if its ears are worn, broken, or elongated. (Figure 12).

Clean, inspect, and lubricate the caliper slides (Figure 13). Replace the caliper if slide wear or damage cannot be cleaned up with emery cloth.

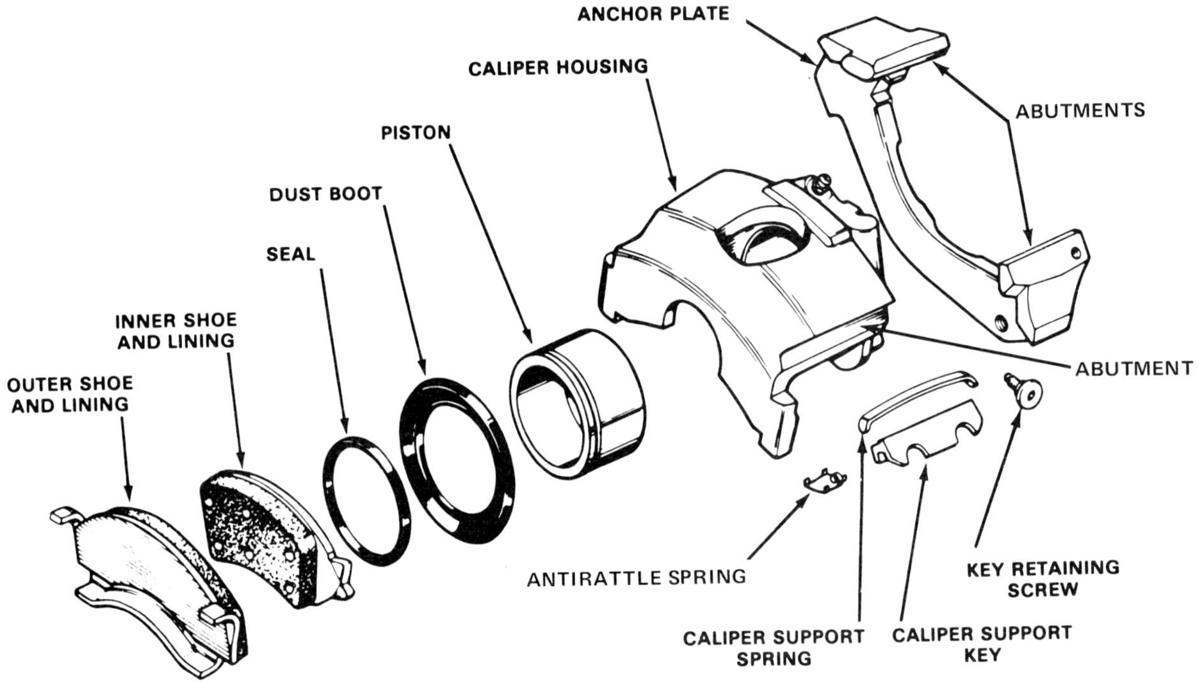

FIGURE 11
A sliding caliper that has been taken apart. The abutments on the caliper and anchor plate must be inspected before it is put back together. (Courtesy of Allied Aftermarket Division, Allied-Signal Corporation)

Pads

Check disc brake pad wear (Figure 14). If the lining has reached minimum thickness (generally 3/32″ or less), the calipers should be rebuilt.

As the pads wear, the piston position moves farther out through the piston seal to make up for pad wear. When new pads are installed, the piston is forced back through the seal. This will result in seal damage if the piston is corroded or pitted. It is good practice to rebuild the caliper so this will not happen.

Disc pad wear sensors are used on many automobiles. They are designed to let the driver know that pad lining has worn to the minimum allowable thickness. Three types of sensors are:

1. **Audible** (Figure 15). A soft metal tab is riveted to the edge of the pad backing plate. This tab is designed to rub on the top part of the rotor when pad thickness reaches a certain wear point. A high-pitched scraping noise is then produced.

2. **Visual.** A sensor probe is recessed in the pad backing plate. If the rotor wears away the face of the probe, the brake warning lamp comes on.

3. **Tactile.** This sensor creates pedal pulsation. A raised bump on the rotor face contacts a section on the pad backing plate when the pad thickness reaches a certain point.

Hardware

Caliper mounting hardware is affected by heat, weather conditions, abrasive particles, and normal wear. Always carefully check the caliper hardware when installing new disc pads. Inspect for corrosion, scoring, alignment, and wear; replace as required. Various forms of disc brake hardware are illustrated in Figure 16.

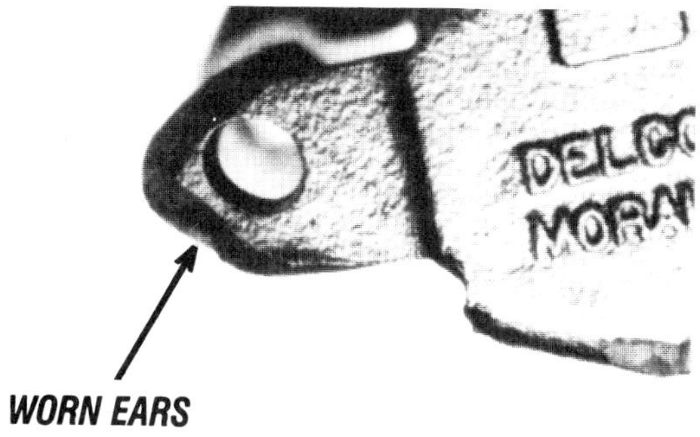

FIGURE 12
Replace caliper if its ears are damaged or worn. (Courtesy of Stuart-Western, Inc.)

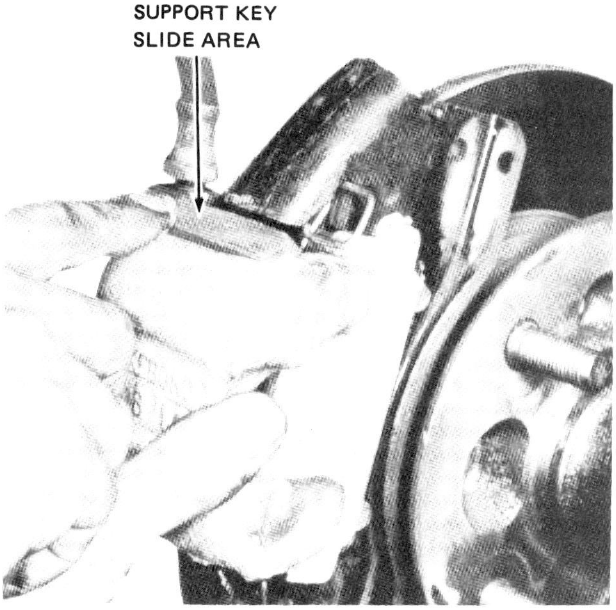

FIGURE 13
Lubricating the caliper support key slide area.

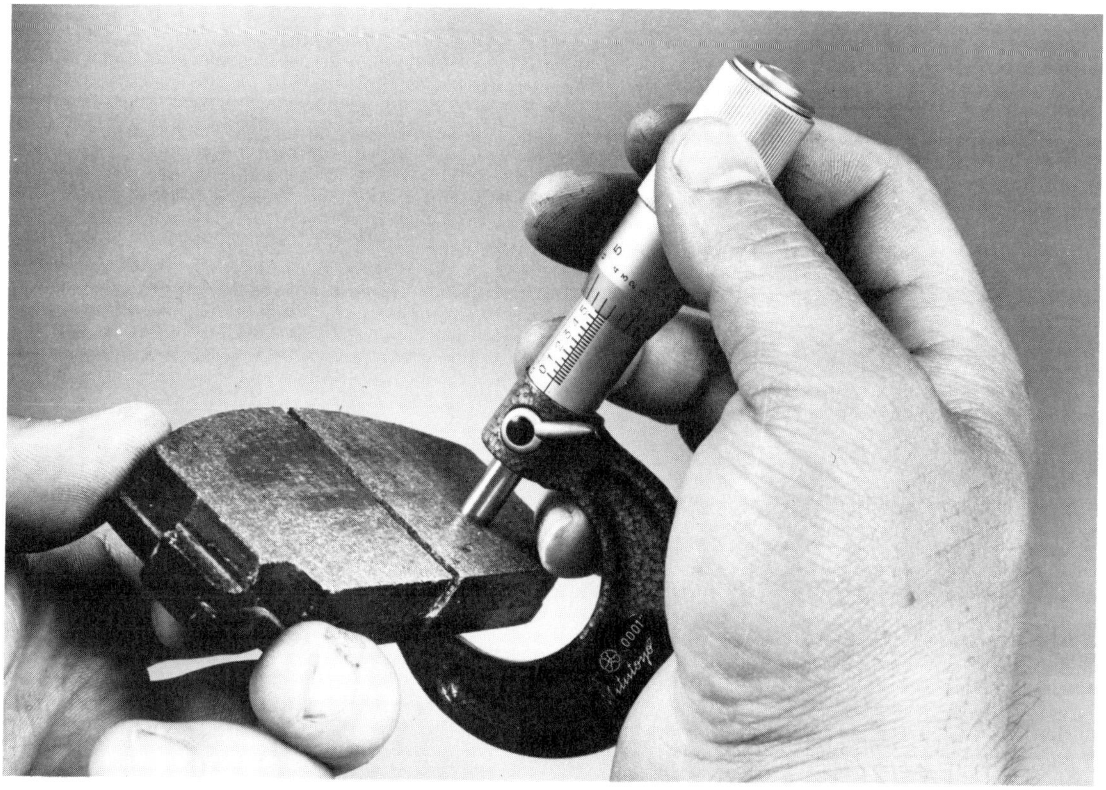

FIGURE 14
Checking disc brake pad wear. What is the reading on this micrometer?

FIGURE 15
Audible disc brake wear indicator. Do not bend the indicator when installing new pads. Some shops do this in order to extend the wear limit. (Courtesy of Pontiac Motor Division, General Motors Corporation)

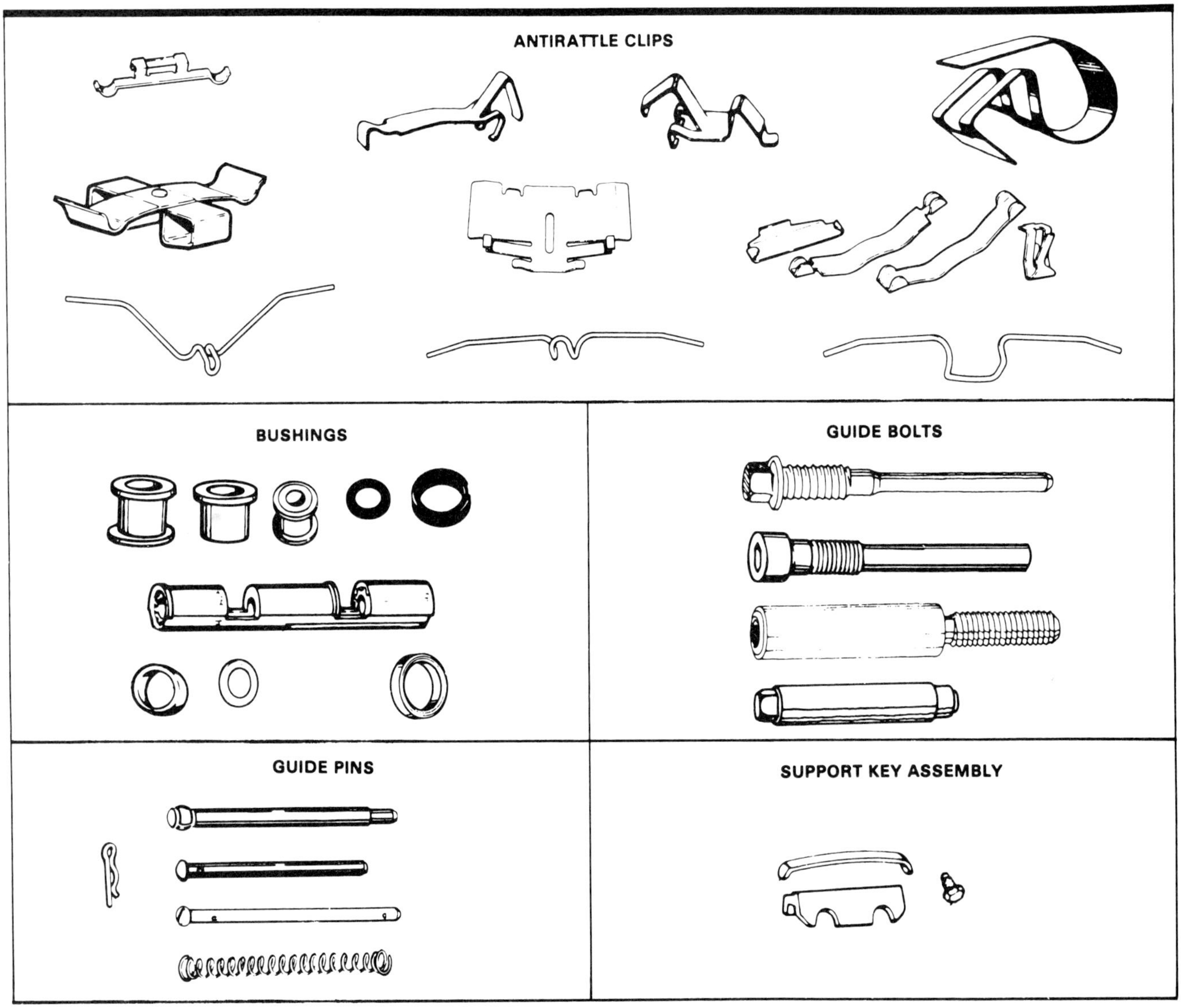

FIGURE 16
Assortment of caliper mounting hardware. (Courtesy of Allied Aftermarket Division, Allied-Signal Corporation)

SHOP TIP

A very common disc brake complaint is "squeal" when braking. This author has had very good results curing "squeal" complaints by using the following procedure:

1. Clean and deburr all metal-to-metal contact surfaces with emery cloth.

2. Coat the backside of the pads with antiseize lubricant (Figure 17).

3. Coat all anchor contact areas with antiseize lubricant (Figure 18).

4. Coat the backside of the caliper piston (Figure 19).

5. Coat the caliper slide areas (including any supports or guides).

FIGURE 18
Coating anchor abutments and antirattle plate with antiseize lubricant.

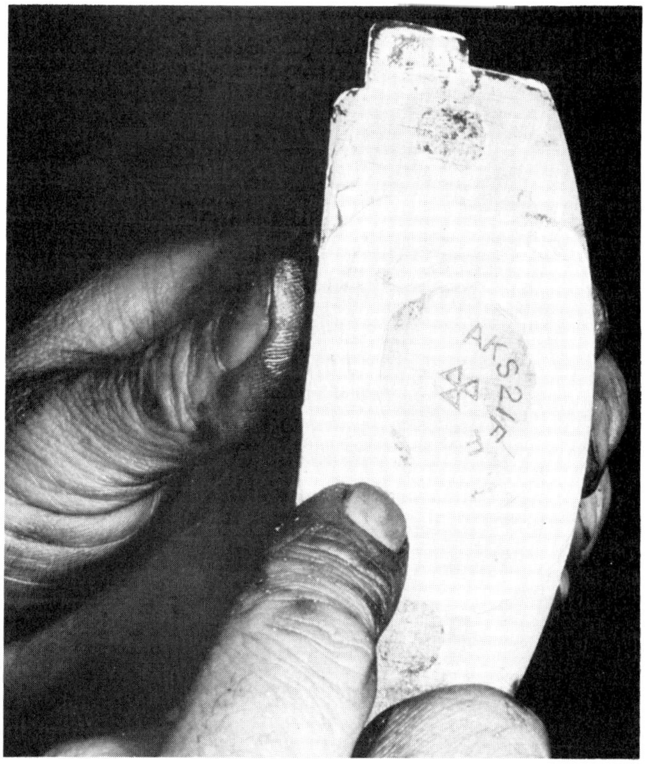

FIGURE 17
Coating backside of pad with antiseize lubricant.

FIGURE 19
Applying antiseize lubricant to piston-to-pad contact surface.

CALIPER OVERHAUL

1. Disconnect the hydraulic brake hose from the caliper.

2. Place the caliper on the bench and apply air pressure gradually to the fluid inlet port in order to force the piston out of the bore (Figure 20).

CAUTION: Use a block of wood or a stack of shop towels to catch the piston when it is forced out of the caliper bore. Keep your hands clear of the piston path.

Internal expanding pliers are sometimes used to remove pistons from caliper bores (Figure 21).

3. Remove the boot from the caliper (Figure 22).

4. Use a small, pointed wooden or plastic stick to work the piston seal out of its groove (Figure 23).

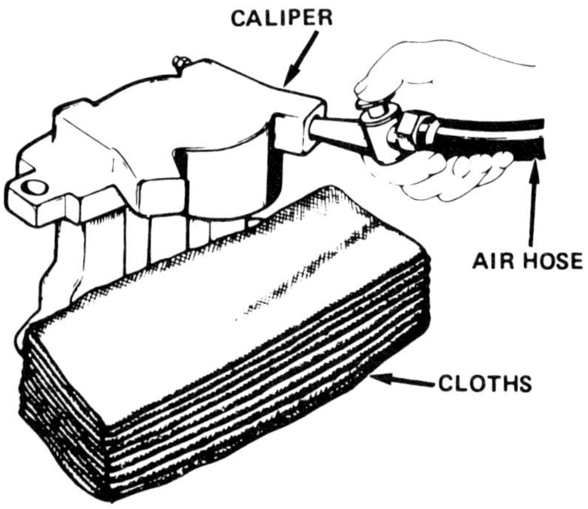

FIGURE 20
Removing caliper piston with air pressure. (Courtesy of Wagner Division, Cooper Industries, Inc.)

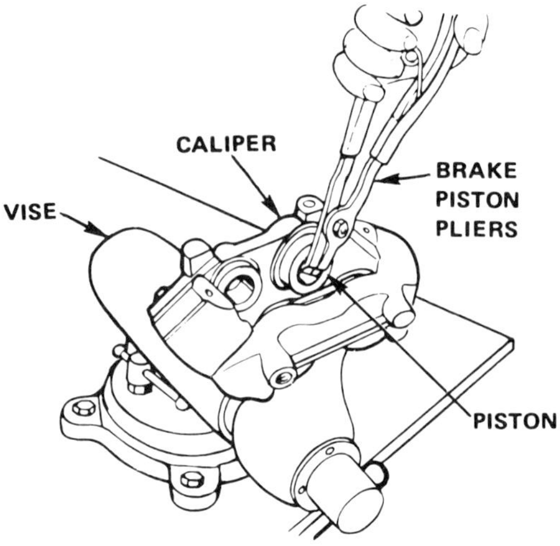

FIGURE 21
Removing caliper piston with internal expanding pliers. (Courtesy of Allied Aftermarket Division, Allied-Signal Corporation)

5. Remove and set aside caliper bushing hardware (if so equipped).

6. Remove the bleeder screw.

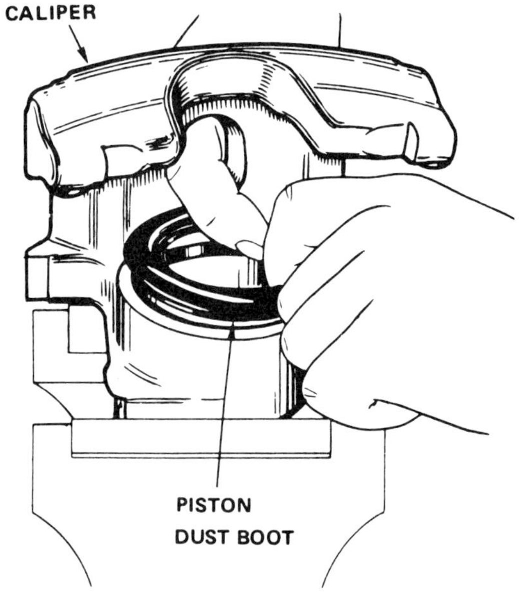

FIGURE 22
A piston dust boot being removed by hand. (Courtesy of Allied Aftermarket Division, Allied-Signal Corporation)

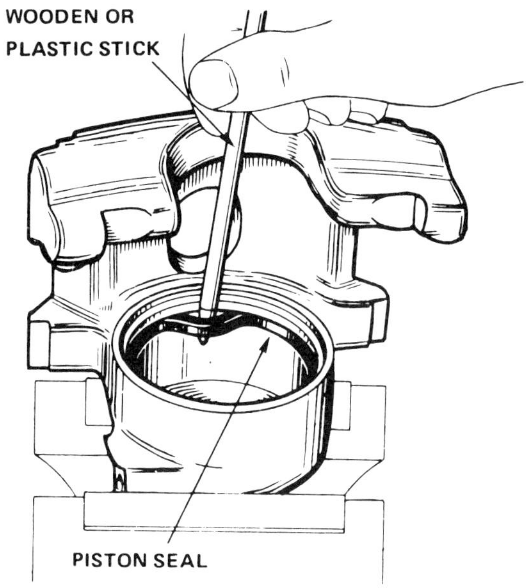

FIGURE 23
Removing a caliper piston seal. A plastic knitting needle is an excellent tool to use for this job. (Courtesy of Allied Aftermarket Division, Allied-Signal Corporation)

FIGURE 24
Brake parts cleaner. This product cleans fast and leaves no residue.

> **SHOP TIP**
> Just before disconnecting the caliper brake hose, many technicians will depress the brake pedal to push the piston out of its bore hydraulically.

Cleaning and Inspection

Clean all parts using denatured alcohol or brake parts cleaner (Figure 24).

Replace steel pistons if the plated surface is worn, rusted, pitted, scored, or corroded. Black stains on the piston are caused by the piston seal and will do no harm.

Inspect phenolic (plastic) pistons for swelling, cracks, chips, or gouges. See Figure 25 (a) and (b). Replace the pistons if any of these conditions is evident.

Inspect the piston bore. Remove any corrosion from the seal groove and boot groove with crocus cloth. Thoroughly clean the caliper after using the crocus cloth.

Assembling

1. Lubricate the new piston seal and caliper bore with clean brake fluid or assembly lubricant (often supplied with the caliper rebuild kit).

2. Install the seal in the groove in the bore. Gently work the seal around the groove, using clean fingers, until it is properly seated. *Be sure the seal is not twisted or rolled.*

3. Install the boot into the bore.

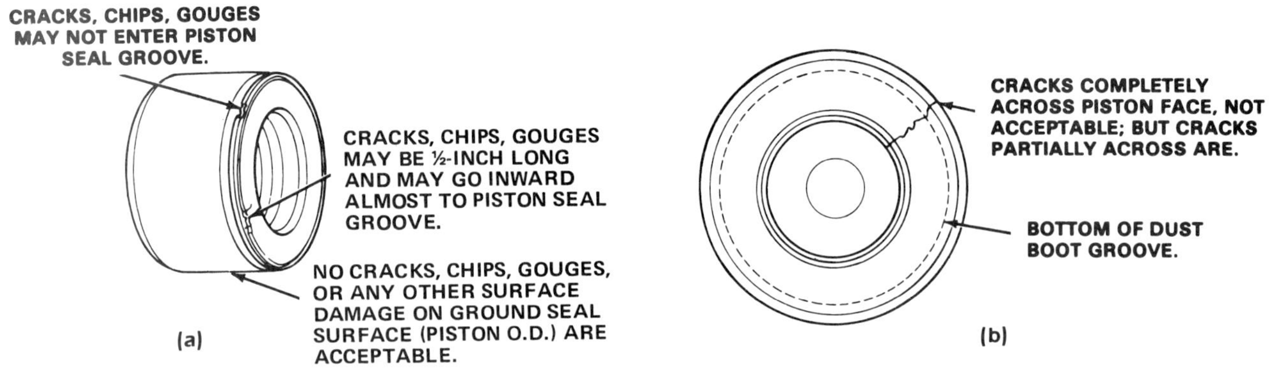

FIGURE 25
(a) Inspecting a phenolic (plastic) piston. (b) This phenolic piston must be replaced. (Courtesy of Allied Aftermarket Division, Allied-Signal Corporation)

4. Coat the piston with lubricant (as specified above). With your fingers spreading the boot, work the piston down the bore until it is bottomed.

NOTE: In some cases, it is easier to push the piston and boot into the bore as one piece (Figure 26). Then use a blunt tool or special driver to seat the boot (Figure 27).

5. Install new guide bushing hardware (if so equipped).

6. Install the bleeder screw.

7. Connect the brake hose to the caliper. It is recommended that a new copper gasket be used (if so equipped). Copper gaskets take a set and may not form a seal if reused.

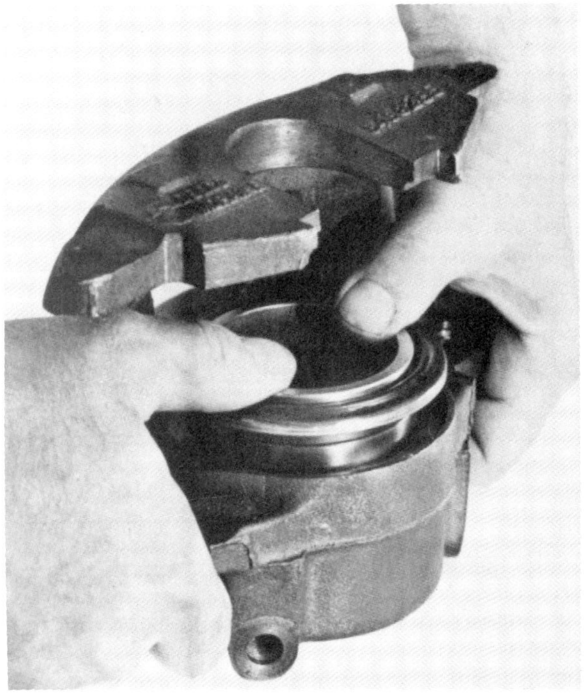

FIGURE 26
Pushing the piston and boot into the caliper bore. (Courtesy of Raybestos-Manhattan, Inc.)

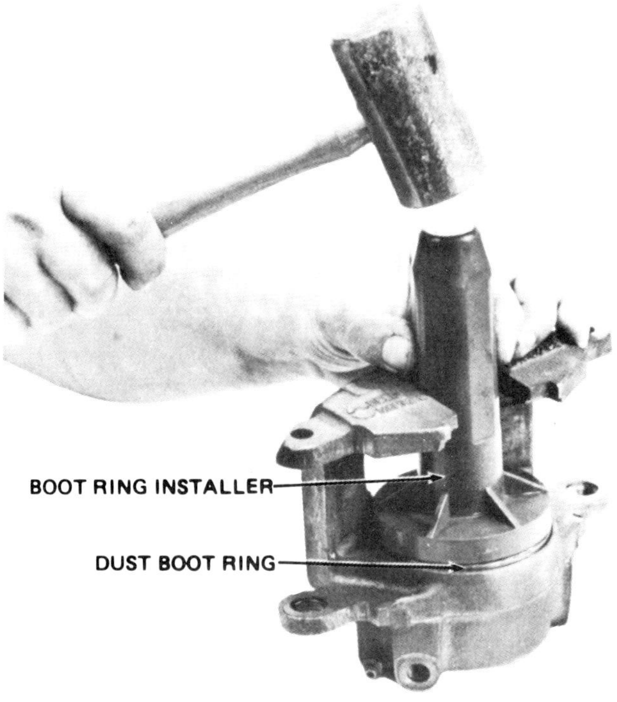

FIGURE 27
Seating the dust boot ring. (Courtesy of Raybestos-Manhattan, Inc.)

INSPECTING THE ROTOR

Look at the rotor carefully for scoring, ridges, and rust (Figure 28).

Light scoring or ridges that go all the way around the rubbing surface of the rotor at the same depth will do no harm (Figure 29). If the scoring or ridges are excessive, the rotor must be machined or replaced.

FIGURE 28
On ventilated rotors, inspect the vents for obstructions, rust, and loose balance weights. Unobstructed vents will allow for proper heat dissipation during braking. Loose weights can produce a rattle noise.

Runout

Excessive lateral runout of the brake rotor will cause a "knocking back" of the piston(s), which will create increased pedal travel and chatter when the brakes are applied.

FIGURE 29
Light scoring on the rotor braking surface will do no harm. If there is deep scoring, the rotor must be machined or replaced.

FIGURE 30
Using a dial indicator with a magnetic base to measure rotor runout.

Before runout is checked, wheel bearing end play must be eliminated by tightening the spindle nut.

CAUTION: Be sure to readjust the wheel bearings when the check is completed.

Mount a dial indicator so that the plunger contacts the rotor near the outer edge (Figure 30). Rotate the rotor by hand and measure the total indicated reading (TIR). If the TIR exceeds specifications (Figure 31), the rotor must be machined or replaced.

Parallelism

Parallelism refers to variations in rotor thickness. The two rubbing surfaces of a rotor must be parallel within specifications. If they are not, front end vibration, a pulsating brake pedal, noise, or grab can occur.

Check parallelism with a micrometer (Figure 32). Measure the thickness with a micrometer at six or more equally spaced points

BRAKE ROTOR			
MIN. THICKNESS		Variation From Parallelism	Runout T.I.R.
Machine To	Discard At		
.980	.965	.0005	.004
.980	.965	.0005	.004
.980	.965	.0005	.004
.980	.965	.0005	.004
.972	.957	.0005	.002
.830	.815	.0005	.002

FIGURE 31
Some typical examples of brake rotor specifications.

approximately 60 degrees apart, preferably at the middle of the braking surface. A rotor with too much variation in thickness will require machining or replacement.

Thickness

Rotors have a specified minimum thickness measurement (Figure 33). Limits are listed for each manufacturer's type in most repair manuals.

FIGURE 32
Rotor parallelism can be checked with a micrometer.

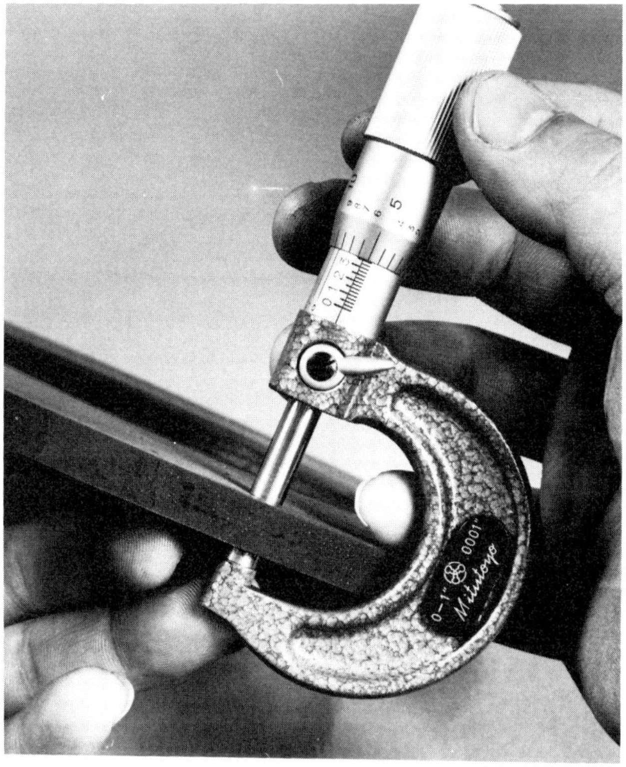

FIGURE 33
Rotors have a minimum thickness dimension. This particular rotor measures .3555" and must be replaced.

MACHINING THE ROTOR

Before machining the rotor, determine whether or not you can clean up the surface and still stay within thickness limits. Most disc brake rotors have a minimum thickness dimension cast into them. This dimension is the minimum wear thickness (often called the *discard thickness*), and not the minimum thickness after machining. The following example illustrates these rotor dimensions:

New rotor thickness (1.390″)

Minimum thickness after refinishing (1.240″)

Minimum wear or discard dimension (1.225″)

WARNING: *Never machine a rotor to the minimum wear or discard dimension. When machining, always wear eye protection.*

Mounting the Rotor for Machining

A disc brake lathe is used to machine ("turn") the rotor (Figure 34). Use whatever cones, spacers, and adapters are necessary to secure the rotor to the lathe arbor (Figure 35 [a] and [b]). A vibration damper should be used on the rotor to prevent "chatter" marks during the machining procedure (Figure 36).

FIGURE 34
Disc brake lathe. This lathe can machine both sides of the rotor at the same time. (Courtesy of Ammco Tools, Inc.)

Rotor Machining Tips

1. The best type of lathe is one that machines both sides of the rotor at the same time. This produces the least distortion during machining, because there is equal cutting pressure on both sides of the rotor.

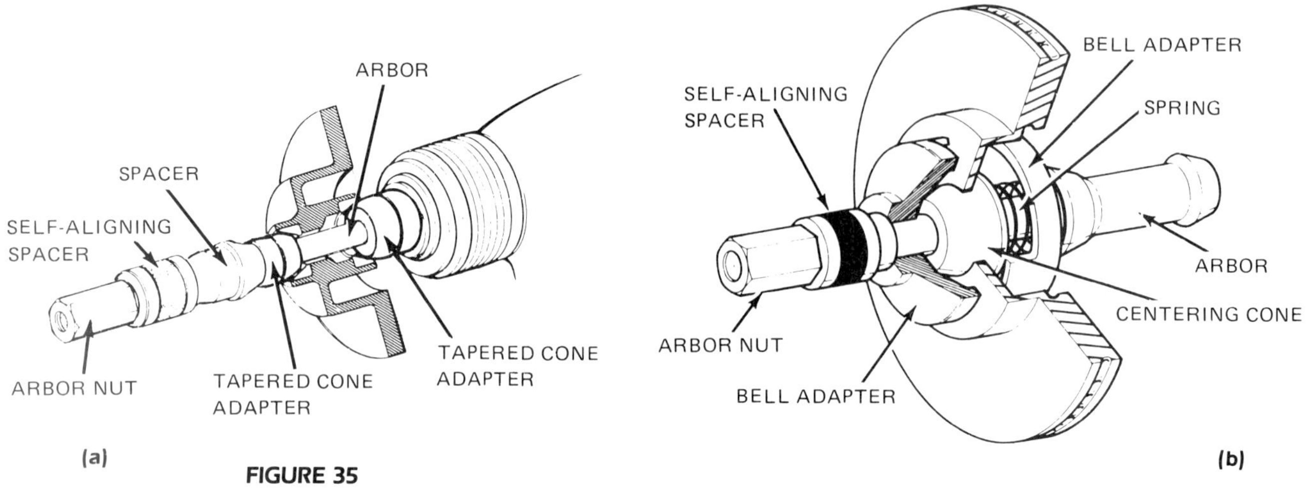

FIGURE 35
(a) A **hubbed** rotor mounting set-up. (b) A **hubless** rotor mounting set-up. (Courtesy of Ammco Tools, Inc.)

2. Always cut *both* sides of a rotor before removing it from the lathe. This will ensure that both sides of the rotor are parallel.

3. After the rotor has been machined, both outside edges are usually razor sharp. While the rotor is still revolving in the lathe, take a mill-smooth file and put a 1/32″ chamfer on these two edges. This will help prevent chipping the pads during caliper installation. Also, on some cars, it will help prevent cutting the piston boots when installing the caliper.

CAUTION: *Be careful that the file does not "catch" in the rotor.*

4. Some import and domestic manufacturers recommend that front wheel drive rotors be machined while on the vehicle (Figure 37).

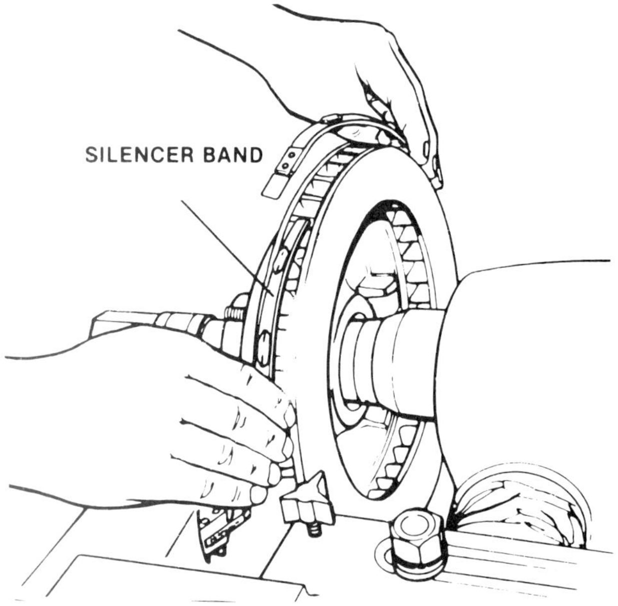

FIGURE 36
Attaching a silencer band (vibration damper) to the rotor. This will prevent "chatter" marks when machining. (Courtesy of Ammco Tools, Inc.)

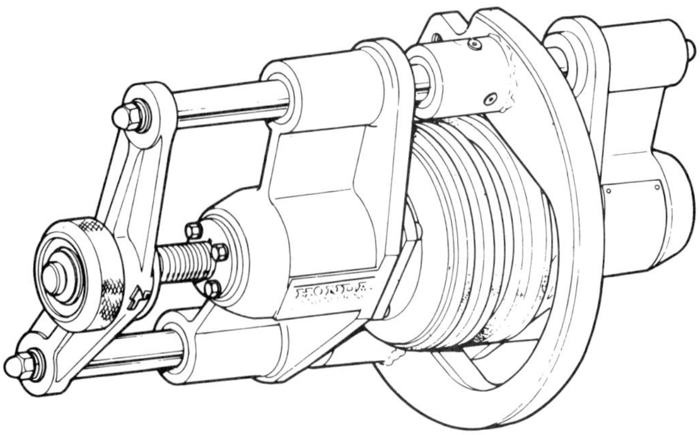

FIGURE 37
An on-the-car rotor grinder. (Courtesy of American Honda Motor Company, Inc.)

NOTE: This procedure will cure vibration problems caused by the caliper mounting surfaces being manufactured untrue to the rotor.

5. The friction surfaces of a newly refaced rotor should be given a nondirectional swirl (Figure 38). This helps to seat the pads and prevent squeal.

SHOP TIP

A rotor can be given a quick swirl finish by using a sanding pad chucked in a 1/4″ drill motor. Revolve the pad against the rotor as it rotates in the lathe.

6. Always clean all braking surfaces of the rotor after machining. Use a clean rag saturated in denatured alcohol or brake parts cleaner. This will remove the fine bits of metal left from machining.

7. With fixed caliper rotors, equal amounts of metal should be removed from both sides of the rotor.

FIGURE 38

A refaced rotor should be given a nondirectional swirl finish. This helps pad seating and can prevent "squeal."

8. When turning rotors, it is difficult to keep metal shavings out of the hub. To cure the problem, cut the bottom out of a paper cup and place it over the hub when mounting the rotor.

SHOP TIP

Steam clean the grease out of hubs before turning a rotor or drum. After machining, blow out any chips left in the hub. WEAR EYE PROTECTION!

PAD APPLICATION

Semimetallic disc brake pads were introduced as original equipment (OE) in the mid-1970s because the asbestos braking material used in automobiles until then was found to have cancer-causing properties. An added benefit of the semimetallic pads is their more favorable performance under hot braking conditions.

Related to this is the down-sizing of today's automobiles, which includes the brake system. Because of the reduced size of the disc brake rotor and the change from vented to solid rotors on most cars (especially front wheel drives), there is much more heat created at the rotor. Semimetallic pads will withstand this increased heat better than asbestos.

Semimetallic pads are not 100 percent metal. They are made of steel wool particles and different resins blended together. Semimetallics have a lower coefficient of friction than asbestos. Most asbestos pads on the market today are edgebranded with an "FF" or "FG." Most semimetallics are marked "EE." Semimetallic material is found in complete sets or in hybrid sets (the inner or outer pad is semimetallic, and the other pad is organic).

PASSENGER CARS AND TRUCKS

Year and Model		FMSI Relined Sets			
		Front	Size	Rear	Size
ALFA ROMEO					
82	Spider Veloce	734-D45+		735-D30	
82-81	GTV6	734-D45+		735-D30	
81-77	Sprint Veloce	736-D31+		7057-D127	
	Spider Veloce	734-D45+		735-D30	
82-81	Alfetta	736-D31+		7057-D127	
AMERICAN MOTORS					
Concord					
83-82	Exc. S/W	7122A-D203●		433	9 x 2
	S/W	7122A-D203●		267	10 x 1-3/4
81-80	4 Cyl.	7026-D90●		433	9 x 2
	6 Cyl.	7027-D91●		433	9 x 2
	S/W	7027-D91●		267	10 x 1-3/4
Eagle					
85-82	All	7122A-D203●		267	10 x 1-3/4

● OEM is Semi-Metallic + Semi-Metallic Pads Available

FIGURE 39
Sample brake catalog listing. (Courtesy of Stuart-Western, Inc.)

NOTE: OE semimetallic applications should be followed. The use of asbestos pads in semimetallic applications can result in short pad life. Always refer to an appropriate brake catalog if there is any question regarding correct application (Figure 39).

Some disc brake pads have a material identification code stamped on the backside (Figure 40).

Semimetallic Installation and Break-In

There are several areas that need extra attention when installing semimetallic disc brake pads.

1. The caliper hardware should be carefully inspected. Replace it if it is not in perfect condition. Failure to do so can result in a misaligned or "hanging" caliper.

2. New caliper boots and seals should be installed (especially if the car has high mileage and has never had this done).

3. When the rotors are turned, use a slow cross feed rate of .002" per revolution. Then give the rotor a nondirectional swirl finish.

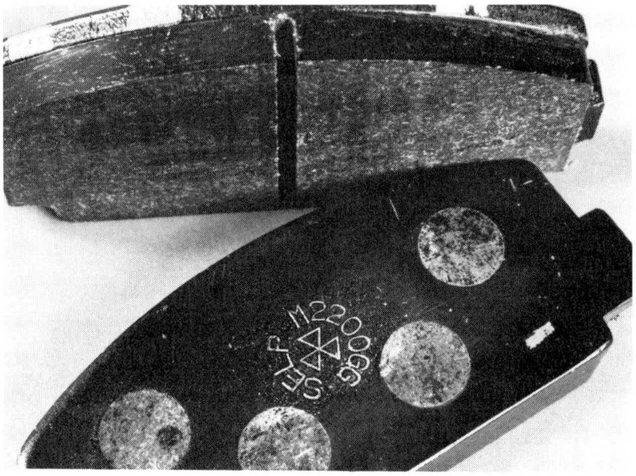

FIGURE 40
Using the wrong pad material can cause noise and/or premature wear. Note the material code.

4. After installation, road test the car, making 20-25 stops with *light pedal* application. Allow time between stops for cooling. This process will burnish the semimetallic friction surface to the rotor and achieve full surface-to-surface contact.

CAUTION: Do not make hard stops or apply extreme pressure during semimetallic break-in. This can result in glazing and noise problems.

5. When upgrading from asbestos to semimetallic pads, it is recommended that semimetallic pads be installed only on vehicles with power brakes.

SHOP TIP
A problem associated with semimetallics is the tendency of some pads to rust to the rotors. If rusting is severe, the pad material can tear loose.

CALIPER INSTALLATION

As the pads, caliper, and related hardware are installed, apply a thin film of antiseize lubricant to all metal-to-metal contact points.

The step-by-step installation procedure is as follows:

1. Install the inboard pad along with any support hardware (Figure 41). The pad must lie flat against the piston (Figure 42).

2. Position the outer brake pad in the caliper. On some calipers, it is necessary to crimp the outer pad ears (Figure 43). The pad must fit snugly to the caliper. Install any required hardware.

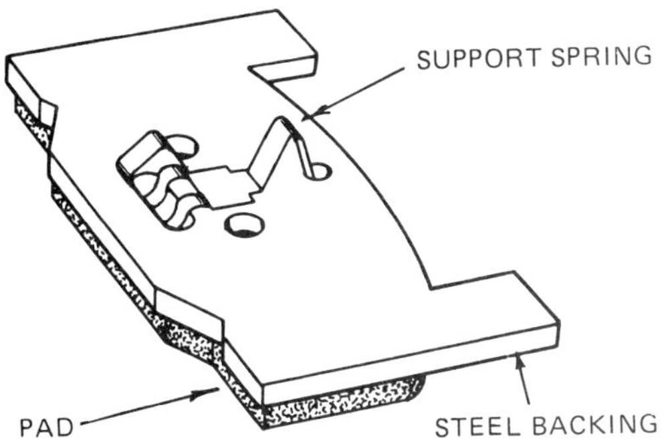

FIGURE 41
Placement of pad supporting spring. (Courtesy of Allied Aftermarket Division, Allied-Signal Corporation)

FIGURE 42
Replacing the inboard pad. (Courtesy of Raybestos-Manhattan, Inc.)

FIGURE 43
Crimping outer shoe ears so the pad fits snugly to the caliper. (Courtesy of Raybestos-Manhattan, Inc.)

3. The caliper may now be positioned on the anchor plate. Slowly slide the caliper assembly over the rotor and onto the machined abutments of the anchor plate.

4. Complete the installation by installing any remaining hardware. Use a torque wrench to tighten all caliper mounting bolts (Figure 44). Specifications can be found in most service manuals.

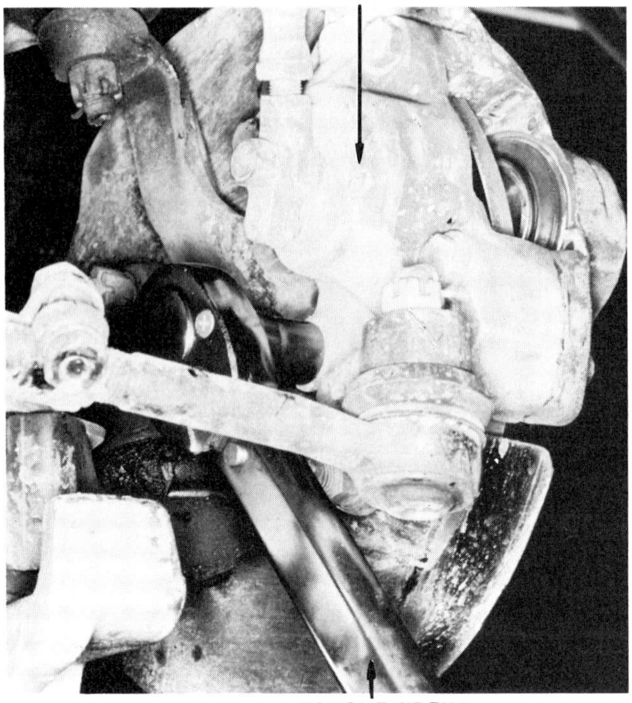

FIGURE 44
Caliper mounting bolts should be tightened with a torque wrench.

Caliper Alignment

Certain fixed calipers must be aligned when they are installed. The clearance between the rotor friction face and the caliper pad mounting surfaces must be within .010″ on both sides. If the clearance varies more than .010″, add or remove shims as required (Figure 45).

CALIPERS WITH PARKING BRAKES

Rear disc brakes are being used on a number of imported and domestic vehicles. These rear calipers are similar to the front calipers except for an added parking brake and adjuster mechanism (Figures 46 and 47). This adjuster mechanism is actuated by the parking brake cable.

Special service tools are available for servicing this type of disc brake mechanism (Figures 48 and 49).

Installing, Bleeding, and Adjusting Delco-Moraine Rear Disc Calipers

1. After removing the wheels, install a lug nut on each rotor to prevent the rotor from falling.

2. Remove the hydraulic lines and actuator levers from the calipers.

3. Remove the caliper mounting bolts and the calipers.

4. Mount the new calipers; right hand (RH) caliper to right hand side; left hand (LH) caliper to left hand side. Install the mounting bolts and torque them to manufacturer's specifications.

FIGURE 45
Certain fixed calipers must be aligned when installed. This is done by using mounting bracket shims. (Courtesy of Allied Aftermarket Division, Allied-Signal Corporation)

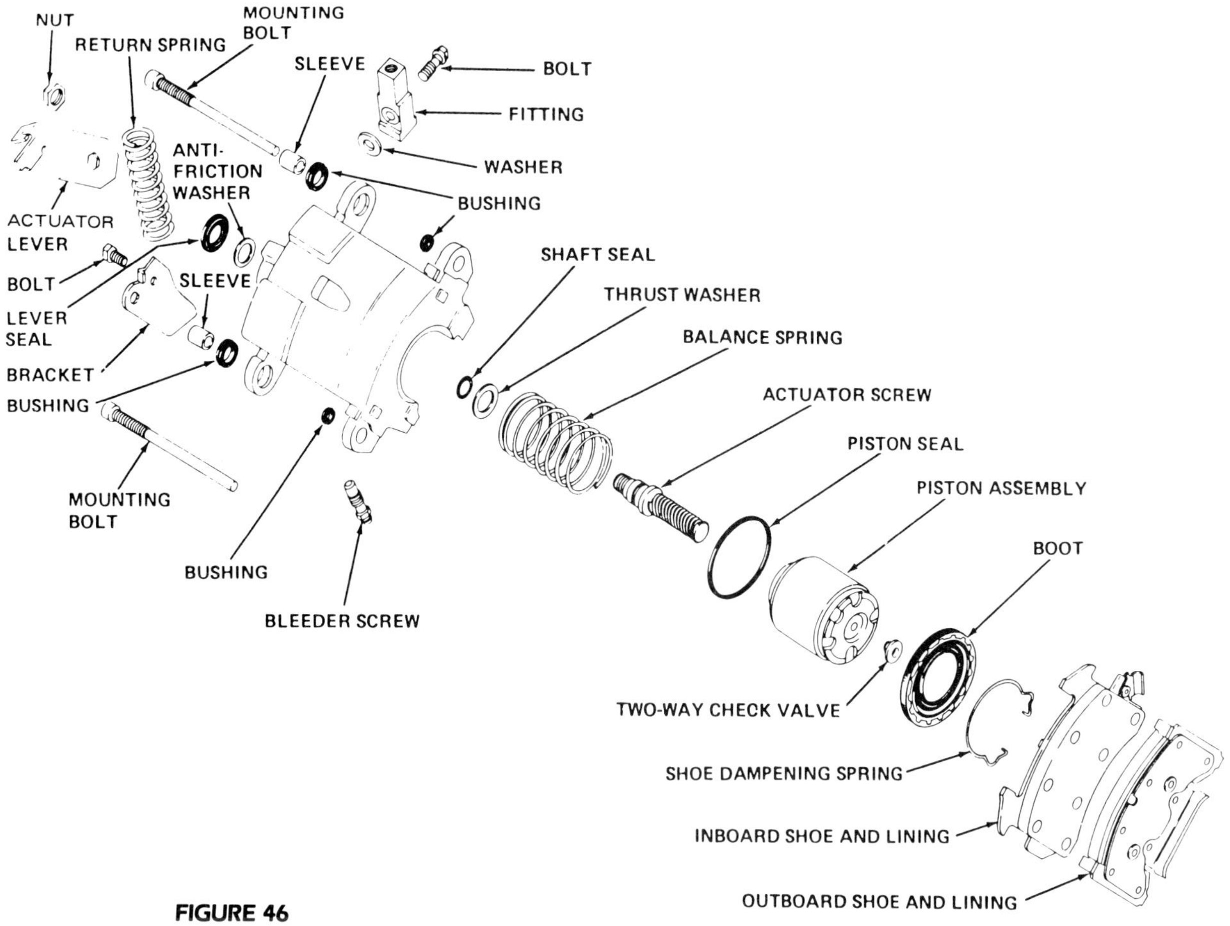

FIGURE 46

Exploded view of a Delco-Moraine rear caliper with an integral parking brake. Locate the two-way check valve. This part prevents air from entering into the piston assembly. (Courtesy of Buick Motor Division, General Motors Corporation)

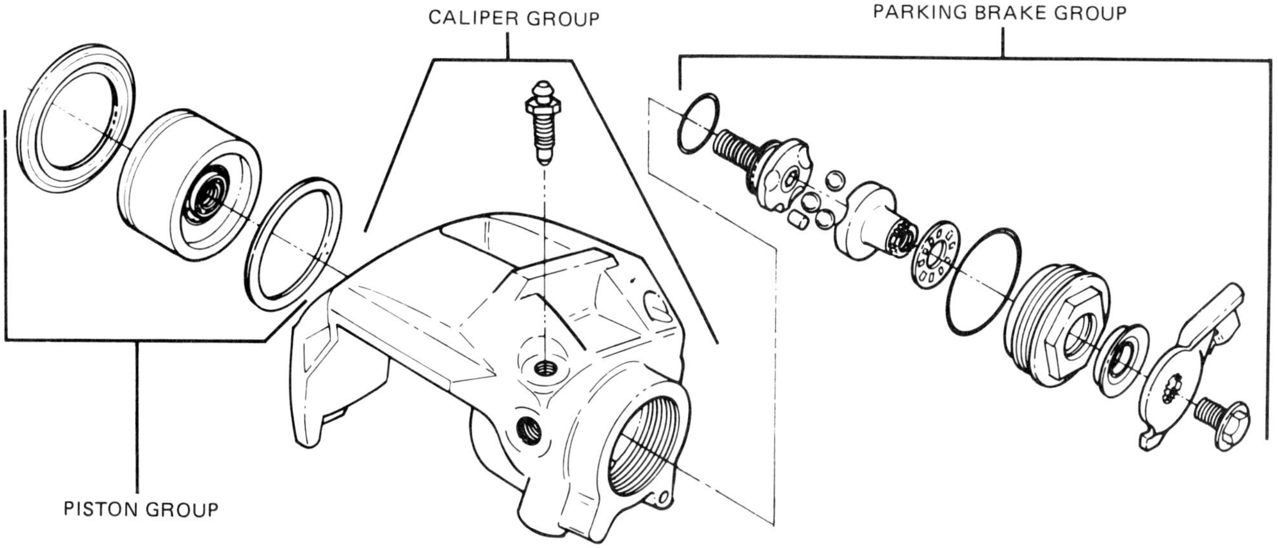

FIGURE 47

Exploded view of a Ford rear caliper. The parking brake has a mechanism that adjusts pad-to-rotor clearance, if needed, whenever the parking brake is used. (Courtesy of Ford Parts and Service Division)

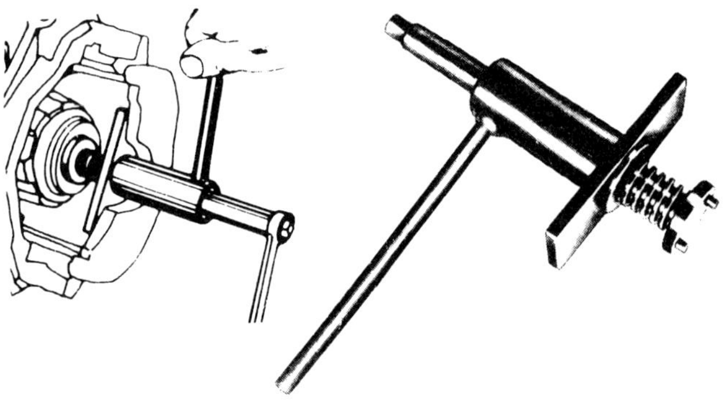

FIGURE 48

Rear disc brake piston adjuster. This is a necessary tool for installing new rear disc pads on Ford Motor Company automobiles. The tool makes it easier to turn the piston back into the caliper bore, which is required due to the parking brake mechanism. (Courtesy of Specialty Products Company)

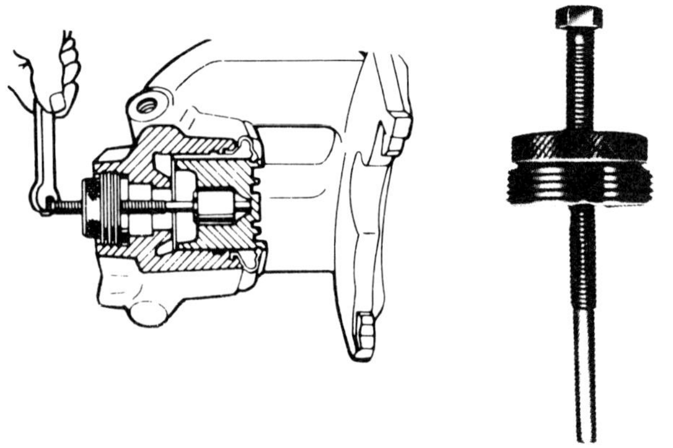

FIGURE 49

Rear disc brake caliper piston remover. This tool prevents damage to the parking brake adjuster mechanism (located in the piston) during overhaul. The piston can be ruined by blowing it out with compressed air. Designed for use on 1975 through current Ford and Mercury automobiles. (Courtesy of Specialty Products Company)

5. Connect the hydraulic lines. Bleed the system with a pressure bleeder at *20 to 25 psi*, until clear fluid is present.

6. Pump the brake pedal several times.

7. Adjust the caliper at the actuator screw (lead screw) until the pads are tight against the rotor.

8. Back the lead screw off 1/16" to align the lever against the caliper stop.

NOTE: *If the lead screw requires more than a 1/16" turn, look for these problems:*
a. Emergency brake cable adjusted wrong prior to this service.
b. Incorrect rotor thickness.

9. Install the RH actuator lever against the caliper stop on the RH caliper.

10. Install the LH actuator lever against the caliper stop on the LH caliper.

11. Check for firm brake pedal and parking brake operation. *The pads will adjust mechanically through the parking brake operation.*

SHOP TIP

A low brake pedal complaint on GM vehicles can be caused by a faulty two-way check valve (refer back to Figure 46). This valve prevents air from getting into the piston assembly. The solution is piston replacement.

Kelsey-Hayes Rear Disc Calipers

An exploded view of the late design Kelsey-Hayes rear disc caliper is shown in Figure 50. The rear rotors have curved fins in the direction of forward rotation, and the left and right are not interchangeable (Figure 51).

FINAL CHECKS AND ADJUSTMENTS

1. Fill the master cylinder reservoirs, and bleed the hydraulic system (Figure 52).

2. Seat the pads by pumping the brake pedal several times. This will adjust the pad-to-rotor clearance.

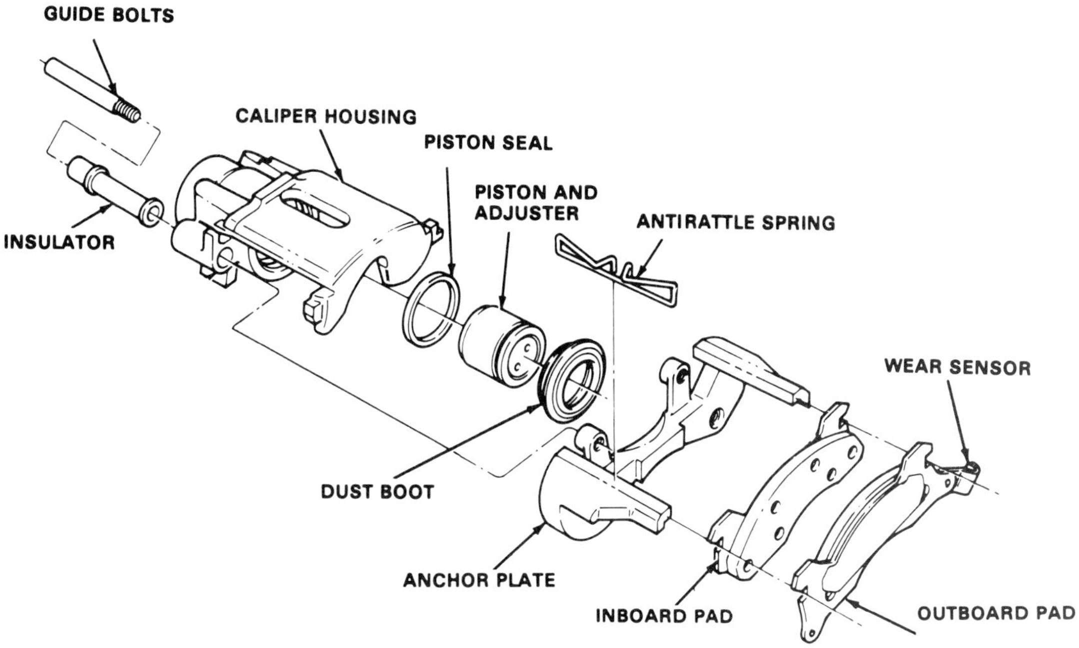

FIGURE 50
Kelsey-Hayes floating rear caliper. (Courtesy of Allied Aftermarket Division, Allied-Signal Corporation)

WARNING: Never move a car until a firm brake pedal is obtained!

3. Maintain firm pedal pressure and inspect the caliper for leaks.

4. Reinstall the wheel, using a criss-cross (star) tightening sequence (Figure 53). A torque wrench should be used. Impact wrenches can distort rotors and create runout problems.

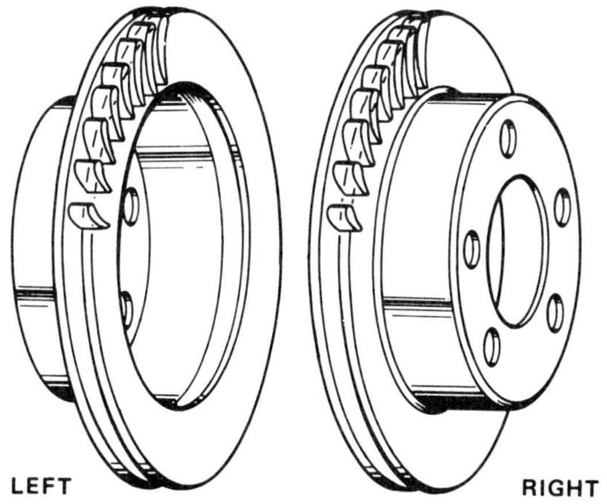

FIGURE 51
Some rotors have directional cooling fins and are not interchangeable left to right. (Courtesy of Allied Aftermarket Division, Allied-Signal Corporation)

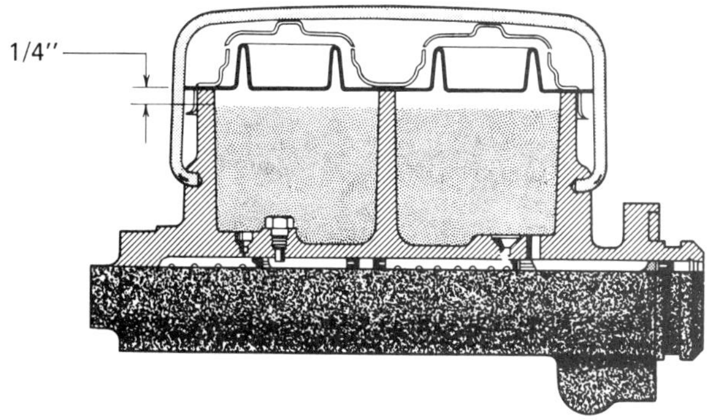

FIGURE 52
Master cylinder brake fluid level should be a specified distance from the top of the reservoir—usually about 1/4". (Courtesy of Chevrolet Motor Division, General Motors Corporation)

FIGURE 53
Wheel lug nut tightening sequence. (Courtesy of Chrysler Corporation)

Road test the vehicle, making several stops. Pulling to either side is normal during the first one or two applications, while the pads are completing the seating process. After one or two stops, there should be no more tendency to pull.

If pulling or any other abnormal conditions continue, check the troubleshooting chart in this section for possible causes.

> **SHOP TIP**
> Special split dies are available to repair damaged rotor or drum studs on the car (Figure 54).

FIGURE 54
These split dies can be used to quickly repair damaged rotor or drum stud threads. (Courtesy of Heli-Coil Products, Division of Mite Corporation)

TRUE CASE HISTORY

Vehicle	Mazda 626
Engine	Four-cylinder, 2000 cubic centimeters
Mileage	7,800. Problem since vehicle was new.
Problem	Front disc brakes would make a "groaning" noise when starting out after a stop. Customer was told that this noise was normal and would eventually go away.
Cause	Calipers hanging up on the slides. This was making the pads drag on the rotors up to about 10 mph.
Cure	Lubricated all metal-to-metal contact areas on the calipers and pads with antiseize compound.

STUDY QUESTIONS

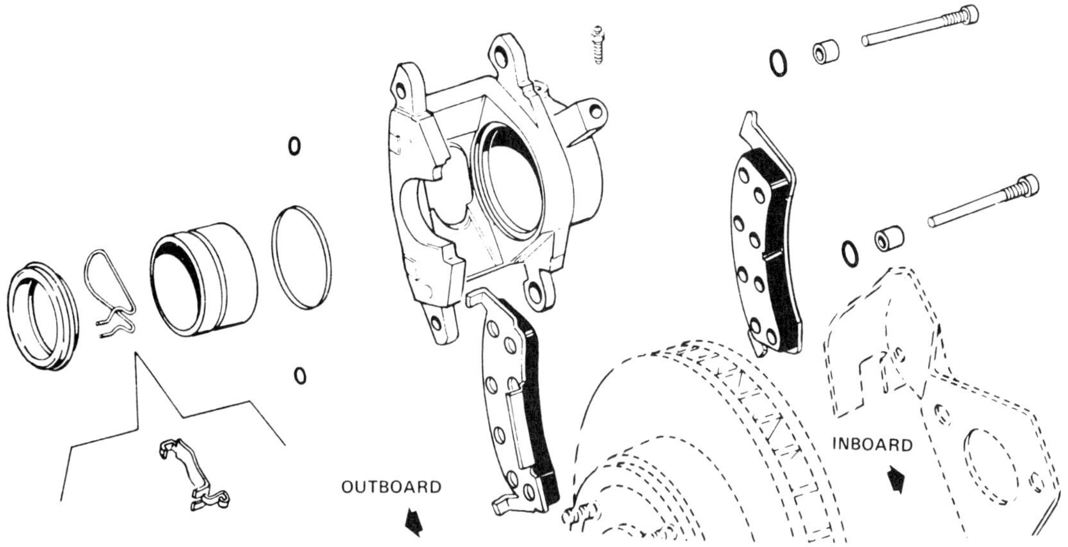

1. An automobile has the above type of disc brake on the front wheels. The inboard pad is badly worn. The outboard pad shows barely any wear.

   Technician A says that too much rotor runout could be the cause.

   Technician B says that a defective piston seal could be the cause.

   Who is right?
   a. A only **c.** Both A and B
   b. B only **d.** Neither A nor B

2. After new front pads have been installed and the calipers rebuilt, excessive pedal effort is needed to stop.

   Technician A says that this could be caused by the wrong pad material.

   Technician B says that this could be caused by air in the system.

   Who is right?
   a. A only **c.** Both A and B
   b. B only **d.** Neither A nor B

3. **Technician A** says to replace the piston seal when rebuilding a caliper.

   Technician B says to replace the guide pins when rebuilding a caliper.

   Who is right?
   a. A only **c.** Both A and B
   b. B only **d.** Neither A nor B

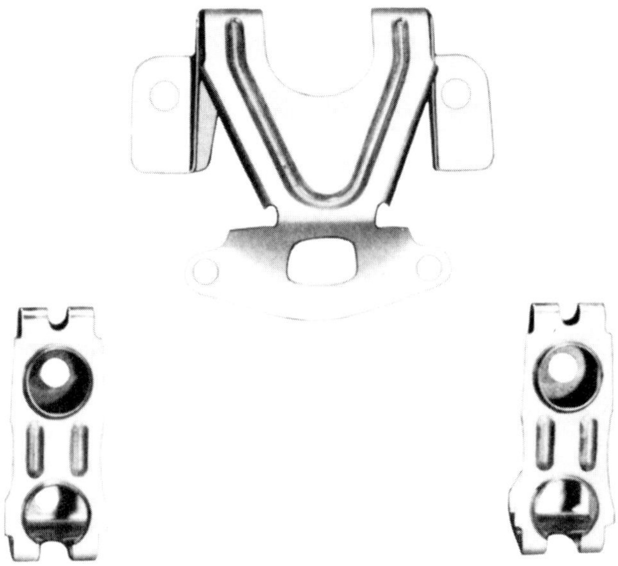

4. The parts shown above are defective.

   Technician A says that this can cause brake pull.

   Technician B says that this can cause excessive up-and-down caliper movement on bumpy roads.

   Who is right?
   a. A only
   b. B only
   c. Both A and B
   d. Neither A nor B

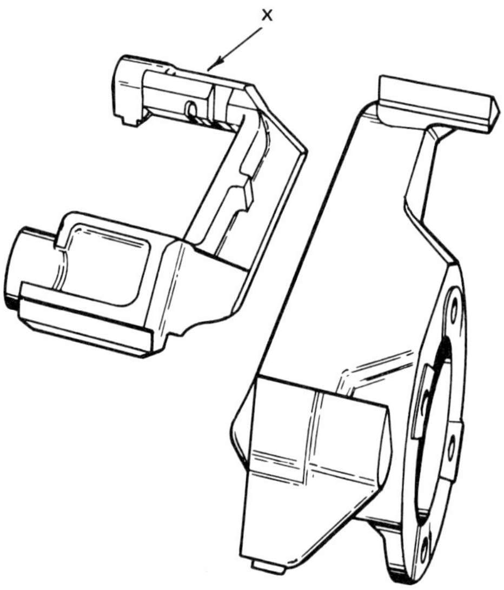

5. The area marked X on the above brake part is badly rusted.

   Technician A says that this could cause noise when braking.

   Technician B says that this could cause tapered pad wear.

   Who is right?
   a. A only
   b. B only
   c. Both A and B
   d. Neither A nor B

6. **Technician A** says that rotor runout can be checked with a micrometer.

   Technician B says that the caliper must be removed when checking rotor runout.

   Who is right?
   a. A only **c.** Both A and B
   b. B only **d.** Neither A nor B

7. **Technician A** says that a rotor should have a nondirectional swirl finish.

   Technician B says that vibration should be dampened when machining a rotor.

   Who is right?
   a. A only **c.** Both A and B
   b. B only **d.** Neither A nor B

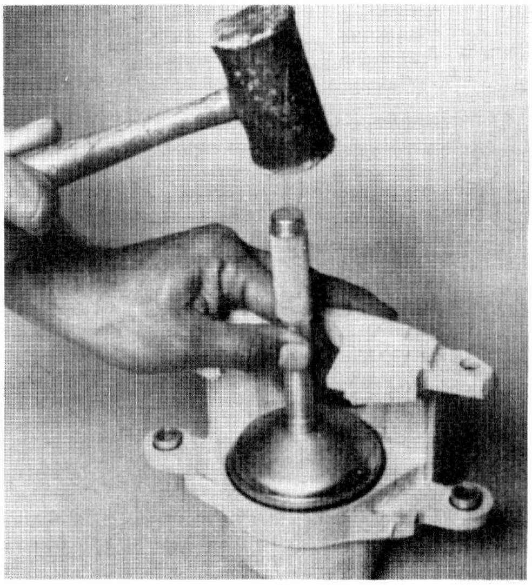

8. What is the technician doing in the above picture?
   a. Installing the seal
   b. Installing the boot
   c. Installing the piston
   d. Installing the guide bushing

YEAR AND MODEL		FRONT PADS
FORD		
LTD, LTD II, Crown Victoria, Custom, Galaxie		
85-83	Crown Victoria.	7082A-D199 +
	S/W	7082A-D199 +
	Police	7082-D150●
85-Late 83	LTD	7072B-D237●
	S/W & Police	7072B-D237●
Early 83	LTD Exc. S/W	7072-D152●
	S/W	7072-D152●
Escort		
85-Late 83	Coupe	7106A-D257●
	Sedan, S/W	7106A-D257●
EXP		
84	w/SVO	—
85-Late 83	w/o SVO	7106A-D257●
Fairmont		
Late 83	9" Brk.	7072B-D237●
	10" Brk	7072B-D237●
Mustang		
85	w/SVO	7082-D200●
84	w/SVO	7082A-D199●
85-Late 83	4 Cyl.	7072B-D237 +
	4 Cyl. Turbo 6 & 8 Cyl. .	7072B-D237 +
Tempo		
85-84	All	7106A-D257●
Thunderbird		
85-84	All	7072B-D237●
83	All	7072B-D237●
FORD TRUCK		
Bronco		
85-84	All	788-D50●
Bronco II		
85-84	All	7125A-D249●
Courier		
83-79	All	7009-D76 +
	or	488
Ranger		
85-84	All	7125A-D249●

● OEM is Semi-Metallic + Semi-Metallic Pads Available

9. A section from a disc brake pad catalog is shown above.

   Technician A says that semimetallic pads are available for all the vehicles listed.

   Technician B says that semimetallic pads are original equipment on all the vehicles listed.

   Who is right?
   a. A only **c.** Both A and B
   b. B only **d.** Neither A nor B

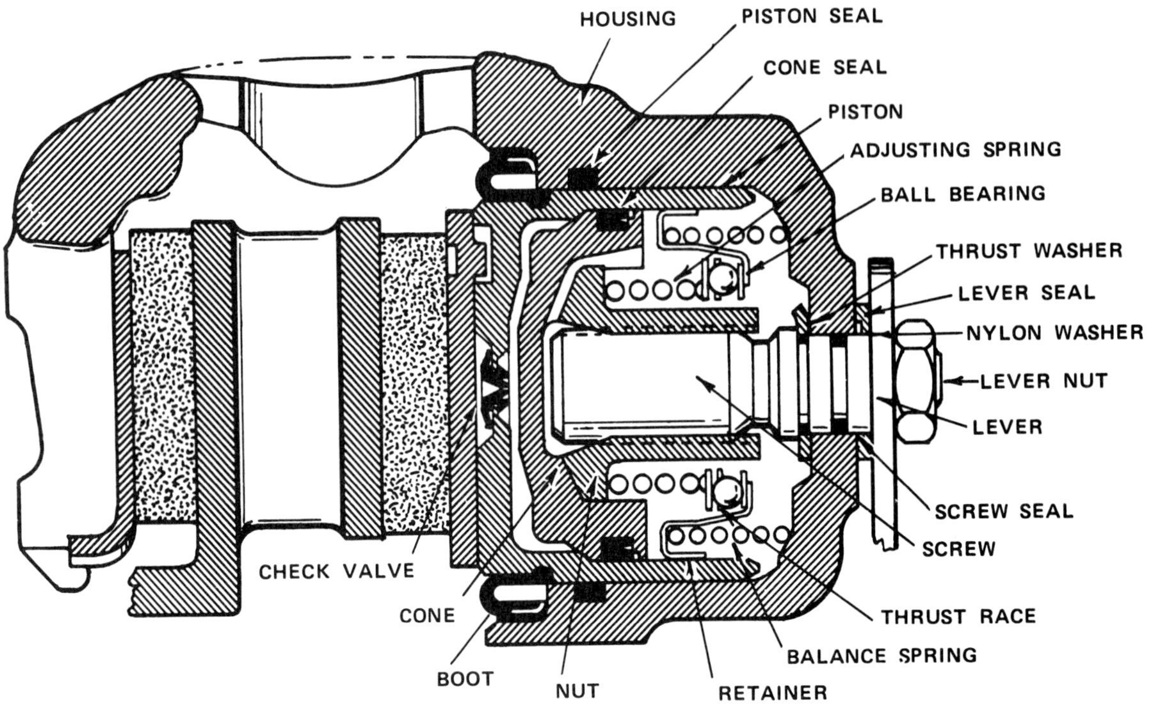

10. Refer to the above brake caliper cross-section illustration.

Technician A says that the check valve prevents air from entering the piston assembly.

Technician B says that the housing contains an integral parking brake assembly.

Who is right?

a. A only
b. B only
c. Both A and B
d. Neither A nor B

11. **Technician A** says that a star pattern tightening sequence should be used when tightening lug nuts.

Technician B says that using an impact wrench to tighten lug nuts can distort the rotor.

Who is right?

a. A only
b. B only
c. Both A and B
d. Neither A nor B

12. New front disc brake pads are going to be installed on a car.

Technician A says that it may be necessary to bottom the caliper pistons.

Technician B says that some brake fluid should be removed from the master cylinder.

Who is right?

a. A only
b. B only
c. Both A and B
d. Neither A nor B

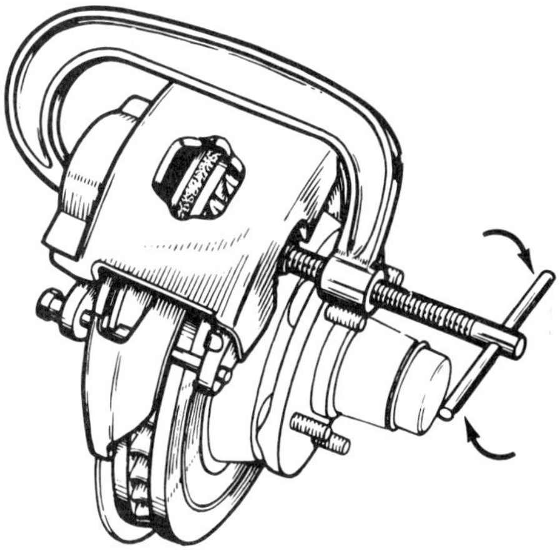

13. The above set-up shows the
   a. Pads being seated
   b. Outboard pads being clinched to the caliper
   c. Caliper mounting pin being pressed out
   d. Caliper piston being pushed into its bore

14. Brake fluid is being added to a master cylinder that has plastic reservoirs.

    Technician A says to add brake fluid until both reservoirs are filled to the top.

    Technician B says that heavy-duty brake fluid will dissolve plastic reservoirs.

    Who is right?
    a. A only
    b. B only
    c. Both A and B
    d. Neither A nor B

Section D

POWER ASSIST UNITS DIAGNOSIS AND REPAIR

After studying this section, and with hands-on practice on a "live" vehicle, you should be able to

- Test pedal free travel with and without engine running to check power booster operation.
- Check vacuum supply (manifold or auxiliary pump) to vacuum type power booster with a vacuum gauge.
- Inspect the vacuum type power booster unit for vacuum leaks; inspect the check valve for proper operation; repair or replace parts as necessary.
- Inspect and test hydro-boost system and accumulator for leaks and proper operation; repair and replace parts as necessary.

POWER BRAKES

Power brakes on passenger cars are used in conjunction with the basic hydraulic braking system. All of the systems produced today employ some sort of an assist (booster) unit, which acts like an extra foot on the brake pedal. This reduces the amount of braking effort the driver must use.

There are two types of power brake systems:

1. Those that use air pressure and a vacuum to create the boost force (Figure 1).

2. Those that use hydraulic pressure from the power steering pump (Figures 2 and 3).

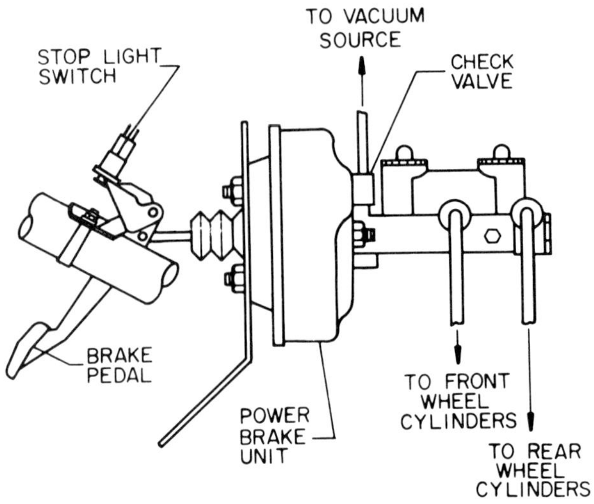

FIGURE 1
Typical power brake unit installation.

Vacuum Boost Assist

Power brake vacuum booster units come in different shapes, but they all work basically the same way. There is a large diaphragm (or piston) in the power section of the booster unit (Figure 4). On most cars, both sides of the diaphragm are under vacuum. When the brake pedal is depressed, atmospheric pressure is let in to one side of the diaphragm, forcing it forward to press against the master cylinder push rod (Figure 5).

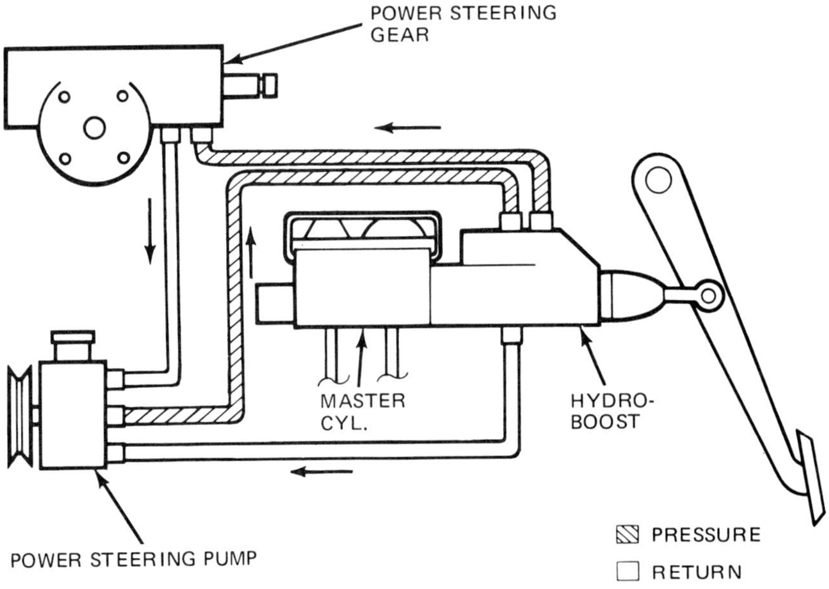

FIGURE 2
Fluid flow in a hydro-boost system. (Courtesy of Ford Motor Company)

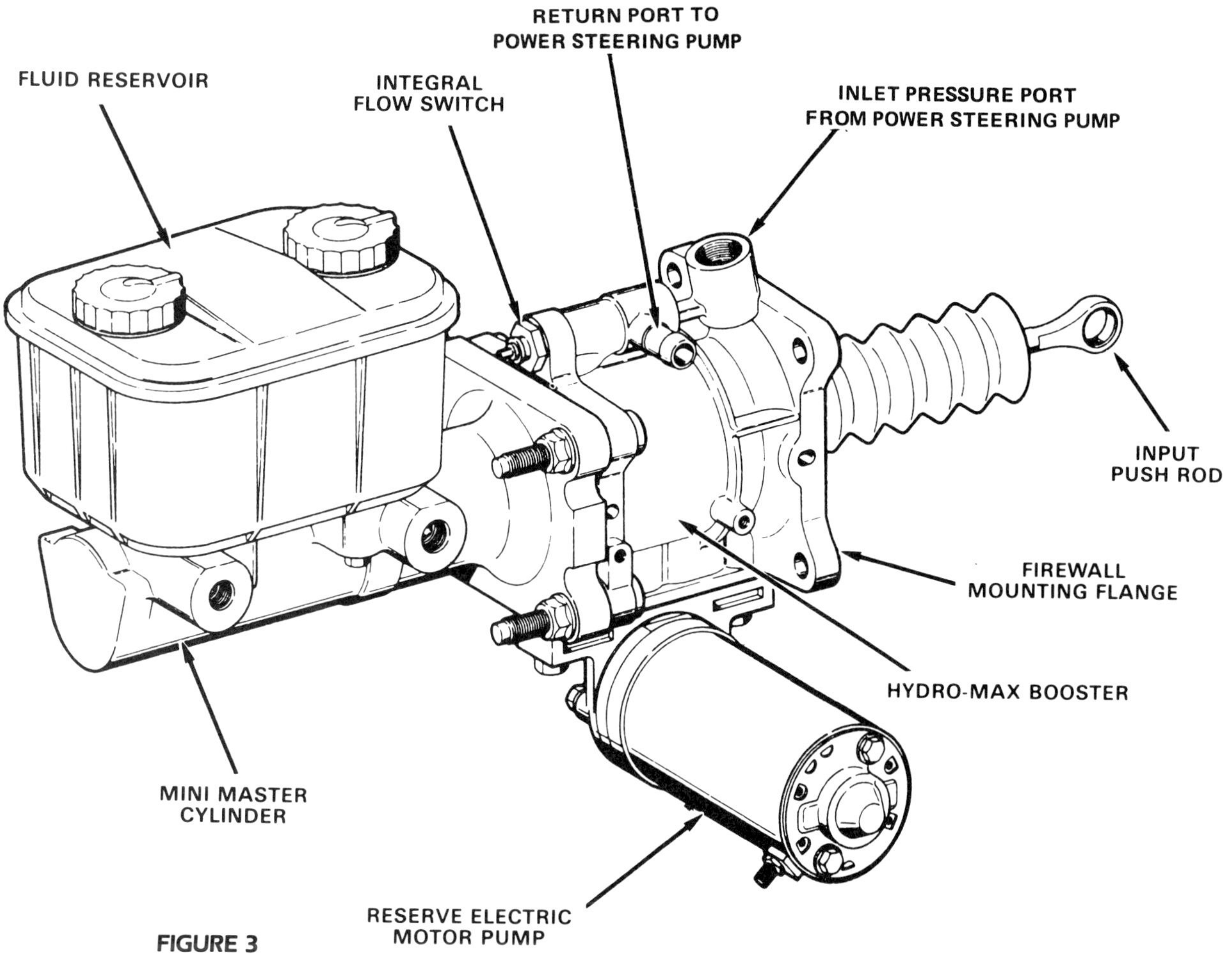

FIGURE 3

The hydro-max is a power steering pump-powered booster which, together with a mini master cylinder, provides power assist for applying the brakes. A reserve electric motor pump provides a power source for the hydraulic booster if normal flow from the power steering pump is interrupted. (Courtesy of Allied Aftermarket Division, Allied-Signal Corporation)

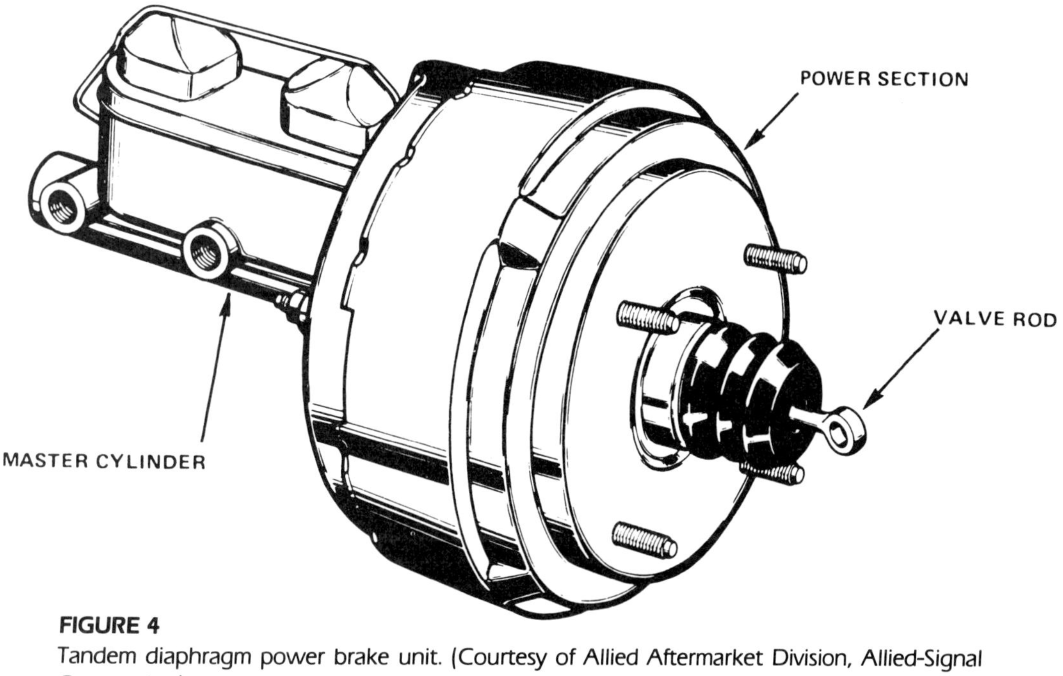

FIGURE 4

Tandem diaphragm power brake unit. (Courtesy of Allied Aftermarket Division, Allied-Signal Corporation)

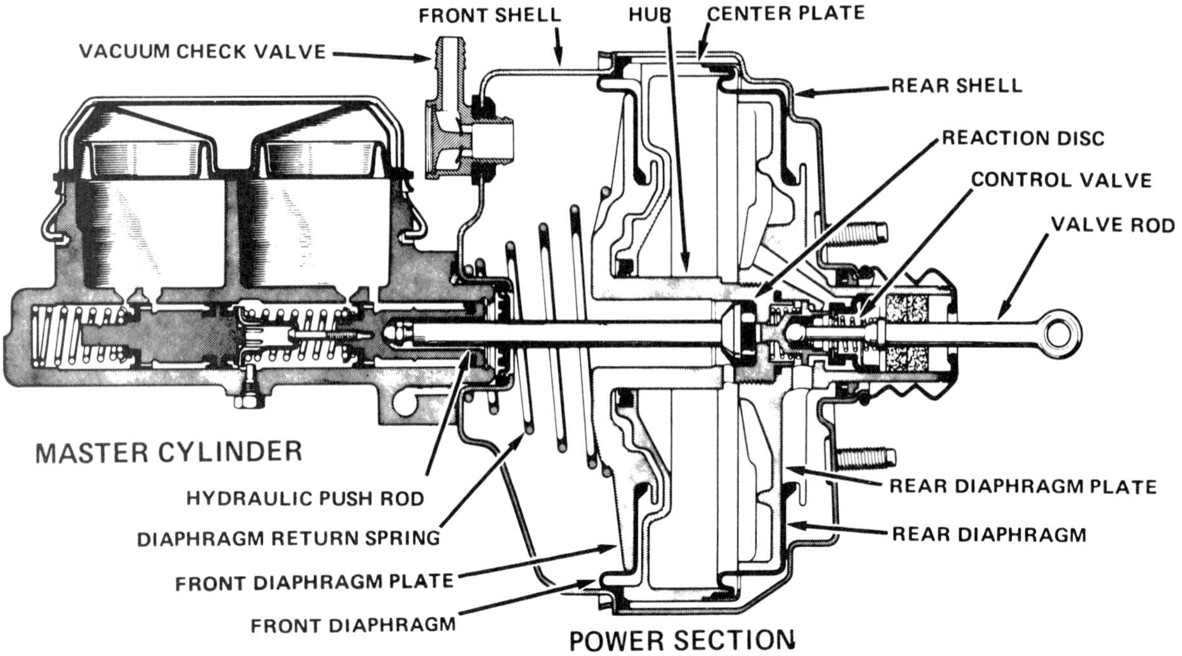

FIGURE 5

Cross-section of tandem diaphragm power brake unit. (Courtesy of Allied Aftermarket Division, Allied-Signal Corporation)

Hydraulic Assist

The hydraulic power assist was developed when diesel engines were introduced and engine vacuum began to be used more and more to operate different parts of the automobile. It works by pressurized fluid from the power steering pump and fits in the same place as a vacuum booster (Figure 6).

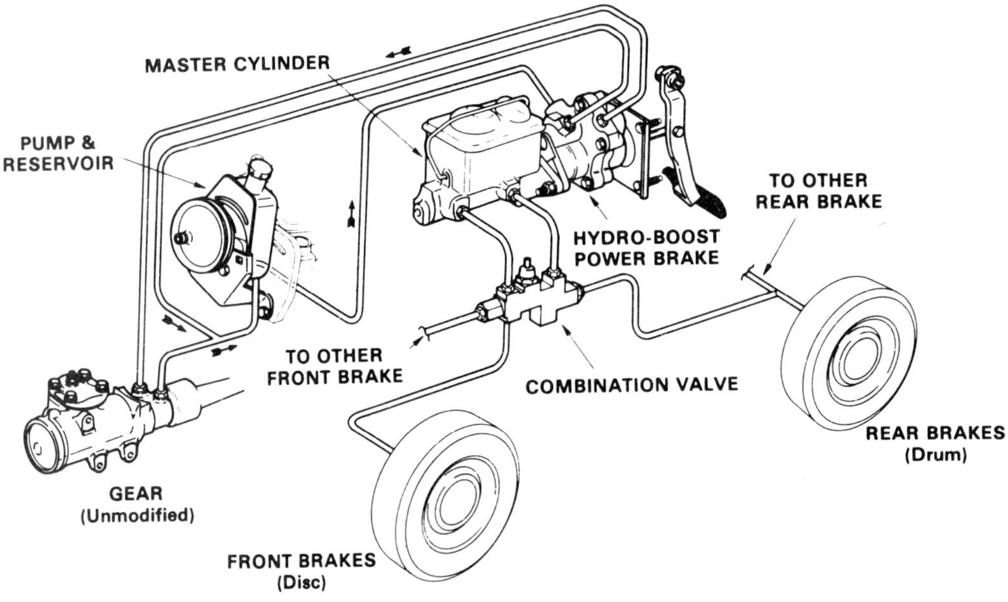

FIGURE 6

Schematic diagram of hydro-boost system. (Courtesy of Allied Aftermarket Division, Allied-Signal Corporation)

BRAKE PEDAL FREE PLAY

In any brake system, the brake pedal must have the proper amount of free play. Without this free play, the master cylinder piston assembly is pushed inward (forward) and the brakes are constantly applied.

Brake pedal free play can be checked as follows:

1. Stop the engine and release the vacuum stored in the brake booster by pumping the brake pedal several times.

NOTE: If a vacuum remains in the booster, the correct free play cannot be determined.

2. Lightly push the brake pedal with your finger until it meets resistance, and measure the pedal stroke (Figure 7).

3. If free play is not within specification, adjust it according to the manufacturer's procedure.

VACUUM BOOSTER OPERATION CHECKS

An operation check should be made whenever you think that a vacuum booster may not be working properly, or any time a new or rebuilt booster has been installed.

Vacuum System Test

With the engine stopped, step on the brake pedal several times. Then, hold foot pressure on the brake pedal and start the engine. *If at this time the pedal moves slightly downward, the vacuum system is working properly.* If the brake pedal does not "fall away" during this test and the complaint is a hard pedal, check the vacuum. You should have 17 to 20 inches of vacuum (Figure 8). If not, check the booster rubber hose and tubing to make sure the passages are clear. Check the operation of the vacuum pump (if there is one) by connecting a vacuum gauge to the outlet port of the pump. Test the vacuum check valve (refer to the next section).

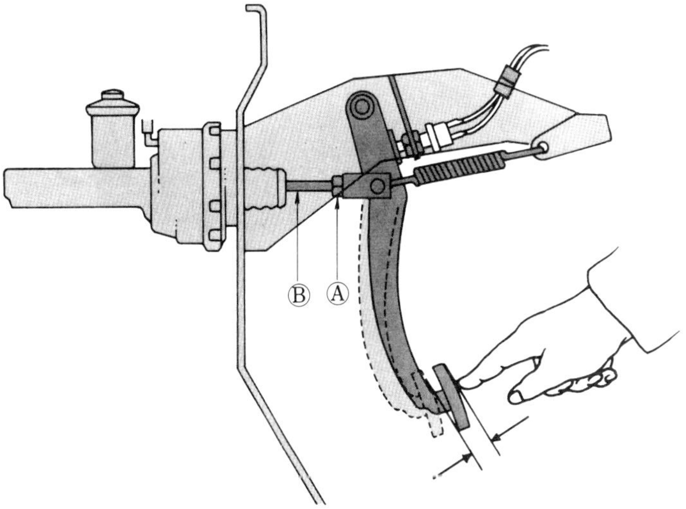

FIGURE 7
Checking brake pedal free play. Adjustment on this vehicle is made by loosening lock nut A, and turning push rod B. (Courtesy of Toyota Motor Sales, U.S.A., Inc.)

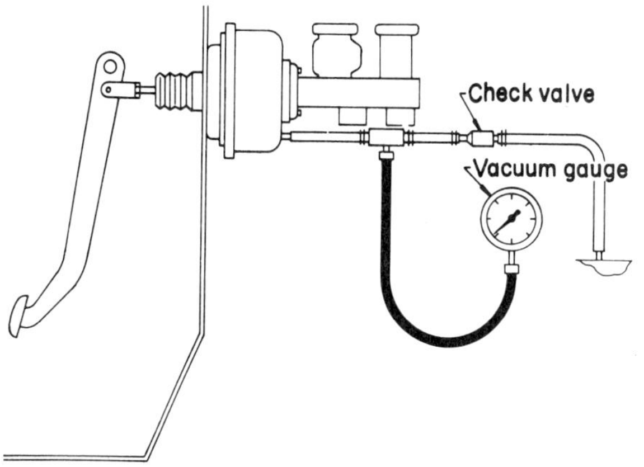

FIGURE 8
A vacuum gauge inserted in the vacuum line between the check valve and the power brake unit provides a quick method for finding system leaks.

Vacuum Check Valve Test

If the power brake vacuum check valve is defective (Figure 9), the power assist portion of the braking system will not work. Only the mechanical/hydraulic portion remains. The check valve can be tested as follows:

1. Run the engine at 2000 rpm for several seconds.

2. Release the accelerator and turn off the engine.

3. Wait 90 seconds and apply the brakes. Two or more applications should be power assisted.

4. If the brake applications are not power assisted, disconnect the hose at the check valve.

5. Blow into the check valve. If air passes through it, the valve is defective.

WARNING: A defective vacuum check valve can allow oil from the crankcase ventilation system to enter the booster unit. This will ruin the diaphragm and seals.

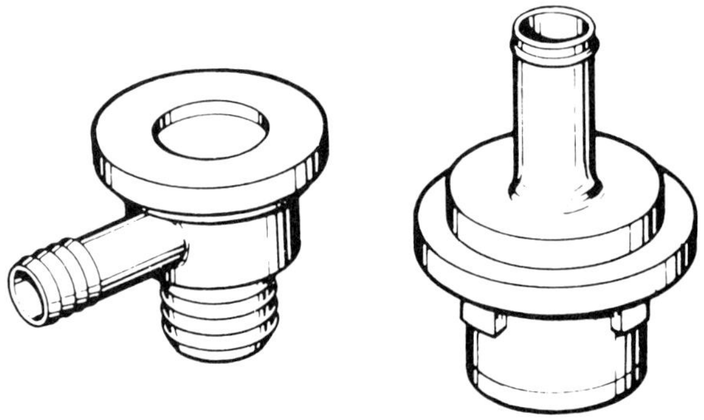

FIGURE 9
Typical power brake unit check valves. (Courtesy of Raymark Industries, Inc.)

Air Tightness Test

Start the engine and shut it off after running it for 1 or 2 minutes. Step on the brake pedal several times with normal braking pressure. If the pedal goes down normally the first time but gradually becomes higher with each following depression, the booster is working properly. If the pedal height remains unchanged, the booster is defective.

Air Tightness Under Load Test

With the engine running, step on the brake pedal and then stop the engine. Hold the pedal down for 30 seconds. If the pedal height does not change, the booster is not leaking. If the pedal rises, the booster is defective.

SHOP TIP
All power brake installations should have the vacuum supply line going up as high as possible from the intake manifold before going down to the power unit (Figure 10). This will help prevent raw gas fumes from getting into the unit and causing the rubber diaphragm to be damaged and fail. It will also prevent a possible explosion in the vacuum section of the power brake unit if the engine backfires.

Brake Drag Test

Raise both front wheels off the floor. Push the disc brake pistons back into the cylinder bores so that the wheels will rotate freely. Spin the wheels and start the engine. Allow the engine to run at idle speed with no foot pressure on the brake pedal. When the wheels come to a stop, check them for brake drag. If drag is noticed, the power brake booster is defective or the push rod is adjusted wrong (Figure 11). This can cause severe pad and rotor wear (Figure 12).

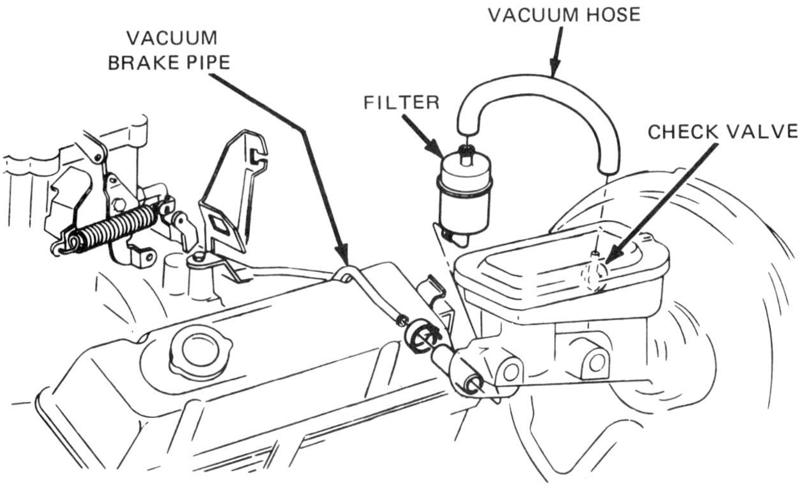

FIGURE 10

This power brake unit uses a filter and a vacuum hose that is mounted high in order to prevent the entry of gas fumes. (Courtesy of Pontiac Motor Division, General Motors Corporation)

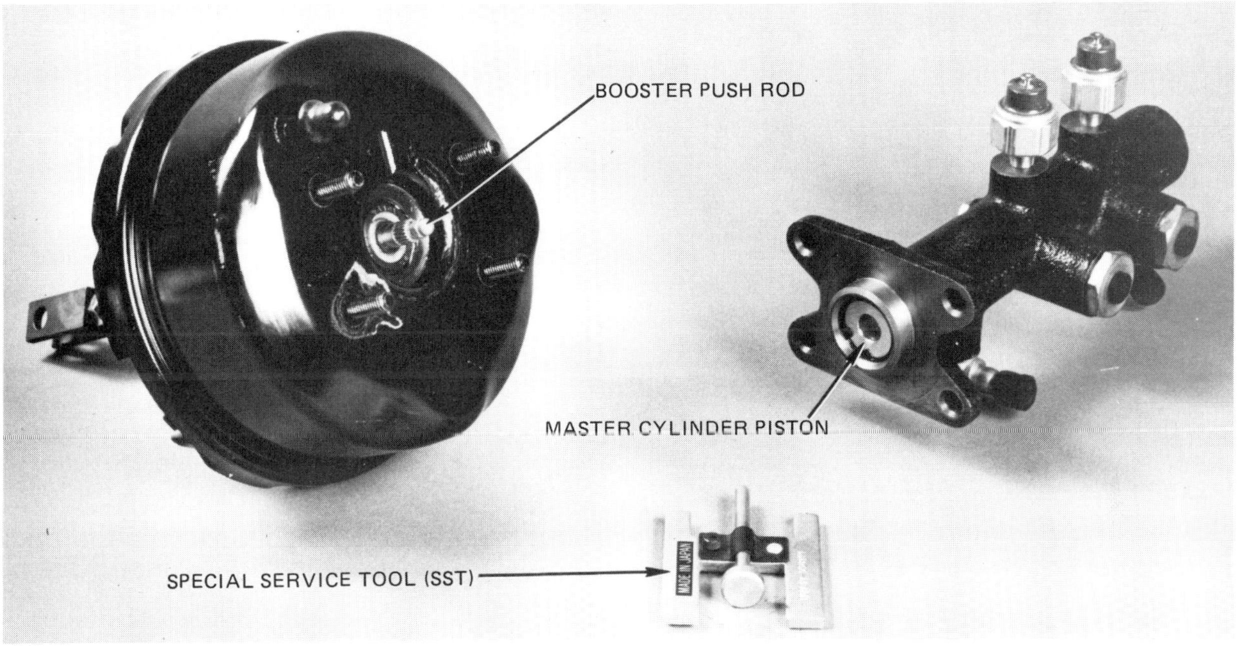

FIGURE 11

Special service tool (SST) is used to measure the clearance between the master cylinder piston and the booster push rod.

FIGURE 12

An improper booster push rod adjustment caused this to happen.

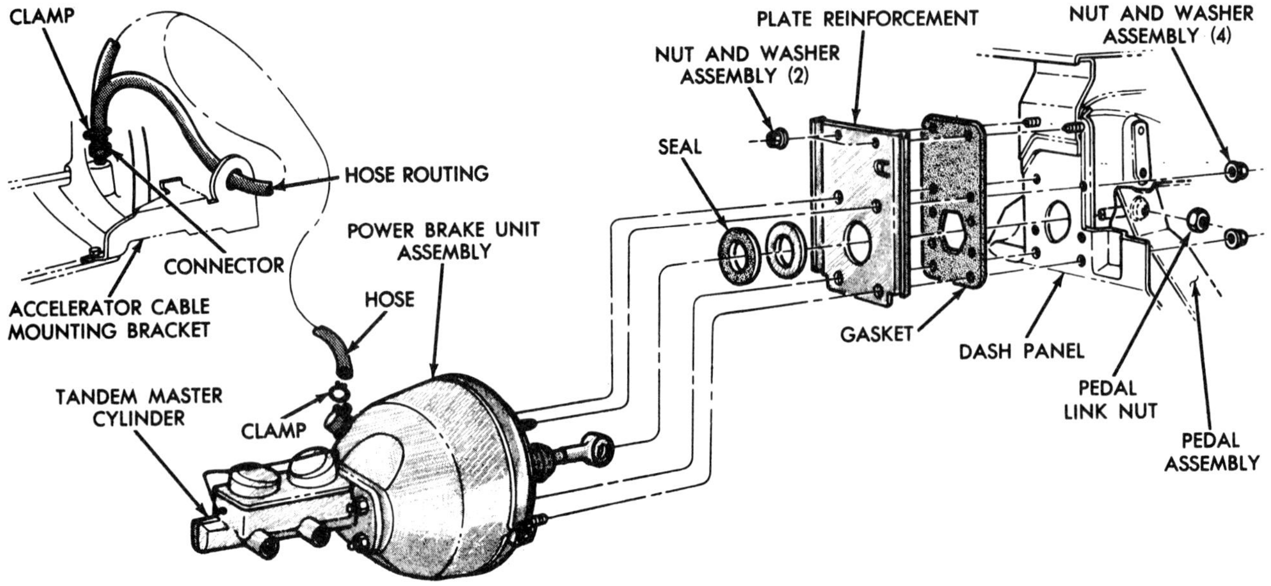

FIGURE 13
Power brake assembly mounting details. (Courtesy of Chrysler Corporation)

Hidden Brake Fluid Loss

Sometimes a vehicle equipped with power brakes has a brake fluid loss problem without showing any sign of an external fluid leak. This can be diagnosed by removing the vacuum hose from the intake manifold to the power brake unit (Figure 13). If the inside of the hose is wet from brake fluid, the master cylinder secondary cup is at fault. The master cylinder should be replaced or rebuilt.

SHOP TIP

Here is a quick check for proper vacuum booster operation on a vehicle that uses a hydraulic stop light switch (Figure 14). Begin by removing the stop light switch and installing a pressure gauge. Take a reading with your foot on the brake. Keeping the same pedal height, start the engine and take another reading. There should be a substantial pressure increase in the second reading.

VACUUM POWER BRAKE TROUBLESHOOTING

Refer to Chart 1 for other vacuum power brake problems.

FIGURE 14
Hydraulic stop light switch.

HYDRO-BOOST TROUBLESHOOTING

A hydro-boost unit differs from a vacuum brake booster not only in the source of power, but in the fact that it is also a part of another vehicle subsystem (the power steering system). Therefore, problems in the steering system may affect the hydro-boost unit. See Figure 15.

CHART 1 TROUBLESHOOTING CHART FOR VACUUM BOOSTER POWER BRAKES

Hard Pedal

1. Faulty vacuum check valve
2. Collapsed or leaking vacuum hose
3. Plugged vacuum chamber
4. Leaking vacuum chamber (can also cause rough engine idle)
5. Brake pedal linkage binding
6. Plugged power brake air filter
7. Low intake manifold vacuum

Pedal Fails (or Is Slow) to Release

1. Excessive seal friction in the power brake unit

Sensitive Brakes

1. Reaction disc leakage, or sticky valve action in the booster unit
2. Reaction disc accidentally pulled out when the master cylinder was replaced
3. Binding seals in the booster unit

Pedal Vibrates and Chatters

1. Binding inside power brake unit
2. Improperly adjusted brake pedal free play

Hydro-Boost Basic Test

With the engine off, pump the brake pedal several times to get rid of all the accumulator reserve pressure. Then, with medium pressure on the brake pedal, start the engine. The pedal should "fall away" slightly and push back. If it does not, the unit may be defective. However, before replacing it, check the probable causes listed in Chart 2.

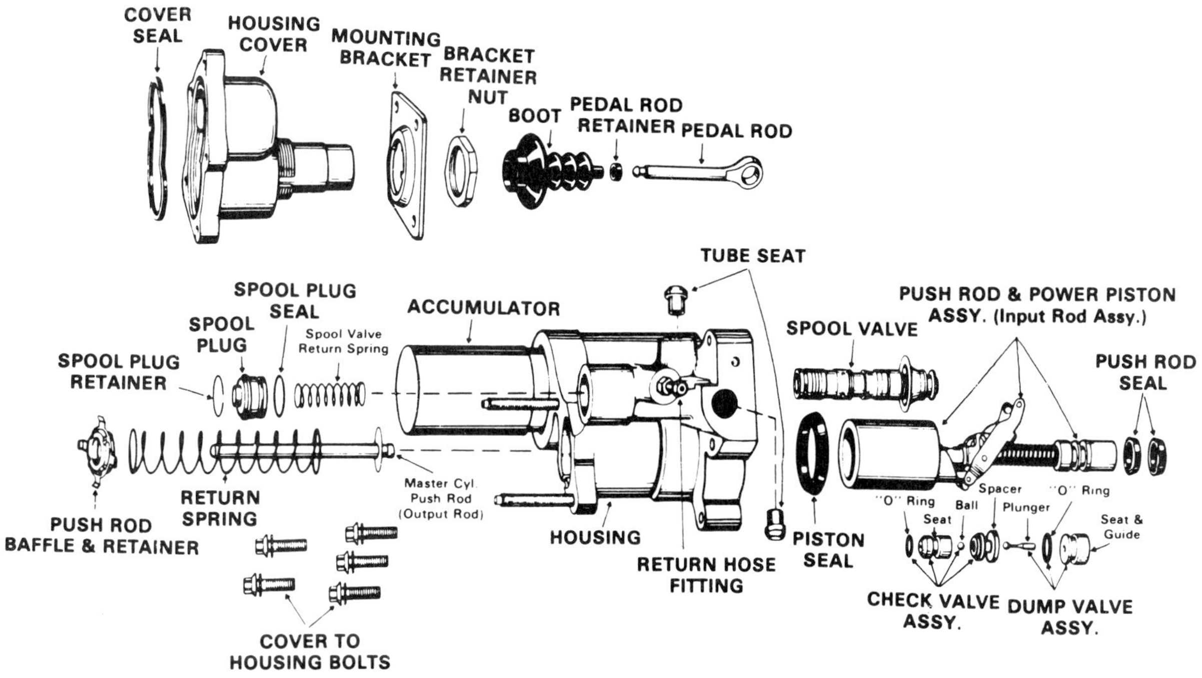

FIGURE 15

Exploded view of a type 1 hydro-boost unit. The accumulator provides reserve stopping capacity if power steering pump flow is interrupted. (Courtesy of Allied Aftermarket Division, Allied-Signal Corporation)

CAUTION: Before removing a hydro-boost unit, pump the brake pedal several times to get rid of any built-up pressure in the accumulator. Otherwise, when the lines are "cracked" for removal, pressurized brake fluid can spray out.

Accumulator Test

With the engine running, turn the steering wheel until the wheels lightly touch their stops. Hold the wheels there for 5 seconds. Release the steering wheel and turn off the engine. Apply and release the brake pedal. You should have several power assisted applications. Recharge the accumulator by again running the engine and holding the wheels at their stops. *You should hear a light hissing sound as the accumulator is charging.* Turn off the engine and wait 1 hour. Without starting the engine, you should have several power assisted brake applications. If not, check the probable causes listed in Chart 2.

Seal Leaks

Figure 16 shows a hydro-boost unit and points out possible fluid leakage areas.

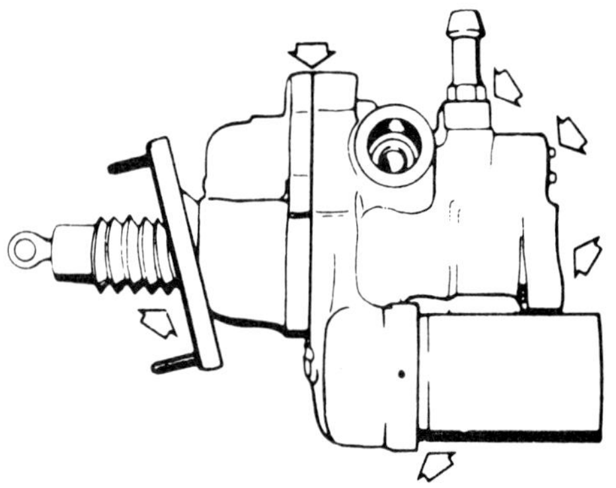

FIGURE 16
Possible seal leakage points on a type 1 hydro-boost unit (Courtesy of Wagner Division, Cooper Industries, Inc.)

HYDRO-BOOST BLEEDING PROCEDURE

Whenever a hydro-boost unit is removed and reinstalled, bleed the steering system. First pour the correct amount of steering fluid into the steering pump. Then start the engine and turn the steering wheel from stop to stop several times. This will usually bleed all air from the booster. Normal driving will remove any remaining trapped air.

CHART 2 TROUBLESHOOTING CHART FOR HYDRO-BOOST POWER BRAKES

Excessive Brake Pedal Effort
1. Loose or broken power steering pump belt
2. Power steering pump reservoir empty
3. Leaks at tube fittings, power steering hoses, booster unit, or accumulator connections
4. Spool valve bore scored or nicked in booster unit
5. Hydraulic lines routed incorrectly
6. Low idle speed

Slow Brake Pedal Return
1. Faulty booster unit spool action
2. Restriction in return line from booster to pump reservoir

"Grabby" Brakes
1. Broken spool return spring inside booster unit
2. Binding pedal linkage

Booster Chatters, Pedal Vibrates
1. Power steering pump belt slips
2. Low fluid level in power steering pump reservoir
3. Contaminated power steering fluid

NOTE: Do not mix power steering fluid and brake fluid. Seal damage will result.

TRUE CASE HISTORY

Vehicle	Toyota Celica equipped with power brakes.
Engine	Four-cylinder, 18R-C model (1968 cc)
Mileage	87,000
Problem	1. Hard brake pedal 2. Pedal squeak during use
Cause	A dry and worn clevis pin (Figure 17).
Cure	Installing a new clevis pin. Antiseize compound was used to lubricate the surface of the new pin.

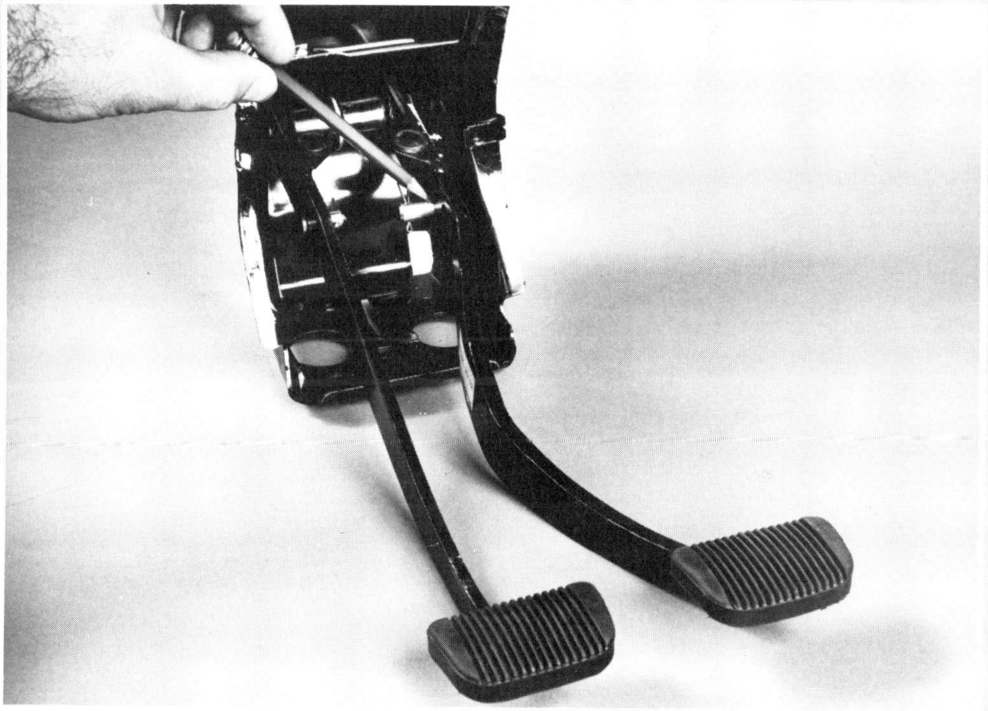

FIGURE 17

STUDY QUESTIONS

1. **Technician A** says that the above adjustment, if incorrect, can cause the brakes to drag.

   Technician B says that the above adjustment, if incorrect, can result in too much pedal travel.

   Who is right?
   a. A only
   b. B only
   c. Both A and B
   d. Neither A nor B

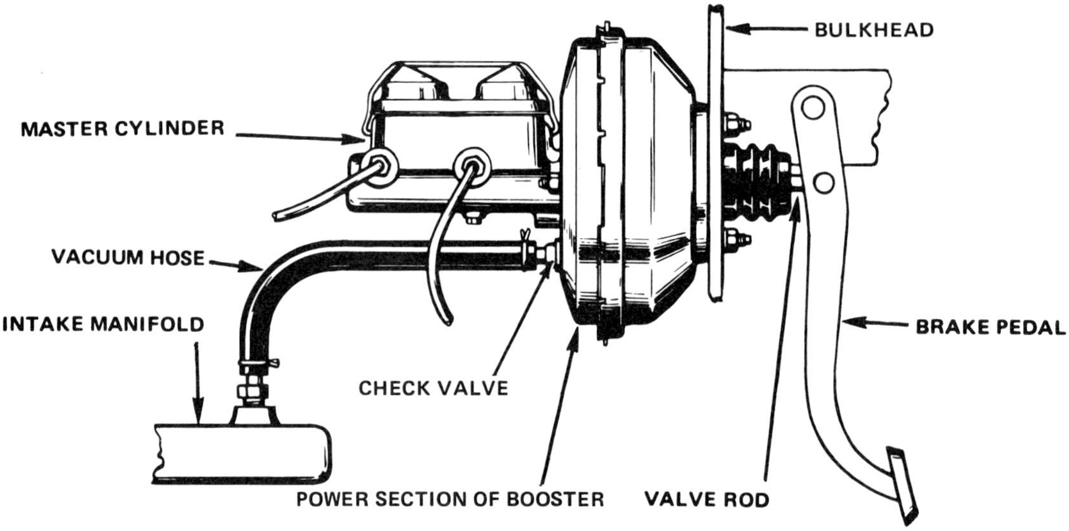

2. A vehicle with the system shown above is losing brake fluid. There is no sign of external leakage. What should a technician do first?
   a. Remove the check valve and measure manifold vacuum
   b. Remove the booster vacuum hose and check inside
   c. Check the brake pedal free travel
   d. Check for a dislodged reaction disc inside the booster

3. With the brakes applied on a car that has a power assisted dual master cylinder, the pedal drops slightly when the engine is started.

   Technician A says that this could be caused by a faulty check valve.

   Technician B says that this could be caused by a faulty master cylinder.

   Who is right?

   a. A only **c.** Both A and B
   b. B only **d.** Neither A nor B

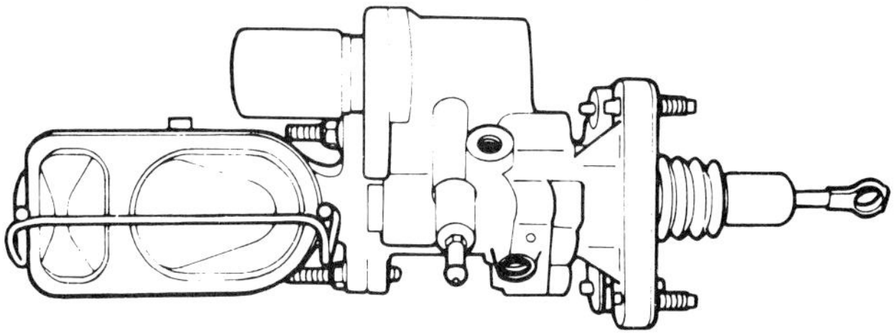

4. The brake pedal chatters on a vehicle with the brake system shown above.

   Technician A says that this could be caused by a slipping power steering belt.

   Technician B says that this could be caused by an outside leak at the accumulator.

   Who is right?

   a. A only **c.** Both A and B
   b. B only **d.** Neither A nor B

Section E

MISCELLANEOUS DIAGNOSIS AND REPAIR

After studying this section, and with hands-on practice on a "live" vehicle, you should be able to perform the following tasks:

- Diagnose wheel bearing noises, wheel shimmy, and vibration problems; determine needed repairs.
- Remove, clean, inspect, repack wheel bearings, or replace wheel bearings and races; replace seals. Adjust wheel bearings in accordance with vehicle manufacturers' specifications.
- Check parking brake system; inspect cables and parts for wear, rusting, binding, and corrosion; clean or replace parts as necessary; lubricate assembly.
- Adjust parking brake assembly; check operation.
- Test parking brake indicator light, switch, and wiring.
- Test, adjust, repair, or replace brake stop light switch and wiring.
- Check operation of antiskid braking systems; adjust or repair according to manufacturers' recommendations.

FRONT WHEEL BEARINGS

The front wheel bearings on automobiles with rear differentials are usually of the tapered roller bearing type (Figure 1).

Front wheel bearing failure can be a serious problem. One type of failure is grease leakage from the wheel hub (Figure 2). This can cause the bearings to overheat and destroy them. In addition, grease can contaminate the brake mechanism and cause grabbing, slipping, or pulling.

Another kind of trouble occurs when the bearings develop pits and fatigue failure because they were not adjusted correctly (Figures 3 and 4).

Loose wheel bearings can cause noise, steering wheel shimmy, cupped or uneven tire wear, vehicle wander and weave, and pulling to one side (Figure 5).

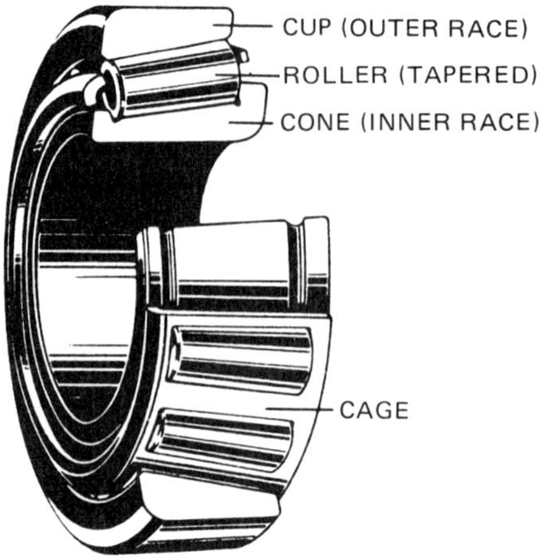

FIGURE 1
Tapered roller bearing. (Courtesy of CR Industries)

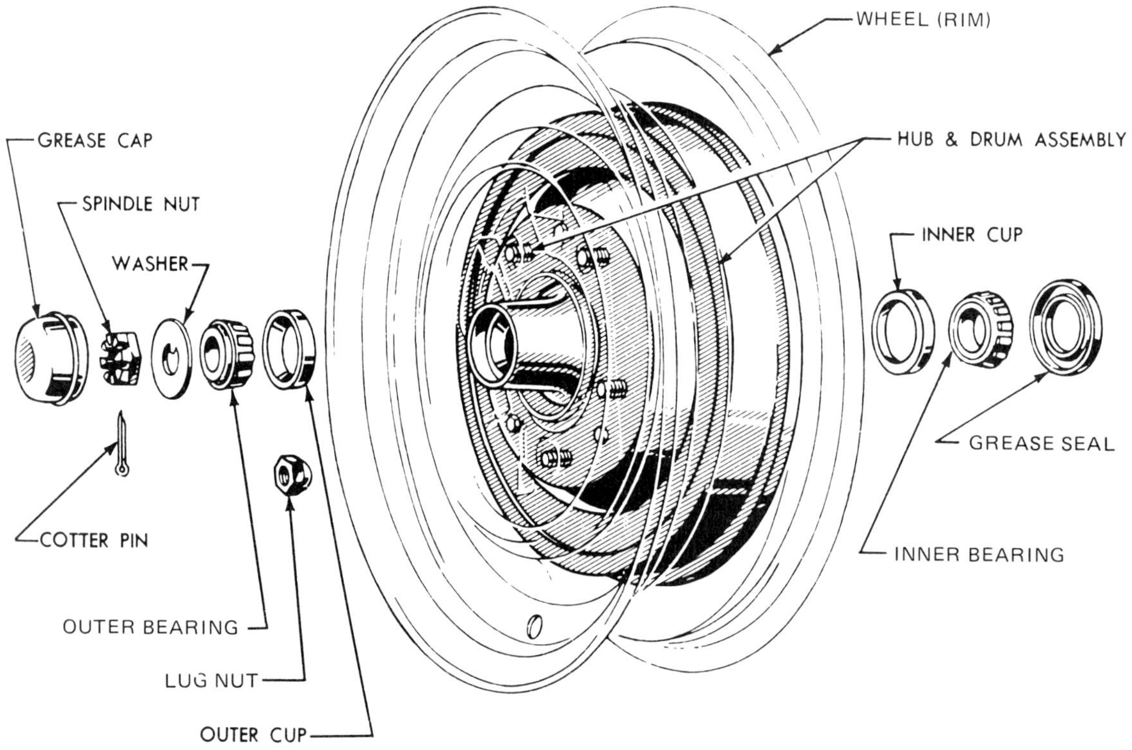

FIGURE 2
Parts of a front wheel assembly. Grease must be kept in the drum (or rotor) hub by the seal. If there is a leak, parts can be ruined. (Courtesy of Ford Parts and Service Division)

FIGURE 3
Wheel bearing failure caused by too tight an adjustment. Notice the chipped small ends of the rollers. (Courtesy of Quaker State Oil Company)

Bearing Checks on the Vehicle

A quick check for bearing roughness can be made by spinning the wheel with one hand and placing your other hand on the bumper. Any roughness felt on the bumper means the bearings must be cleaned and inspected. Bearings that "rumble" or "grind" are worn or pitted and must be replaced.

NOTE: *If the vehicle is equipped with disc brakes, you will usually need to push the pads away from the rotor in order to spin the wheel.*

Bearing looseness can be checked by rocking the wheel with one hand at the bottom of the tire and the other at the top (Figure 6). With the hub cover removed, observe the looseness between the spindle and hub.

COMMENT: *As a rule of thumb, properly adjusted front wheel bearings will allow the wheels to rotate freely with zero (no preload) to .006″ end play (Figure 7).*

Accurate wheel bearing adjustment is the best way to prevent fatigue pitting, flaking, scoring, or brinelling of bearing races and rollers (Figure 8 [a]). See Chart 1 for additional front wheel bearing problems.

FIGURE 4
Metal fatigue produced this wheel bearing cone damage. (Courtesy of Quaker State Oil Company)

FIGURE 5
Cupped tire wear can be caused by loose wheel bearings.

FIGURE 6
Rocking the tire in and out can tell you whether or not the wheel bearings are too loose.

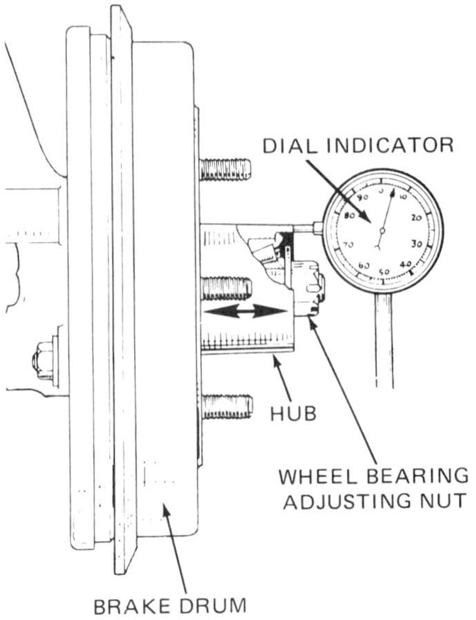

FIGURE 7
A dial indicator can be used to measure bearing end play.

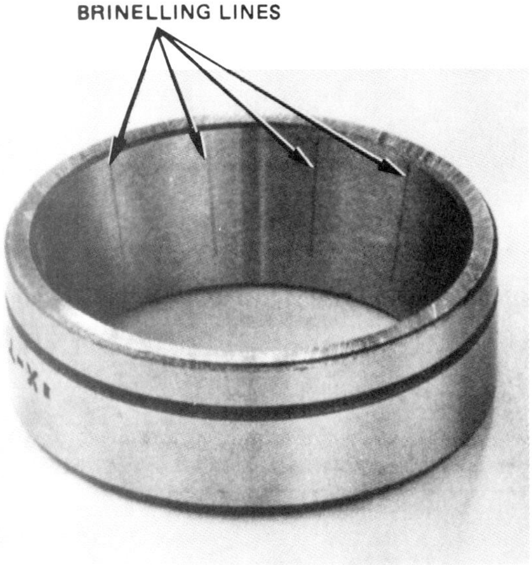

(a)

(b)

FIGURE 8
(a) This cup shows brinelling damage from impact loads. (Courtesy of Quaker State Oil Company) (b) This lining has been soaked with wheel bearing grease and must be replaced. Do not attempt to dry it out and reuse it.

CHART 1 TROUBLESHOOTING CHART FOR FRONT WHEEL BEARINGS

Squeak Noise at Wheel (Brakes Not Applied)
1. Worn, dry, or improperly adjusted wheel bearings
2. Loose lug nuts

Growling, Click, or Rattle
1. Worn, damaged, or dry wheel bearings
2. Loose lug nuts
3. Broken or out-of-position static collector inside grease cap.

Groan, Roughness, or Chatter
1. Worn, dry, or improperly adjusted wheel bearings
2. Loose lug nuts

Brakes Grab or Lock Up When Applied
1. Worn, damaged, or dry wheel bearings

Scraping Noise During a Turn
1. Loose wheel bearings

Lubrication

Front wheel bearing grease should be changed at the manufacturer's recommended intervals. It is very important that all old grease be cleaned out of the wheel hub and bearings before adding fresh grease. Laboratory tests have shown that mixing even small amounts of old, used grease with fresh grease can cause softening and liquifying (Figure 8 [b]). *Use only grease that meets the specifications stated in the manufacturer's service manual.*

SHOP TIP
Wheel bearing lubricant that has been contaminated by water will usually appear milky in color.

Removal (Without Disc Brakes)

1. Raise the vehicle so its front wheels are free of the floor. After lifting the vehicle up, place wheel chocks on both sides of the rear wheels. Then, before working on the vehicle, correctly place jack stands underneath it.

WARNING: Never crawl under a vehicle held up only by a jack!

2. Remove the wheel cover, hub cover, cotter pin, nut lock (if there is one), and spindle nut. See Figure 9.

3. Remove the thrust washer and outer bearing.

4. Carefully slide the wheel, hub, and drum assembly off the spindle. Do not let the inner bearing or seal drag on the spindle threads.

COMMENT: Many shops will try to remove the grease seal at this point. They thread the spindle nut back on and let the seal "catch" on it as the wheel is removed. This practice can bend the wheel bearing cage (Figure 10).

5. Remove the grease seal and the inner bearing (Figure 11).

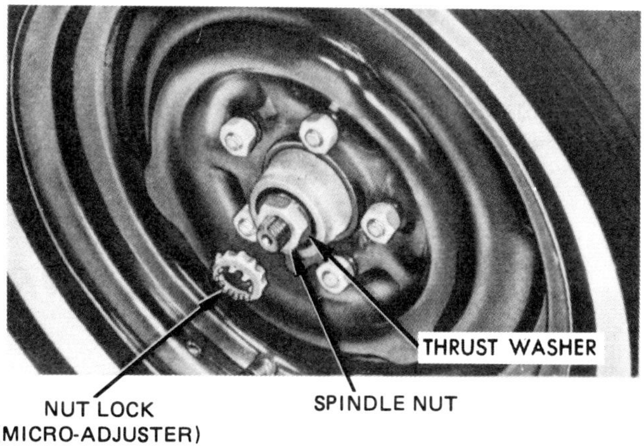

FIGURE 9
A front wheel bearing adjustment mechanism. (Courtesy of Chrysler Corporation)

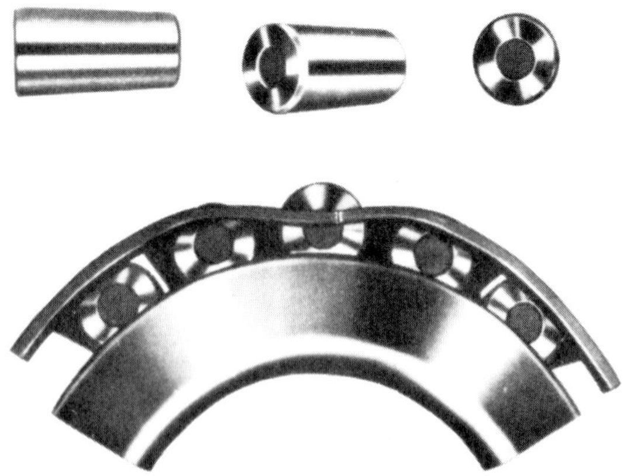

FIGURE 10
Improper bearing removal or handling can bend the bearing cage. This bearing must be replaced. (Courtesy of General Motors Corporation, Product Service Training Division)

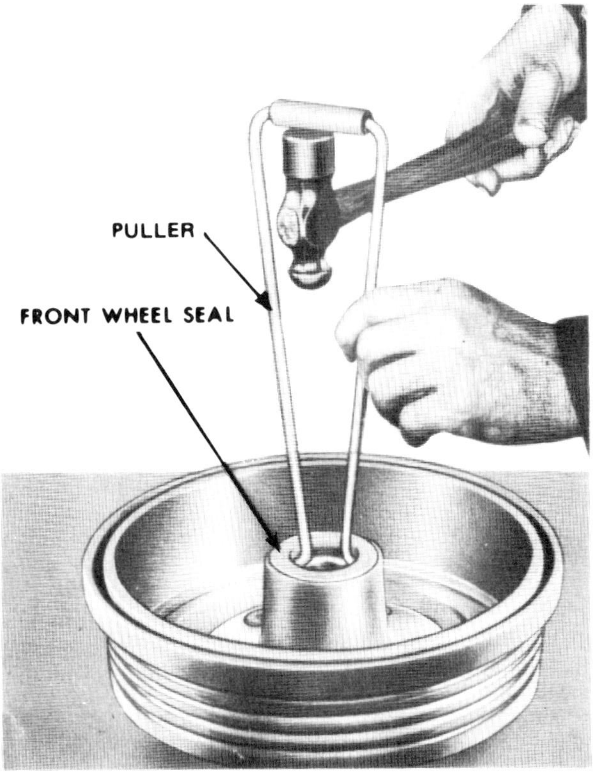

FIGURE 11
This special tool can be used to pull the front wheel seal from a brake drum or rotor. (Courtesy of Ford Parts and Service Division)

Removal (With Disc Brakes)

1. Raise the vehicle so its front wheels are free of the floor.

2. Remove the wheel cover and loosen and remove the lug nuts; remove the wheel and tire assembly.

3. Remove the grease cap, cotter pin, nut lock (if there is one), and spindle nut. See Figure 12.

4. Loosen and remove the caliper mounting bolts.

CAUTION: The brake hose does not need to be disconnected, but the caliper must be supported while disconnected to avoid brake hose damage. The caliper can usually be rested on the "A" arm or suspended by wire.

5. Remove the thrust washer and outer bearing.

6. Carefully slide the wheel hub and rotor assembly off the spindle.

7. Remove the grease seal and the inner bearing (Figures 13 and 14).

FIGURE 12
Prying off the grease cap with special pliers. Channel lock pliers may also be used. (Courtesy of CR Industries)

FIGURE 13
The old seal and inner bearing can be tapped out with a wooden dowel. (Courtesy of CR Industries)

FIGURE 14
This method is often used to remove the grease seal from the hub. (Courtesy of CR Industries)

FIGURE 15
A solvent tank can be used to wash all old grease from bearings. However, dry the bearings thoroughly before repacking with grease. (Courtesy of CR Industries)

Cleaning and Inspection

1. Using clean solvent on a clean rag, remove grease and dirt from the hub cover, spindle nut, thrust washer, hub cavity, and spindle.

2. Wash all old grease from the inner and outer bearings (Figure 15). Dry *thoroughly* using compressed air.

CAUTION: Never spin a bearing with compressed air. This can cause personal injury and bearing damage. Always wear eye protection when using compressed air.

COMMENT: Some shops wash the bearings in hot, soapy water after cleaning them with solvent. They believe this helps the grease stick to the bearing surfaces better.

3. Inspect the cleaned bearing for nicks, cracks, heat discoloration, cage damage, pitted areas, or other defects (Figure 16). If any of these are present, discard the bearing and install a new one along with a new cup. *Never install a new bearing with a used bearing cup, or vice versa.*

FIGURE 16
Excessive heat will cause softening of races and rollers along with heat discoloration. The discoloration can range from faint yellow to dark blue. Replace bearings that have been damaged by overheating. (Courtesy of General Motors Corporation, Product Service Training Division)

SHOP TIP

If the electrical system on a vehicle is not properly grounded to the body frame, it may seek a ground through the drive line and into the wheel bearings. This will "cook" the bearings and cause early failure. If the bearing balls or rollers are black, poor body grounding is the reason.

4. Inspect the bearing and seal surfaces on the spindle for burrs, nicks, or embedded particles or wear (Figure 17). Minor imperfections can be smoothed out with crocus cloth. If a score or groove is deep enough to catch your fingernail, the spindle should be replaced.

FIGURE 17

A badly damaged spindle. Too much heat caused the lubricant to break down, and the bearing cones became welded to the spindle.

SHOP TIP

When inspecting, replacing, or repacking bearings, make sure that the inner cones of the bearings are free to "creep" on the spindle. The bearings are designed to "creep" so that there will be a constantly changing load contact between the cones and the rollers. Polishing and applying a thin layer of grease to the spindle will permit this movement.

5. Inspect the bearing cups for looseness in the hub. The hub, cups, and bearings must be replaced if looseness exists. Look for etching, pitting, or spalling (metal flaking) on cup surfaces. If any of these conditions are present, replace the cup and its bearing (Figure 18 [a] and [b]). *Always replace cups with the correct tools in order to ensure proper cup seating.*

Bearing Repacking/Seal Installation

1. Be sure the new seal is the correct one for the application.

2. By hand, or with a bearing repacker, force grease through the cage and rollers on all surfaces of the bearing (Figures 19 and 20).

3. Fill the hub grease cavity (if there is one) with the recommended wheel bearing grease (Figures 21 and 22).

4. Place the inner side of the drum or rotor face-up and set the inner bearing in the hub.

5. Lightly coat the lip of the new seal with grease.

6. Slide the seal onto the proper driver. *The sealing lip must point toward the bearing.*

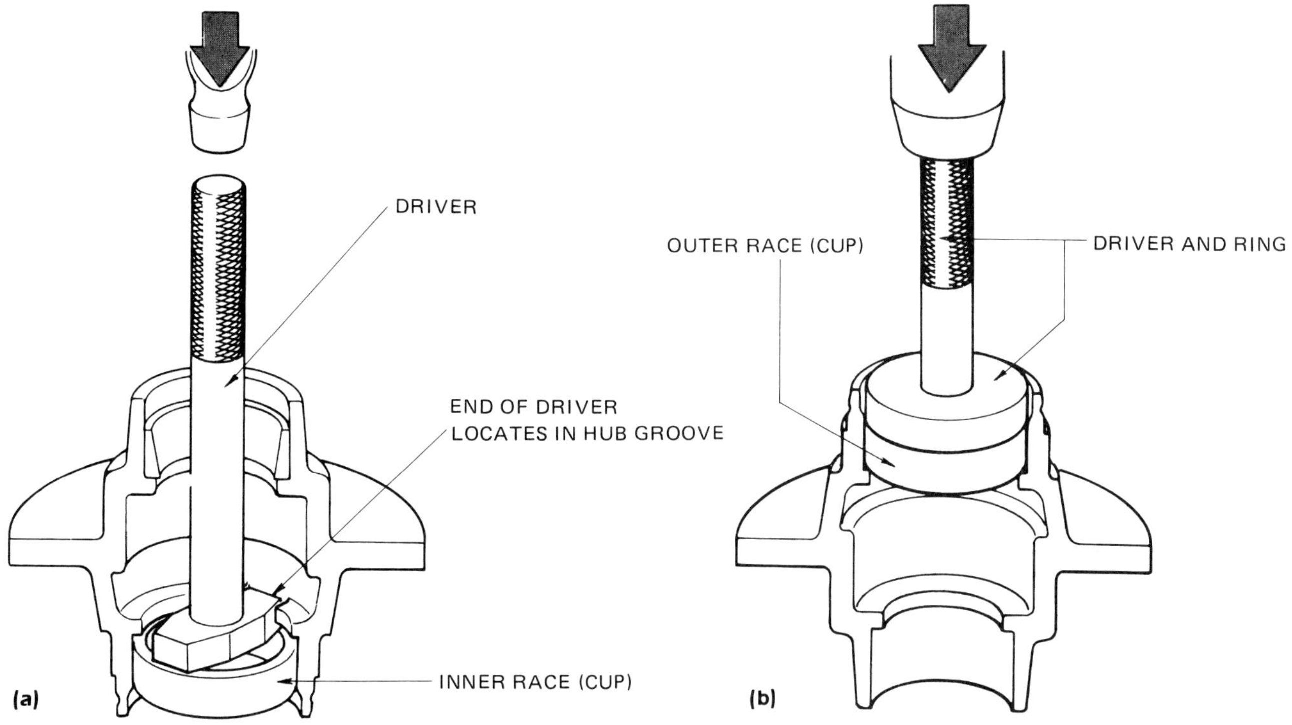

FIGURE 18

(a) Bearing cups are pressed in the hub. This picture shows an inner cup being driven out. (b) Driving in an outer bearing cup. Using the proper tools will help to make sure that a cup is fully bottomed (seated) in the hub. (Courtesy of Toyota Motor Sales, U.S.A., Inc.)

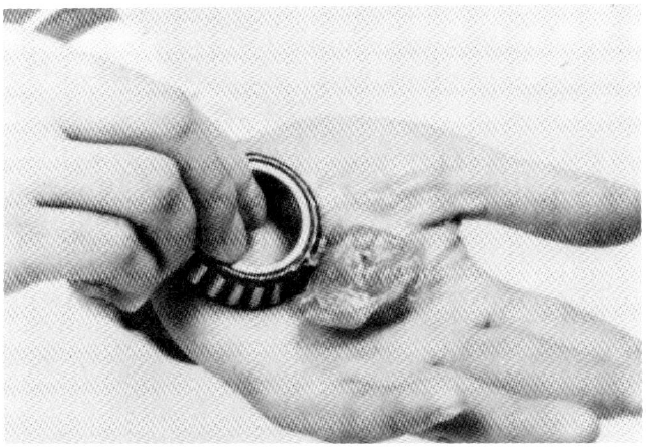

FIGURE 19

Hand packing grease into a bearing. The grease must be forced completely through the rollers and cage. (Courtesy of Toyota Motor Sales, U.S.A., Inc.)

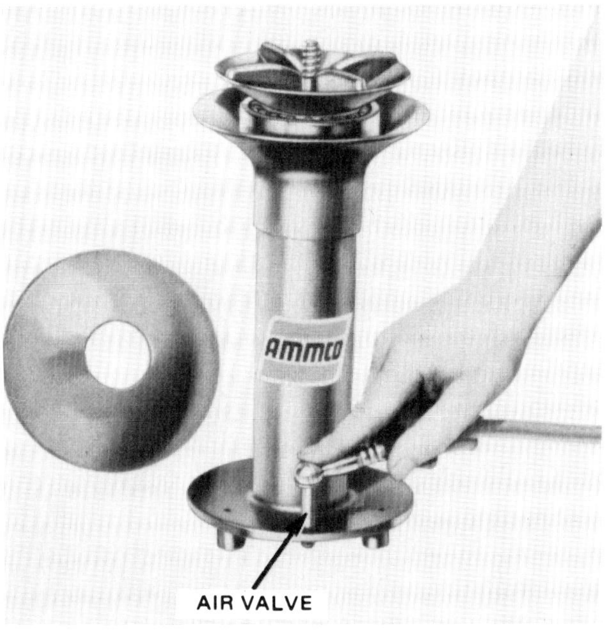

FIGURE 20

Air-operated bearing repacker. (Courtesy of Ammco Tools, Inc.)

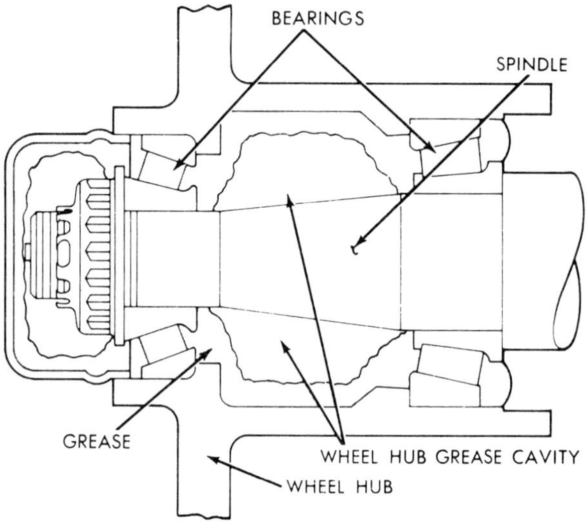

FIGURE 21
Front wheel bearing lubrication. Some hubs are built with a grease cavity. As shown, a small amount of grease is placed inside. This extra reservoir of grease assists lubrication flow through the bearings. DO NOT place grease inside a hub that does not have a grease cavity; the grease seal will leak. (Courtesy of Chrysler Corporation)

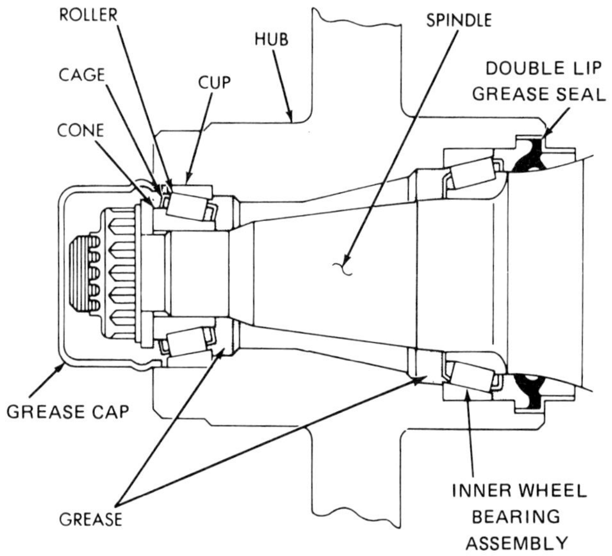

FIGURE 22
Wheel bearing hub built without a grease cavity. (Courtesy of Chrysler Corporation)

7. Position the seal so it starts squarely in the hub without cocking. Tap the driver until the seal is seated (Figure 23). *If the proper driver is not available, use a wood block and hammer to drive in the seal. Never hammer directly on the seal.*

8. Carefully lift and push the drum or rotor assembly onto the spindle. Keep the drum or rotor centered, so the seal is not touched or damaged by the spindle threads. Push the drum or rotor back until the seal and inner bearing are seated on the spindle.

9. Install the outer bearing, washer, and adjusting nut in that order.

10. Replace the caliper on cars that have disc brakes.

11. Replace the wheel. Replace the lug nuts and tighten them evenly.

Bearing Adjustment

12. Rotate the wheel to be sure the brakes are not dragging. Dragging brakes can cause a false adjustment.

13. Adjust the bearing assembly to the manufacturer's recommended setting (Figure 24). If this information is not available, follow these guidelines:

While rotating the wheel, tighten the adjusting nut until there is a slight drag and all bearing surfaces are in contact. Back off the adjusting nut 1/6 to 1/4 turn, or enough so that the wheel rotates easily with zero (no preload) to .006″ end play.

FIGURE 23
Installing a grease seal. Tap the driver until the seal is properly seated in the hub. (Courtesy of CR Industries)

VEHICLE	BRAKE SHOE	BRAKE DRUM		BRAKE PAD	BRAKE ROTOR					CALIPER		WHEEL	WHEEL BEARING SETTING		
		DIAMETER			MIN. THICKNESS										
YEAR, MAKE AND MODEL	* Minimum Lining Thickness	Standard Size	Machine To	Minimum Lining Thickness	Machine To	Discard At	Variation From Paral-lelism	Runout T.I.R.	DESIGN	Mounting Bolts Torque (ft-lbs)	Bridge, Pin or Key Bolts Torque (ft-lbs)	Lugs or Nuts Torque (ft-lbs)	STEP 1 Tighten Spindle Nut (ft-lbs)	STEP 2 Back Off Retorque (in-lbs)	STEP 3 Lock, or Back Off and Lock
GENERAL MOTORS CORP — BUICK															
82 Century, Skyhawk															
w/vented disc	*	7.880	7.899	.030	.830	.815	.0005	.004	40	—	28	100	NON	ADJUSTABLE	
w/solid disc	*	7.880	7.899	.030	.444	.429	.0005	.004	40	—	28	100	NON	ADJUSTABLE	
82 Regal, LeSabre	*	9.500	9.560	.030	.980	.965	.0005	.004	4	—	35	80†	12	Handtight	½ Flat
82-79 Riviera w/o rear disc brakes	*	9.500	9.560	.030	.980	.965	.0005	.004	4	—	35	100	NON	ADJUSTABLE	
w/rear disc brakes	—	—	—	.030	.980	.965	.0005	.004	41	32	30	100	—	—	—
82-79 Electra, Estate Wagon	*	11.000	11.060	.030	.980	.965	.0005	.004	4	—	35	100	12	Handtight	½ Flat
82-80 Skylark	*	7.880	7.899	.030	.830	.815	.0005	.003	40	—	28	103	NON	ADJUSTABLE	
81-79 Century, Regal, LeSabre	*	9.500	9.560	.030	.980	.965	.0005	.004	4	—	35	80†	12	Handtight	½ Flat
80-76 Skyhawk	*	9.500	9.560	.030	.830	.815	.0005	.005	15	—	—	80	12	Handtight	½ Flat
79 Skylark	*	9.500	9.560	.030	.980	.965	.0005	.004	4	—	35	80	12	Handtight	½ Flat
78-77 Electra, Estate Wagon, Riviera	*	11.000	11.060	.030	.980	.965	.0005	.004	4	—	35	80■	12	Handtight	½ Flat
78-77 Century, LeSabre, Regal, Skylark	*	9.500	9.560	.030	.980	.965	.0005	.004	4	—	35	80■	12	Handtight	½ Flat
76-75 Riviera, Electra, Estate Wagon, LeSabre, Custom	*	12.000	12.060	.030	1.230	1.215	.0005	.005	4	—	30-40	90	19	132	1/6 Turn
76-75 Apollo, Skylark	*	9.500	9.560	.030	.980	.965	.0005	.004	4	—	35	70	19	132	1/6 Turn
75 Century, Regal	*	9.500	9.560	.030	.980	.965	.0005	.004	4	—	35	70	19	132	1/6 Turn
75 Skyhawk	*	9.000	9.060	.030	.455	.440	.0005	.005	15	—	35	70	19	132	1/6 Turn
74-72 Electra, Custom, LeSabre, Centurion, Riviera, Wildcat	*	11.000	11.060	.030	1.230	1.215	.0005	.005	4	—	30-40	75	12	Handtight	½ Flat
74-73 Estate Wagon	*	12.000	12.060	.030	1.230	1.215	.0005	.005	4	—	30-40	75	12	Handtight	½ Flat
74-73 Century, Regal	*	9.500	9.560	.030	.980	.965	.0005	.004	4	—	35	65-70	19	132	1/6 Turn
72 GS, Gran Sport, Skylark, Sports Wagon	*	9.500	9.560	.030	.980	.965	.0005	.004	4	—	35	65-70■	19	132	1/6 Turn

*.030″ over rivet head, if bonded lining use .062″ ■ w/½″ stud 100 ft/lbs. † w/Aluminum whls. LeSabre 90 ft/lbs., Regal 100 ft/lbs.

FIGURE 24

Brake specification charts are widely available. They generally contain factory wheel bearing setting data. (Courtesy of Ammco Tools, Inc.)

14. Put in a new cotter pin and bend it. If a static collector is used in the hub cover, make sure the cotter pin is not in the way. *The cotter pin must be tight.* Replace the hub cover.

15. Torque the lug nuts to the manufacturer's recommended specifications.

16. Replace the wheel cover and lower the car.

SHOP TIP

Reusing old cotter pins can lead to serious trouble. Old cotter pins that have been bent and rebent are very apt to break or shear off. Pieces of broken cotter pins have often caused badly damaged bearings and have even caused the wheels to seize up. A broken cotter pin can also allow the spindle nut to work loose or tighten, either of which is dangerous.

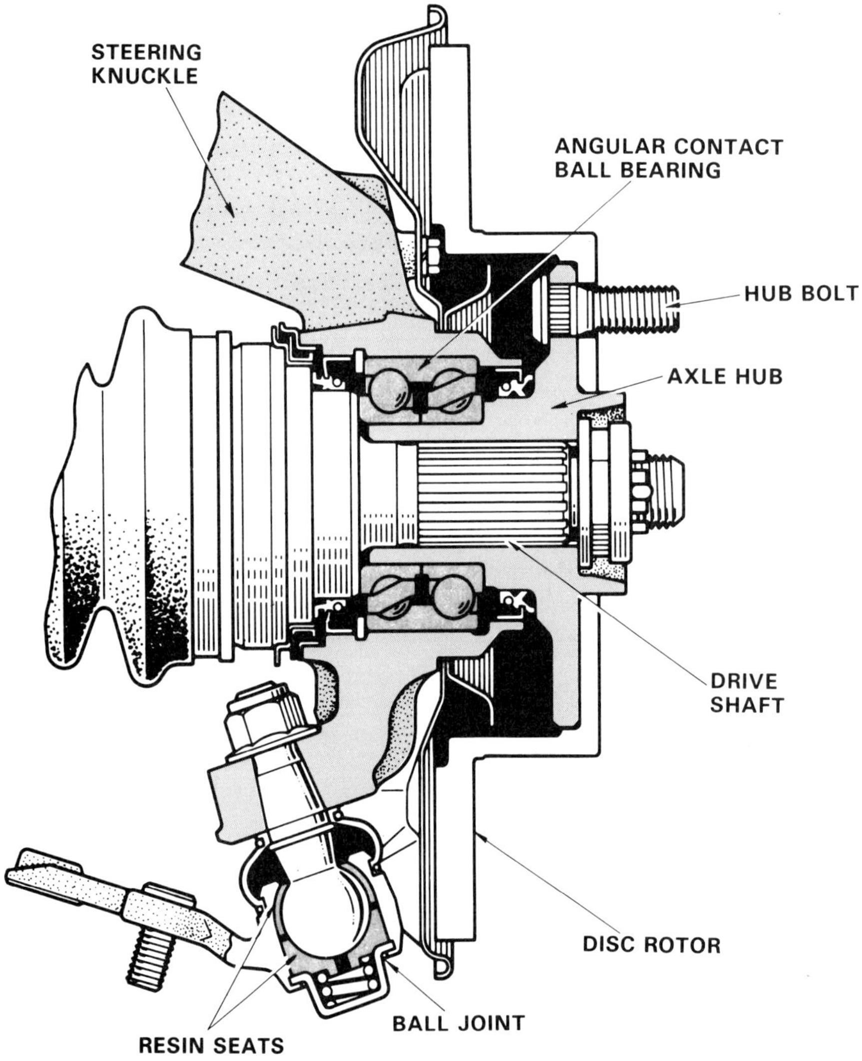

FIGURE 25
This late model front wheel drive assembly has permanently sealed, nonadjustable ball bearings. (Courtesy of Toyota Motor Sales, U.S.A., Inc.)

FRONT WHEEL DRIVE BEARINGS AND HUB

Most front wheel drive automobiles are designed with permanently sealed front wheel bearings that require no adjustment or periodic repacking (Figure 25). Special tools are often required to remove and install the hub (Figure 26).

COMMENT: The rear wheel bearings on some front wheel drive automobiles require periodic repacking. Procedures are similar to those used with front wheel bearings on rear wheel drive cars.

FIGURE 26
Pulling a front wheel drive rotor and hub.

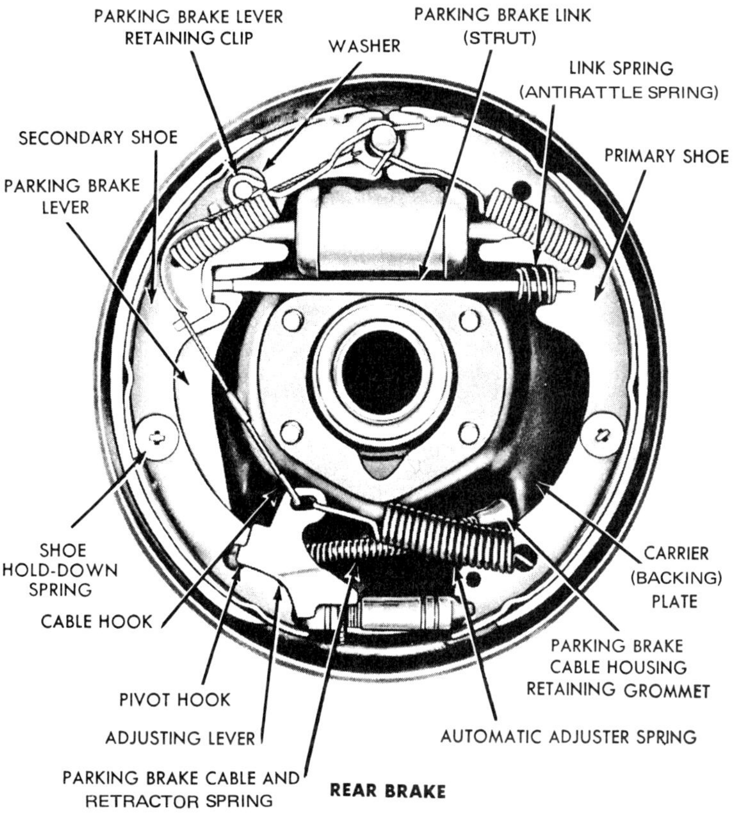

FIGURE 27
Bendix-design right rear brake assembly. Locate the parking brake cable, lever, and strut. (Courtesy of Ford Parts and Service Division)

FIGURE 28
Another rear drum type of parking brake. The self-adjuster lever is operated when the parking brake is applied. (Courtesy of American Honda Motor Company, Inc.)

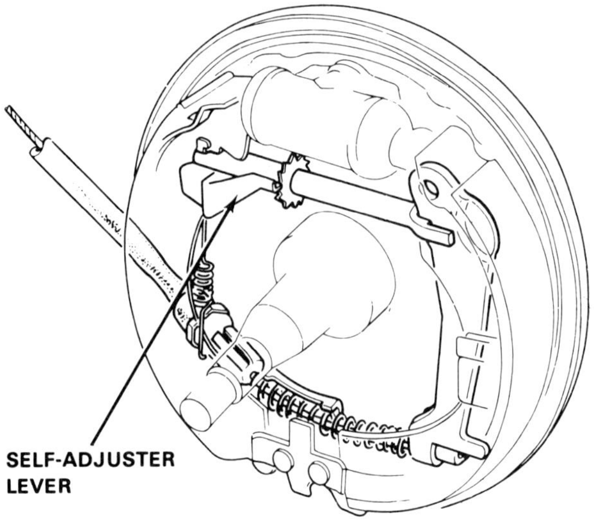

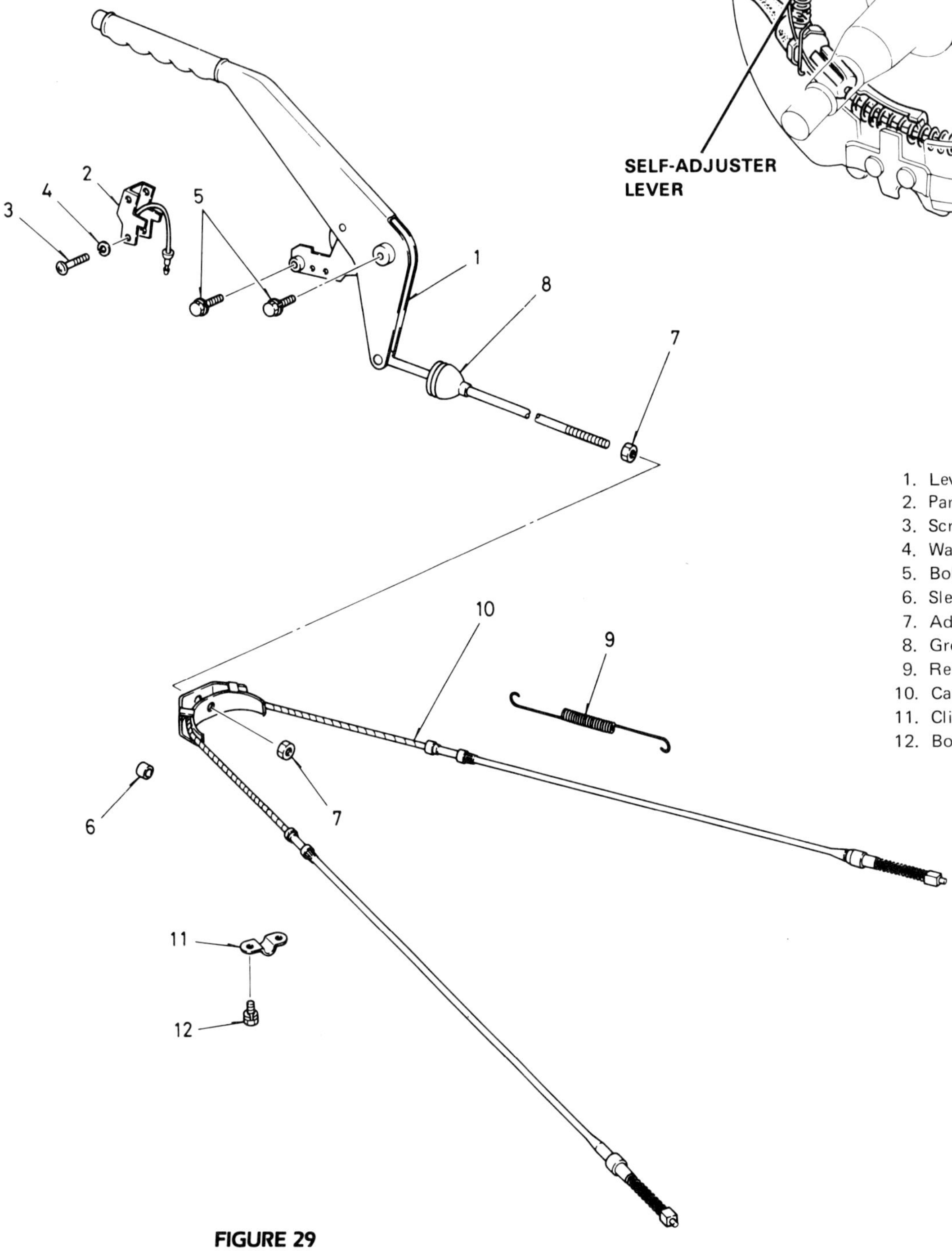

1. Lever arm (release handle)
2. Parking brake switch
3. Screw
4. Washer
5. Bolts
6. Sleeve
7. Adjusting nuts at equalizer
8. Grommet
9. Return spring
10. Cable
11. Clip (cable to floor)
12. Bolt

FIGURE 29
A typical parking brake lever and cable arrangement. (Courtesy of Isuzu Motors Limited)

PARKING BRAKE SYSTEM

The parking brake is a hand- or foot-operated mechanical brake designed to hold the vehicle while it is parked.

Today, several types of parking brakes are used. They include the following:

1. The integral type, which is part of the rear wheel drum brakes (Figures 27 and 28). The brake shoes are mechanically operated by a lever and strut connected to a flexible steel cable. This cable is joined to a front brake cable and equalizer extending to a parking brake pedal or release handle (Figure 29).

2. The auxiliary drum type (Figure 30). This arrangement is found on some GM and Chrysler Corporation cars that use disc brakes on the rear.

3. The rear parking brake (Figure 31). Here, the rotor and pads of the rear disc brake are also used for the parking brake.

COMMENT: On some imported cars, the disc parking brake holds the front wheels.

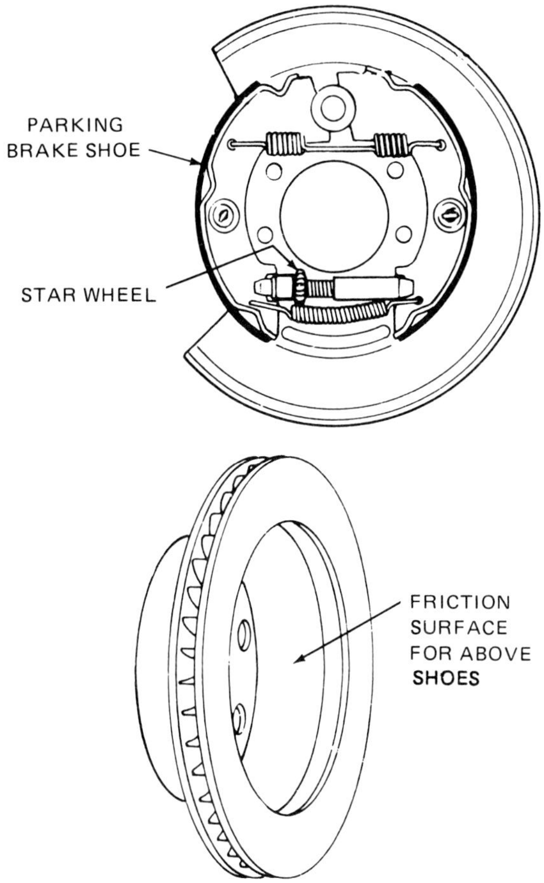

FIGURE 30
These parking brake shoes are cable-operated. Adjustment is made by turning the star wheel. (Courtesy of Allied Aftermarket Division, Allied-Signal Corporation)

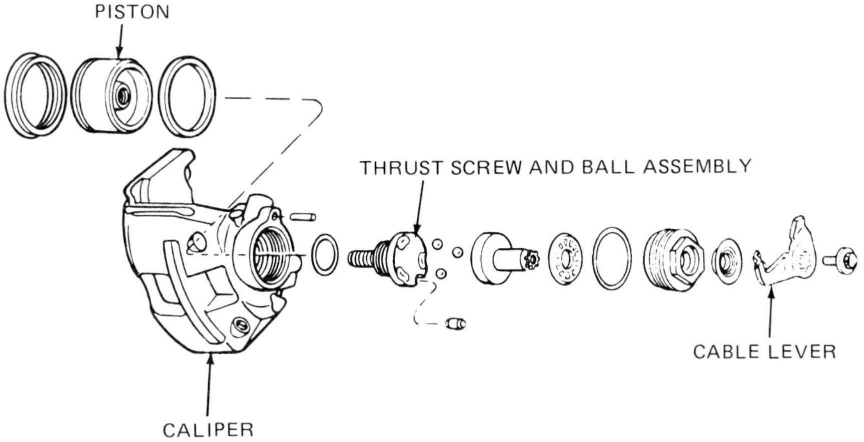

FIGURE 31
A rear disc parking brake. Rotation of the cable lever moves the caliper piston and applies the pads. (Courtesy of EIS Division, Parker Hannifin Corporation)

SHOP TIP

On many rear disc parking brake assemblies, advise the customer to always use the parking brake when parking. This is because the pads adjust automatically when the parking brake is used.

Troubleshooting

Refer to Chart 2 for parking brake problems.

Adjusting Cable Tension

Sometimes it is necessary to adjust the tension in the parking brake cables. Adjustment location depends on the vehicle model (Figures 32 and 33).

A typical parking brake cable adjustment procedure is as follows:

1. Release the parking brake lever and loosen the cable adjusting nut to make sure the cable is slack.

NOTE: The brake shoes must be properly adjusted before adjusting the parking brake.

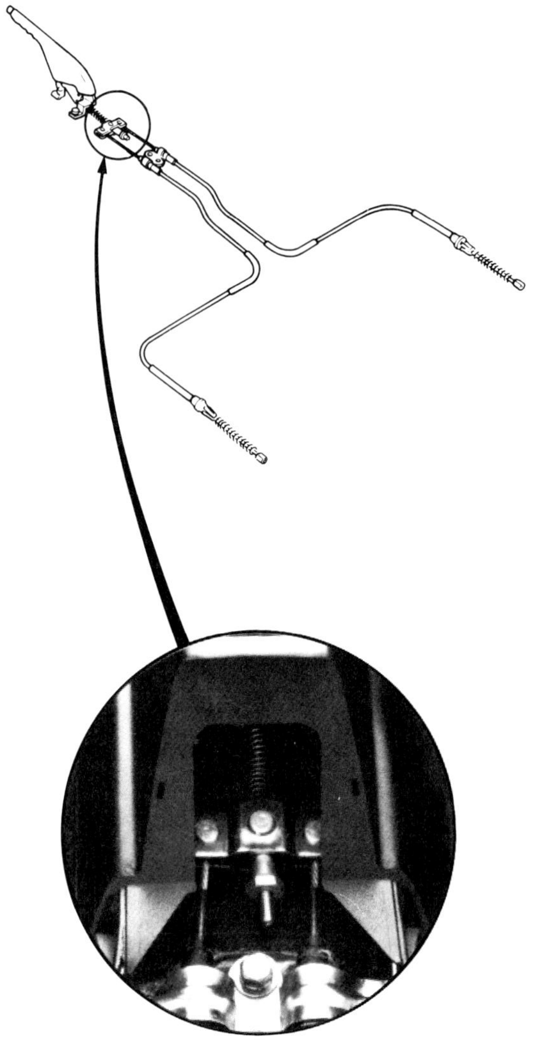

FIGURE 32
The adjuster for the parking brake is often located underneath the car at the cable equalizer yoke. (Courtesy of American Honda Motor Company, Inc.)

CHART 2 TROUBLESHOOTING CHART FOR PARKING BRAKES

Brake Will Not Hold
1. Broken, rusted, or "frozen" brake cable
2. Ratchet or pedal mechanism worn
3. Improperly adjusted brakes or cable
4. Grease- or brake-fluid-soaked brake lining

Dragging Brake
1. Improper cable or brake shoe adjustment
2. Parking brake cable retractor spring weakened by heat
3. Sticking or "frozen" brake cable
4. Bent or rusted cable equalizer
5. Broken brake shoe return spring
6. Broken brake shoe hold-down spring

Brake Does Not Release Properly
1. Binding cable
2. Defective vacuum release diaphragm on pedal mechanism

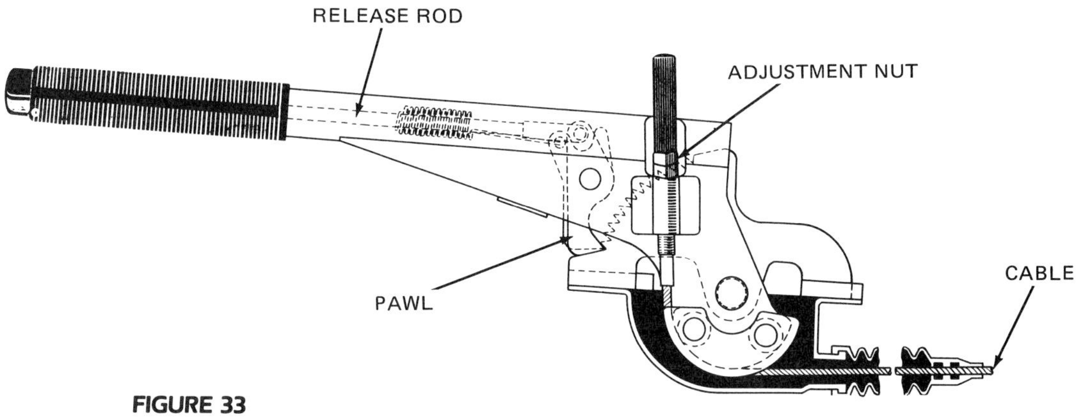

FIGURE 33
Sometimes the parking brake adjustment is made inside the car. (Courtesy of Toyota Motor Sales, U.S.A., Inc.)

2. Clean the cable threads with a wire brush and coat with antiseize lubricant (Figure 34). Lubricate all pivot points (Figure 35).

3. Tighten the cable adjusting nut until a slight drag is felt while rotating the rear wheels.

4. Loosen the cable adjusting nut until both rear wheels can be rotated freely.

5. Back off the cable adjusting nut two full turns.

> **SHOP TIP**
> A seized cable adjusting nut can usually be freed by using an oxyacetylene torch. Heat the nut to a dull red color, then unthread the nut using vise-grip pliers and a box-end wrench. BE SURE TO WEAR APPROVED GOGGLES WHEN USING THE OXYACETYLENE TORCH.

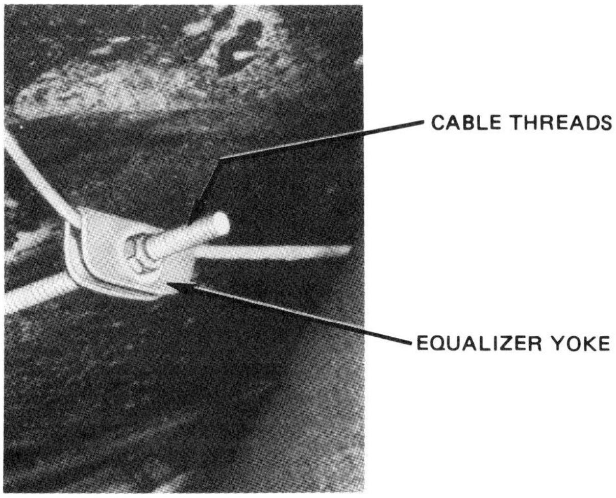

FIGURE 34
Cable threads should be kept clean and lubricated to prevent rusting. (Courtesy of Ford Parts and Service Division)

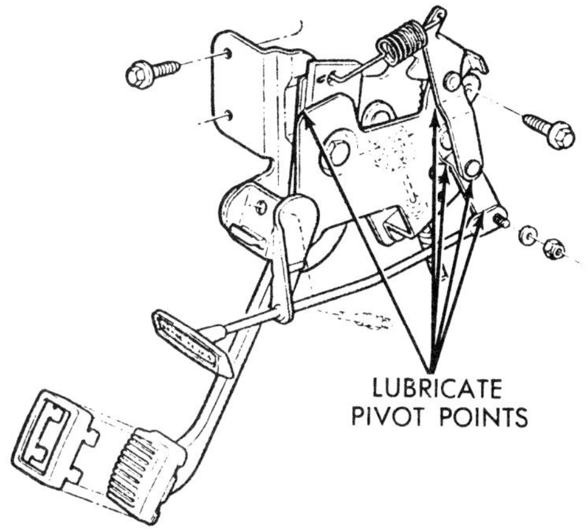

FIGURE 35
Foot-operated parking brake lubrication points. (Courtesy of Chrysler Corporation)

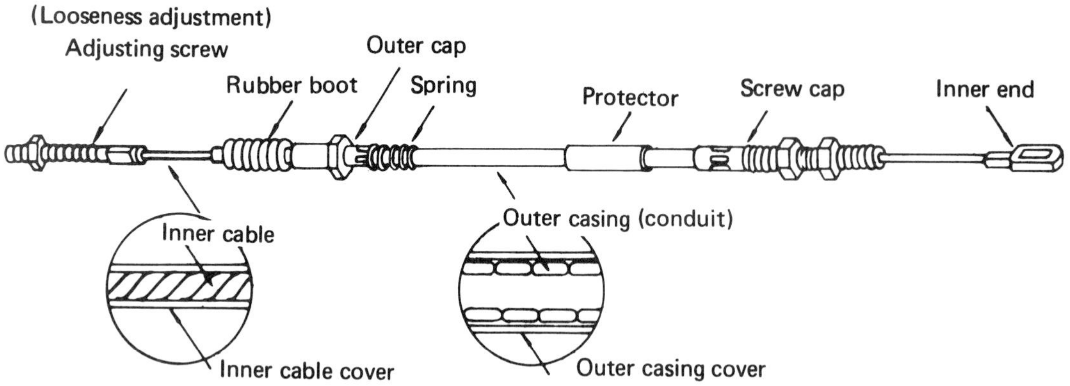

FIGURE 36
Rear parking brake cable and outer casing (conduit).

Replacing Rear Cable

If it becomes necessary to replace the rear parking brake cable (Figure 36), the basic procedure is as follows:

1. Disconnect brake cable from the equalizer.

2. Remove the strut and antirattle spring from the brake shoe assembly. If necessary, refer back to Figure 27.

3. Disconnect the brake cable from the operating lever.

4. Compress any retainers on the end of the brake cable housing and pull the cable from the backing plate (Figure 37).

5. Lubricate the new cable with antiseize lubricant at all metal-to-metal rubbing points.

6. Install a new cable into the backing plate, making sure that it is locked firmly into place.

Indicator Light

On many cars, a parking brake light switch is bolted to the pedal mounting bracket (Figure 38). The switch is turned on by the pedal arm and usually is not adjustable. This switch turns on the parking brake indicator (reminder) light. This light should go off when the parking brake is fully released (Figure 39).

STOP LIGHT SWITCH

The brake stop light switch is an electrical switch designed to cause the stop lights to come on when the brake pedal is depressed.

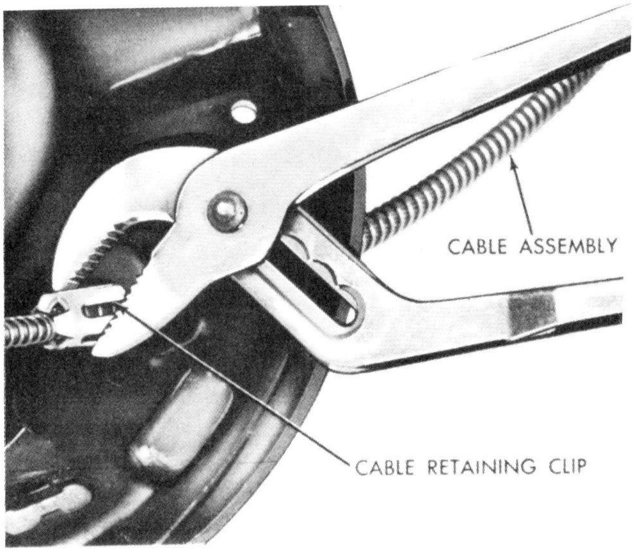

FIGURE 37
Removing brake cable assembly from the backing plate. A box-end wrench slipped over the cable retaining clip can also be used. (Courtesy of Chrysler Corporation)

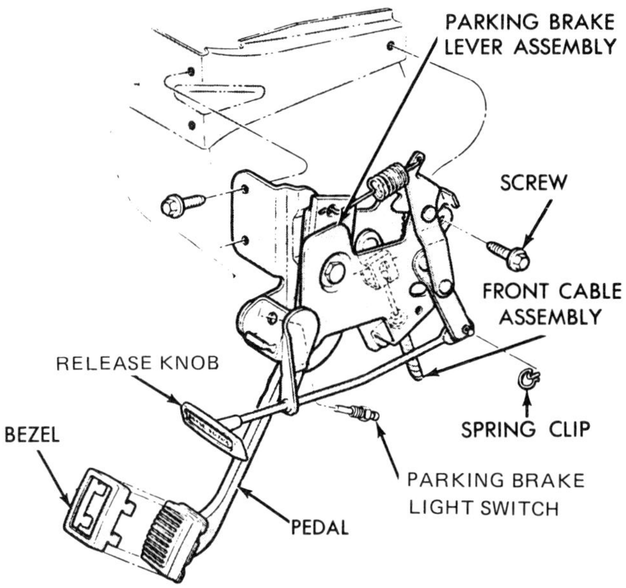

FIGURE 38
A parking brake light switch that is attached to the parking brake pedal. (Courtesy of Chrysler Corporation)

See Figure 40. There are two general types of stop light switches:

1. **Mechanically operated** (Figure 41). These are used on most current model cars.

2. **Hydraulically operated** (Figure 42). These are generally used on older model cars.

Operation

In both types of switches, the circuit through the switch is *open* when the brake pedal is released. When the brake pedal is depressed, the circuit through the switch *closes*.

Servicing

Mechanically operated stop light switches can be damaged by wear, corrosion, or breakage. Hydraulically operated switches can develop leaks.

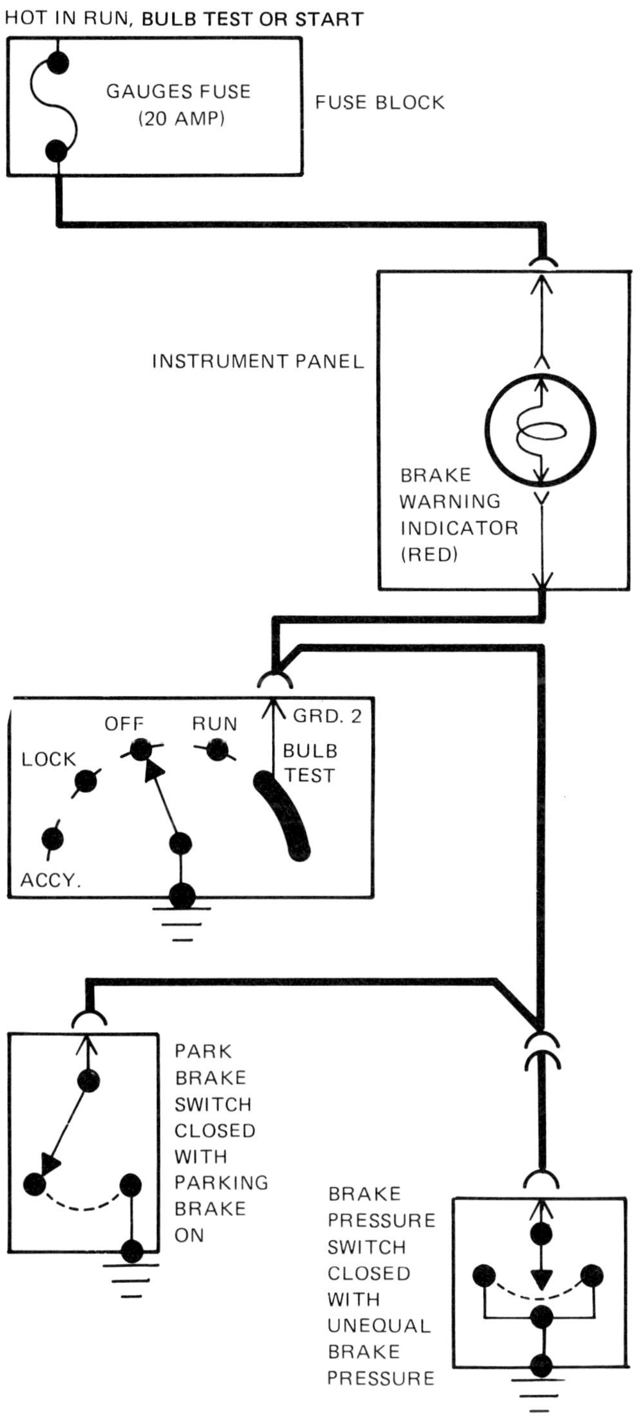

FIGURE 39
Parking brake indicator light and brake warning light circuit.

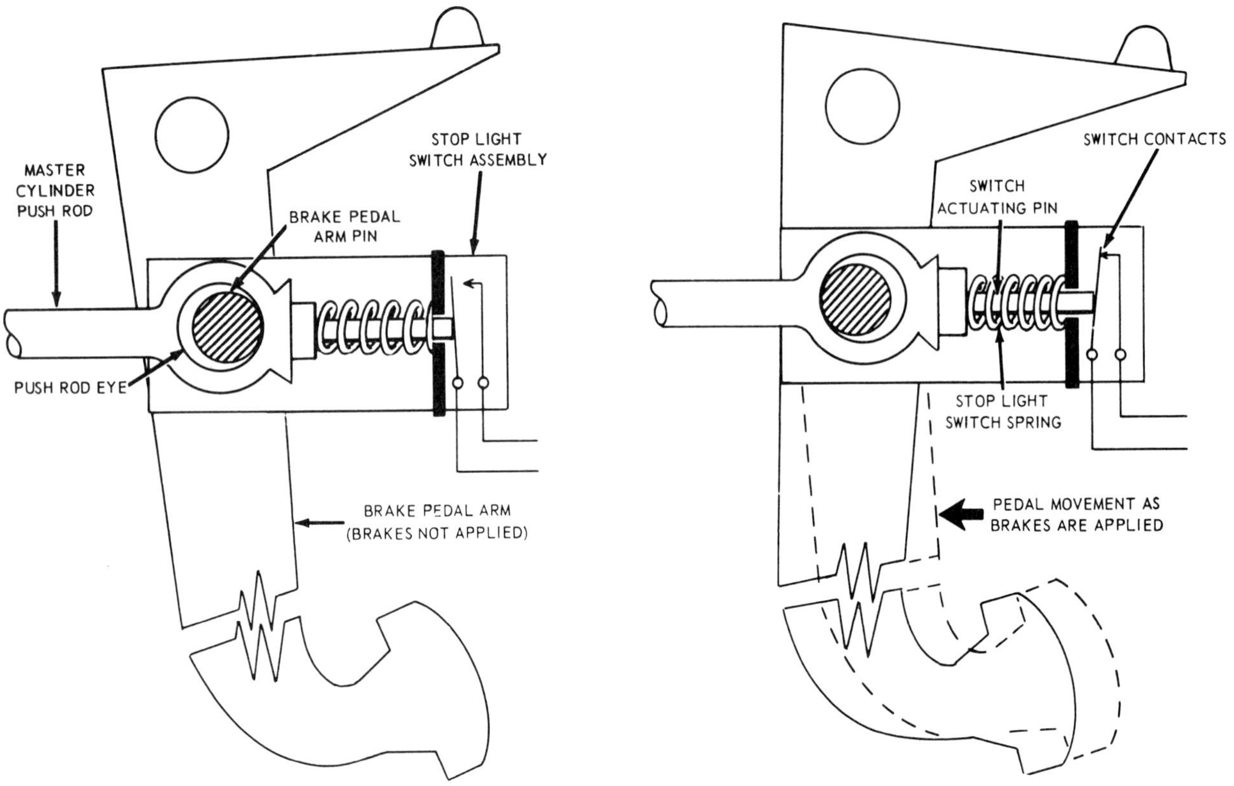

FIGURE 40
Mechanical stop light switch operation. Forward movement of the brake pedal arm pin releases the switch-actuating pin and closes the switch contacts. (Courtesy of Ford Parts and Service Division)

Most mechanical stop light switches are adjustable. The switch should be adjusted so that the stop lights come on when the pedal is pushed in approximately 1/4″.

Before replacing a stop light switch, use a jumper wire to bypass and test the switch. The stop lights will come on if the bulbs, sockets, connectors, and wires have continuity (Figure 43).

If the stop lights come on when a jumper wire is used but do not operate when the brake pedal is pushed down, the switch is faulty.

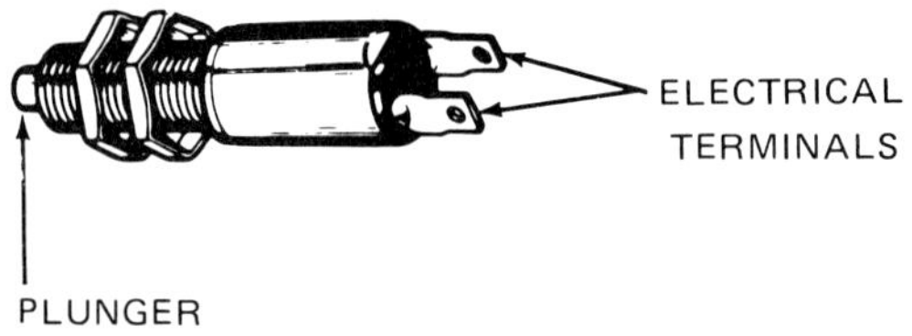

FIGURE 41
A popular type of mechanical stop light switch.

> **SHOP TIP**
> If one or both turn signals will not flash, it may mean that the brake (stop) lights are not working.

FIGURE 42
Hydraulically operated stop light switches are generally found in older cars. However, they are still being used by several automobile manufacturers.

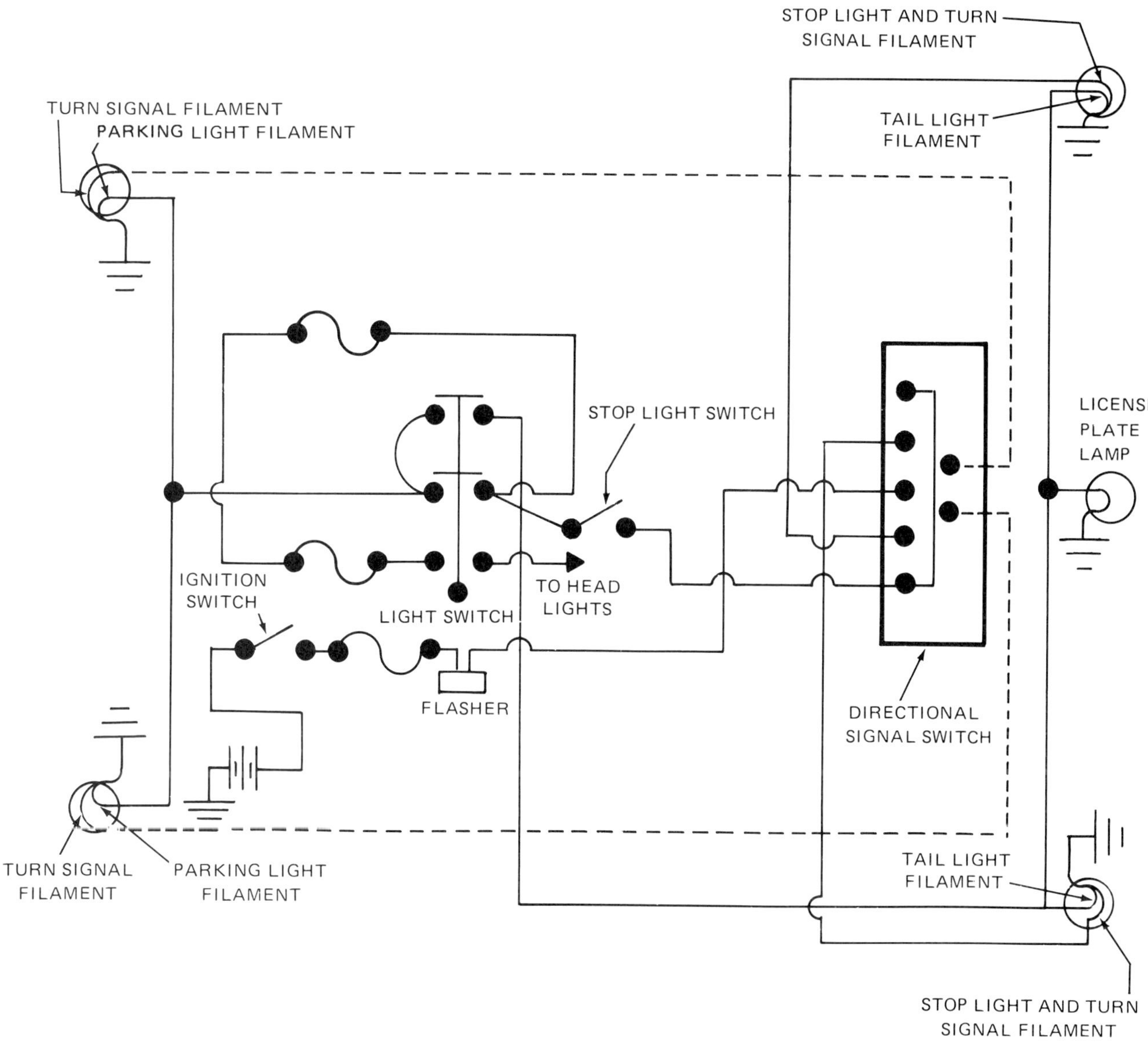

FIGURE 43

Typical stop light and directional signal circuit. Closing the stop light switch allows current to flow to the stop lights through the directional signal switch. A defective directional signal switch can prevent the stop lights from working.

ANTISKID BRAKING SYSTEMS

Antiskid braking systems are designed to prevent the wheels from locking up when the brakes are applied. These systems prevent the skidding caused by a locked wheel, yet maintain the brake pressure needed for maximum stopping effort.

When brake line pressure is not great enough to cause wheel lock-up, the brake system operates normally. However, during hard stops and when braking on wet roads, there can be enough line pressure to cause lock-up. This danger of lock-up is sensed and line pressure is automatically released and applied at a rate of approximately 2 to 4 cycles per second. This cycling continues throughout the braking application until the car is slowed to approximately 5 miles per hour or until the brakes are released by the driver.

On many cars, each time the ignition is turned *on* the antiskid system is operated for one cycle of both release and apply. This cycle may be heard and is normal.

Antiskid systems usually include the following parts:

1. Speed sensors. These are electromagnetic devices that produce an AC voltage with a frequency proportional to rear wheel speed.

2. Control module (logic controller). The control module receives and continuously monitors the voltage signals from the speed sensors. Its basic function is to determine from the voltage signals when wheel lock-up is about to occur and to send an electrical signal to the hydraulic actuator.

3. Hydraulic actuator (pressure modulator). The hydraulic actuator responds to the electrical signal from the control module and modulates line pressure to the brakes.

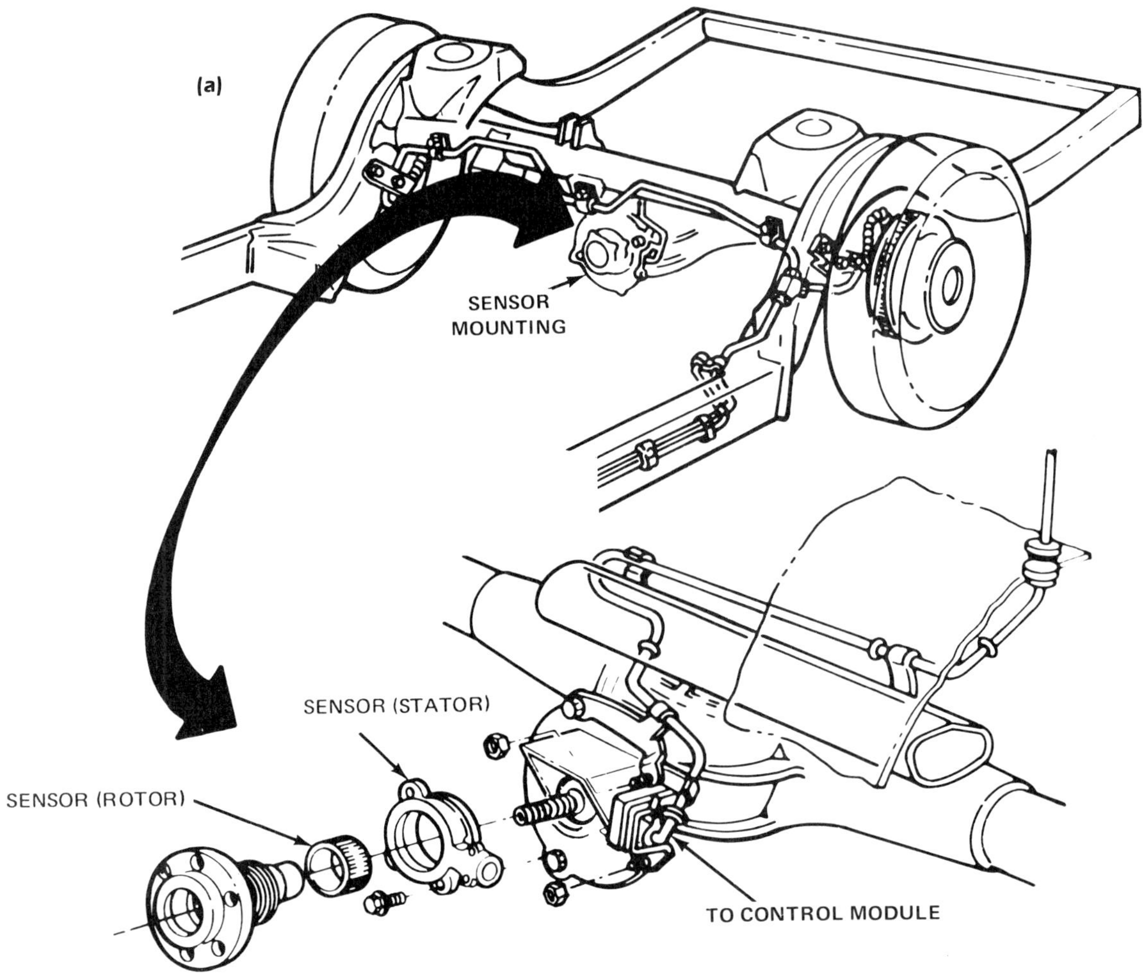

FIGURE 44
(a) This antiskid braking system uses a rear axle drive pinion sensor to report rpm's to the control module. (Courtesy of Ford Parts and Service Division)

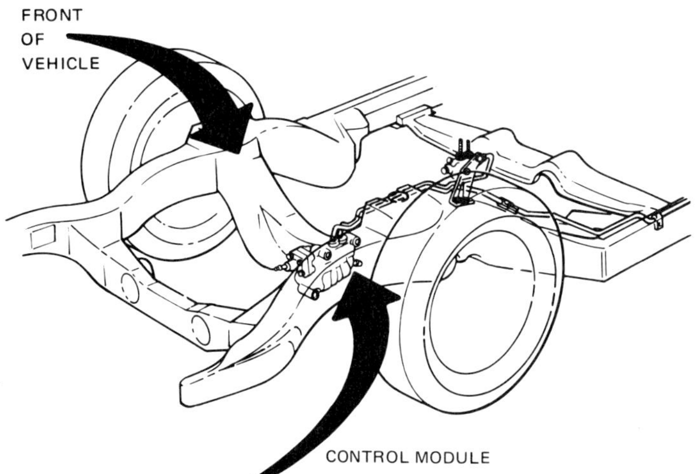

FIGURE 44 (Continued)
(b) Brake line pressure is modulated by a hydraulic actuator. (Courtesy of Ford Parts and Service Division) (c) A left rear wheel speed sensor and wire routing. (Courtesy of Chrysler Corporation)

(b)

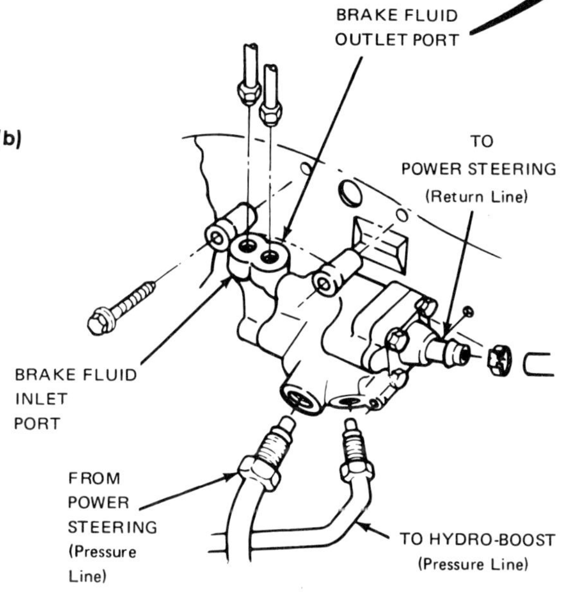

(c)

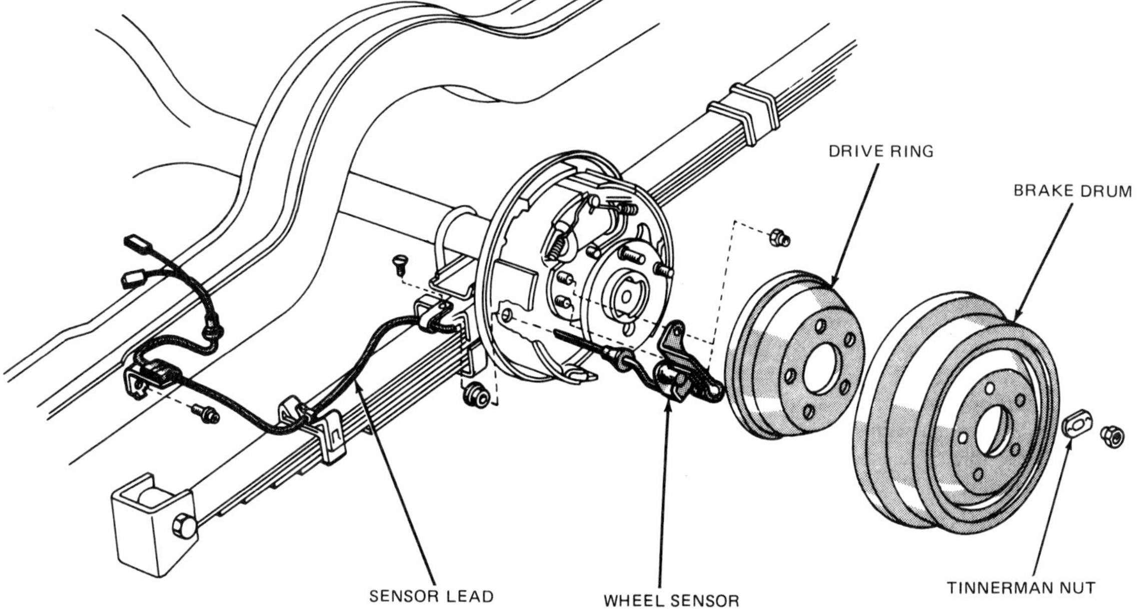

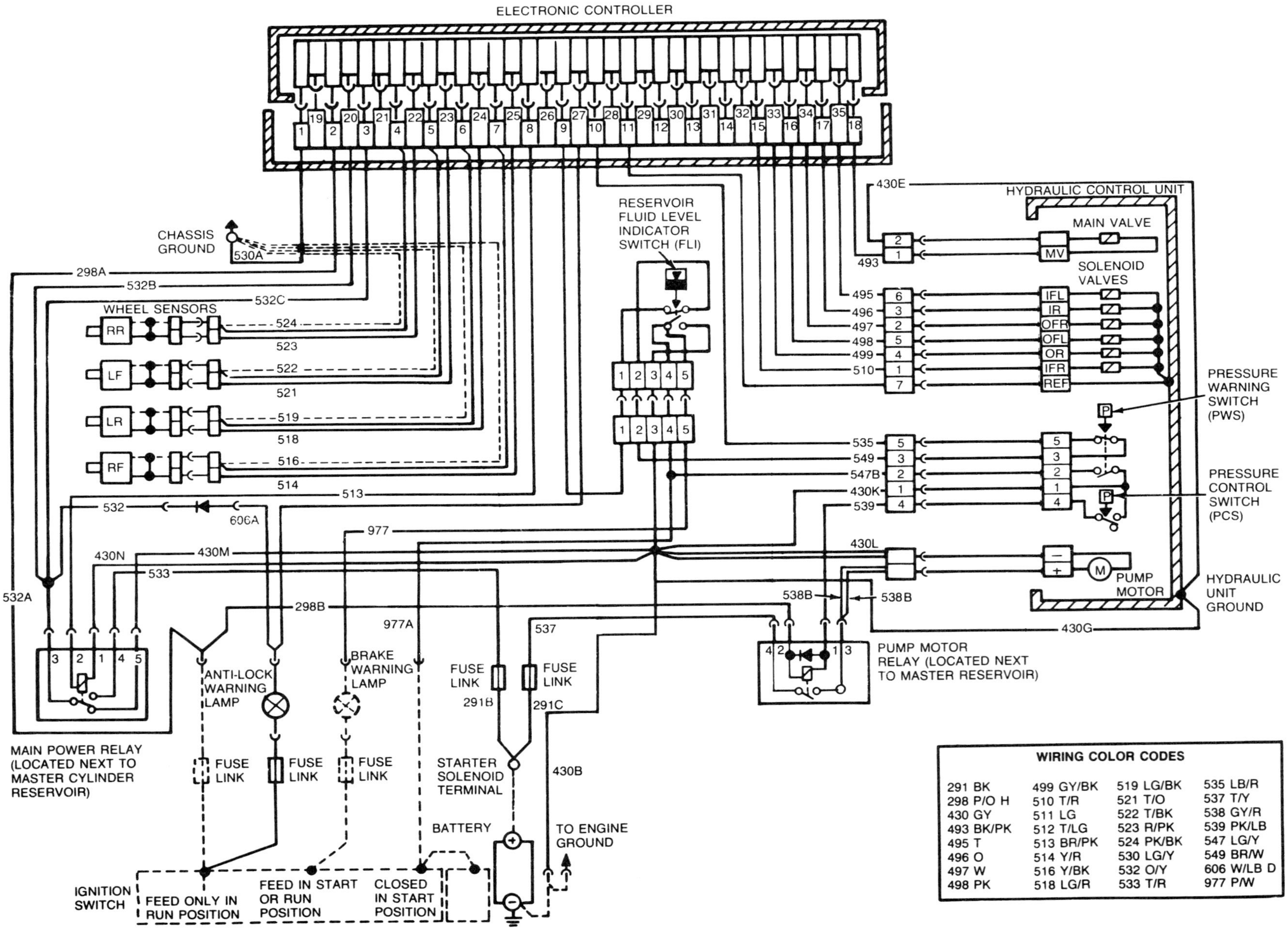

WIRING COLOR CODES			
291 BK	499 GY/BK	519 LG/BK	535 LB/R
298 P/O H	510 T/R	521 T/O	537 T/Y
430 GY	511 LG	522 T/BK	538 GY/R
493 BK/PK	512 T/LG	523 R/PK	539 PK/LB
495 T	513 BR/PK	524 PK/BK	547 LG/Y
496 O	514 Y/R	530 LG/Y	549 BR/W
497 W	516 Y/BK	532 O/Y	606 W/LB D
498 PK	518 LG/R	533 T/R	977 P/W

FIGURE 45

Wiring schematic for a 1986 Ford four-wheel antilock braking system. When troubleshooting, use a volt-ohmmeter (VOM) to check for continuity and resistance values as detailed in the shop manual. (Courtesy of Ford Parts and Service Division)

Figure 44 (a) and (b) show one system. A rear axle drive pinion sensor reports pinion rpm to the control module, which analyzes the input and gives commands to the actuator. Figure 44(c) shows a system that uses wheel sensors.

SERVICING

Antiskid braking systems must be serviced using factory procedures and specifications. An ohmmeter and voltmeter can be used to check out circuitry (Figure 45). The rest of the brake system is conventional, and is serviced as such.

Operational Check

1. Raise the car on a hoist so that all wheels are just above the floor.

2. Place the transmission in neutral and start the engine.

3. Rotate one of the antiskid wheels using a dynamic wheel balancer.

4. Remove the drive motor from the wheel, apply the brake, and use the strobe light to see whether or not the wheel is cycling. If the wheel stops immediately when the brake is applied, the system is not working. Repeat the procedure for the other antiskid wheels.

Repair Precautions

1. On many systems, before the actuator or pressure warning switch is removed (Figure 46), it is mandatory to discharge hydraulic pressure from the system. This is done by turning the ignition *off* and pumping the brake pedal a minimum of twenty times until an increase in pedal force is felt clearly.

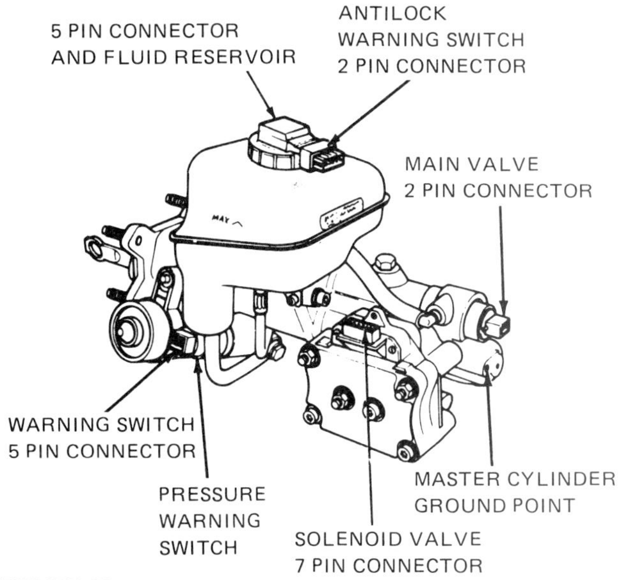

FIGURE 46
Hydraulic control unit (actuator) and pressure warning switch for the system shown in Figure 45. When removing it, do not allow brake fluid to come in contact with any of the electrical connectors. (Courtesy of Ford Parts and Service Division)

2. Correct polarity is very important in an antiskid system. Electrical components are easily damaged if they are connected wrong.

3. Some wheel sensors have an adjustable air gap between the sensor head and the toothed sensor ring (Figure 47).

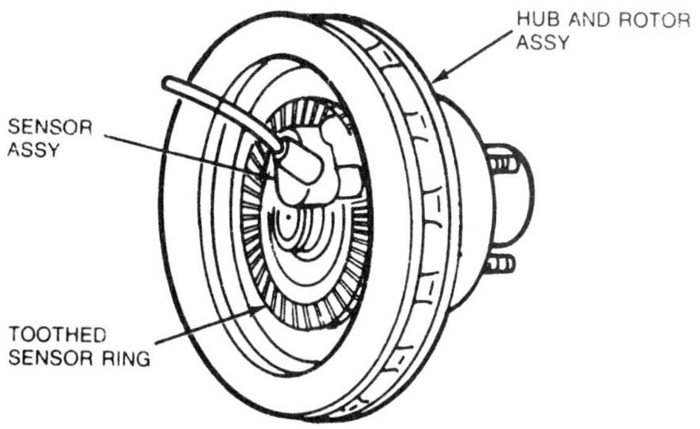

FIGURE 47
Front wheel sensor. The sensor is connected to the logic controller through a wiring harness. If the sensor is to be reused or adjusted, the pole face must be clean of all foreign material. Carefully scrape the pole face with a dull knife or similar tool. (Courtesy of Ford Parts and Service Division)

TRUE CASE HISTORY

Vehicle	Ford LTD equipped with disc brakes.
Engine	Eight-cylinder, 302 CID
Mileage	Unknown
Problem	Low brake pedal would occur after coming out of a quick turn to the right or left. Pedal height returned to normal after the next brake application.
Cause	The front wheel bearings on both sides of the car were adjusted much too loosely. This produced too much rotor thrust when turning, which pushed the caliper pistons further back into their bores. Low brake pedal height then resulted during the next stop.

STUDY QUESTIONS

1. An outer wheel bearing cup is spinning (turning) in the front hub.

   Technician A says to replace the hub.

   Technician B says to inspect the spindle.

   Who is right?
   a. A only **c.** Both A and B
   b. B only **d.** Neither A nor B

2. **Technician A** says that a binding parking brake cable can cause the brakes to drag.

   Technician B says that a weak parking brake retractor spring can cause the brakes to drag.

   Who is right?
   a. A only **c.** Both A and B
   b. B only **d.** Neither A nor B

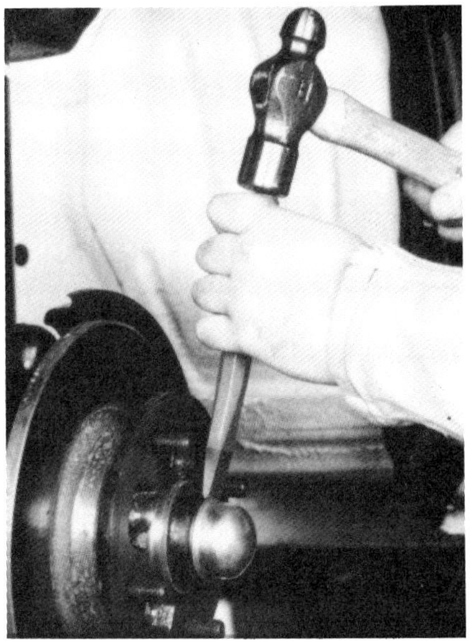

3. The above picture shows a technician
   a. Seating the grease seal
   b. Removing the hub cover
   c. Cutting the bearing retainer
   d. Installing the outer cup

4. **Technician A** says to adjust brake shoes before adjusting the parking brake.

   Technician B says to push the brake pedal down when adjusting the parking brake.

   Who is right?
   a. A only
   b. B only
   c. Both A and B
   d. Neither A nor B

5. The brake warning indicator light on an automobile stays lit while driving.

   Technician A says that this could be caused by a bent brake pedal mounting bracket.

   Technician B says that this could be caused by the parking brake switch being stuck closed.

   Who is right?
   a. A only
   b. B only
   c. Both A and B
   d. Neither A nor B

6. The brake lights (stop lights) on a car do not work.

   Technician A says that this could be caused by a bad directional signal switch.

   Technician B says that this could be caused by a bad ground at the bulb sockets.

   Who is right?
   a. A only
   b. B only
   c. Both A and B
   d. Neither A nor B

7. **Technician A** says that an ohmmeter can be used to check continuity in an antiskid braking system.

   Technician B says that an ohmmeter can be used to check resistance values in an antiskid braking system.

   Who is right?
   a. A only
   b. B only
   c. Both A and B
   d. Neither A nor B

BOOK SUMMARY QUESTIONS

This test will help to further prepare you for the ASE Brake Certification Examination. These questions are based on concepts presented in this book and are similar in content and style to those on the actual test.

1. A power brake booster unit has a vacuum leak. All of the following conditions would exist EXCEPT:
   a. Rough engine idle
   b. Hissing sound
   c. Pedal pulsation
   d. Hard pedal

2. The parking brake pedal on a car does not snap to the OFF position when the release is operated.

   Technician A says that this could be caused by cable bind inside the conduit.

   Technician B says that this could be caused by lack of seepage from the wheel cylinders to lubricate the cable.

   Who is right?
   a. A only **c.** Both A and B
   b. B only **d.** Neither A nor B

3. During the teardown phase of a brake job you discover the front wheel bearings are badly damaged from too much end play. Which brake pedal complaint would this MOST LIKELY have caused?
   a. Spongy
   b. Hard
   c. Rising
   d. Pulsating

4. A set of good brake linings becomes soaked with rear axle lubricant.

   Technician A says that it is okay to wash the lining in solvent and reuse.

   Technician B says that it is okay to dry the lining with an oxyacetylene torch and reuse.

   Who is right?
   a. A only **c.** Both A and B
   b. B only **d.** Neither A nor B

5. A car has drum brakes. Which of these conditions is LEAST LIKELY to cause vibration (chatter) during stopping?
   a. Loose backing plate
   b. Torn wheel cylinder boot
   c. Hard spots on drum
   d. Incorrect toe and heel clearance

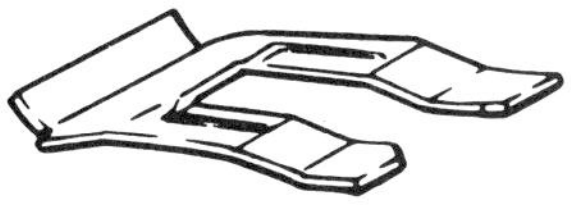

6. Identify the above item.
   a. Caliper retaining clip
   b. Brake hose lock
   c. Brake cable conduit fastener
   d. Disc brake stabilizer plate

7. When "turning" a scored brake drum, the maximum oversize limit for most cars is:
   a. 0.030″
   b. 0.060″
   c. 0.090″
   d. 0.110″

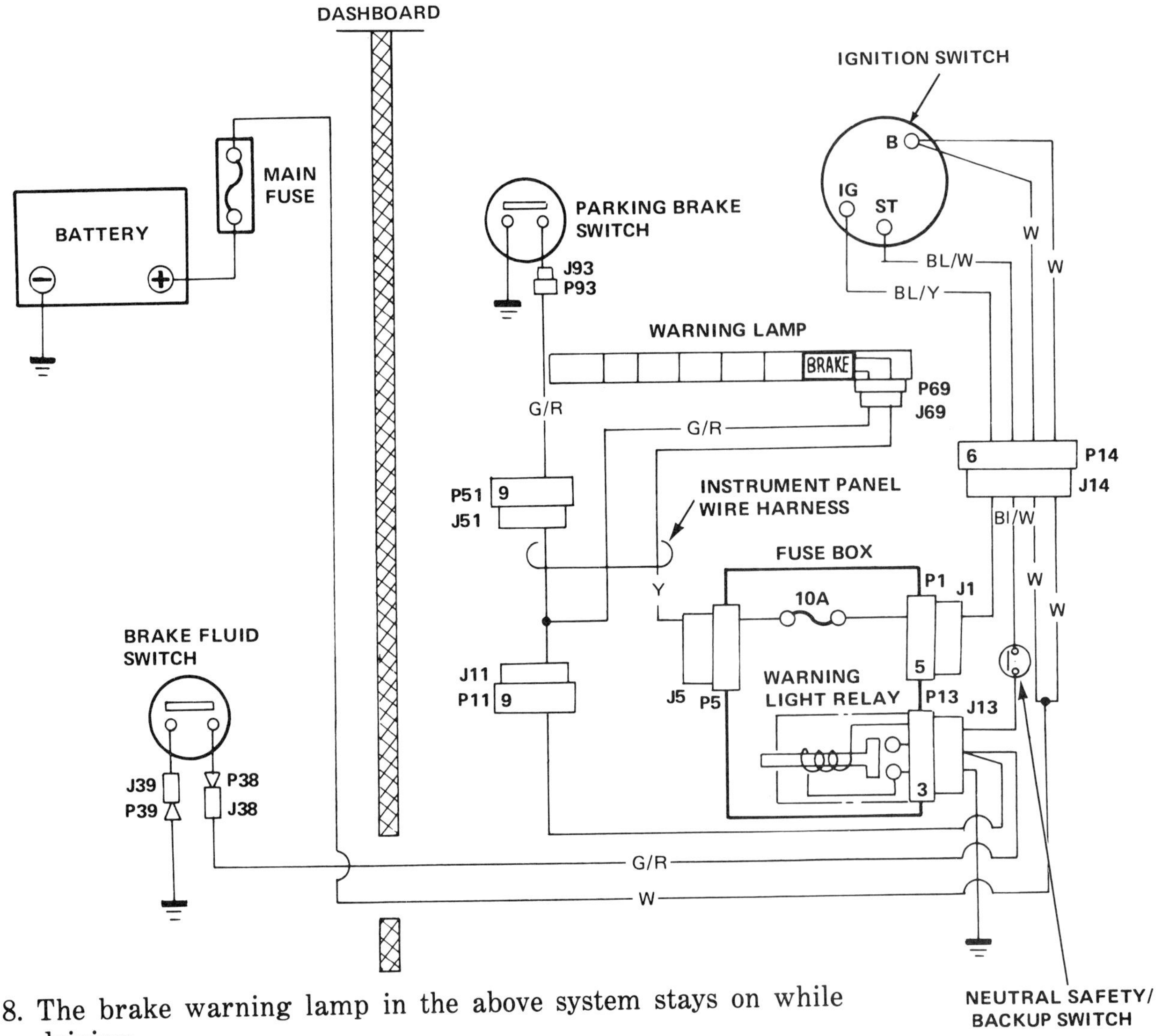

8. The brake warning lamp in the above system stays on while driving.

   Technician A says that this could be caused by a blown main fuse.

   Technician B says that this could be caused by the brake fluid level being too low.

   Who is right?
   - **a.** A only
   - **b.** B only
   - **c.** Both A and B
   - **d.** Neither A nor B

9. On drum brakes, which of the following is the MOST LIKELY cause of a spongy (springy) brake pedal?
   - **a.** Loose wheel cylinder
   - **b.** Air in system
   - **c.** Self-adjuster not working
   - **d.** Weak return springs

10. Which of these should be used to clean the wheel cylinder bore after honing?
    - **a.** Cleaning solvent
    - **b.** An oil-soaked rag
    - **c.** Hot water and soap
    - **d.** Denatured alcohol

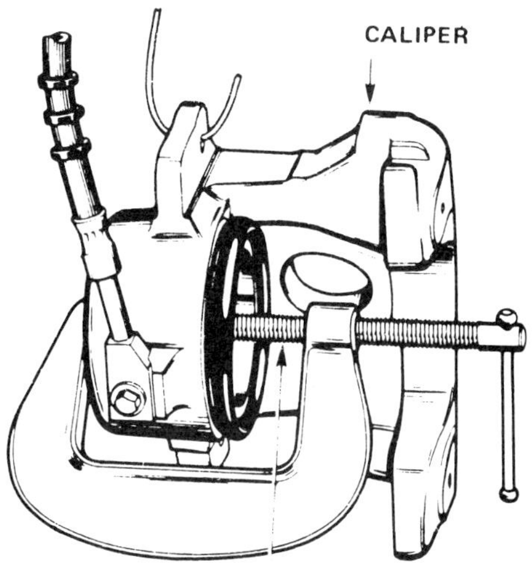

11. The above set-up is being used to
    a. Bottom the piston
    b. Press in a caliper sleeve
    c. Install the boot retainer
    d. Remove a frozen piston

12. The above picture shows a technician
    a. Installing the inner wheel bearing cup
    b. Removing the inner wheel bearing cup
    c. Installing the grease seal
    d. Installing the outer wheel bearing cup

13. The brakes on a car heat up during driving and fail to release.

    Technician A says that this could be caused by the stop light switch being improperly adjusted.

    Technician B says that this could be caused by the metering valve being installed incorrectly.

    Who is right?
    a. A only
    b. B only
    c. Both A and B
    d. Neither A nor B

14. When taking a caliper apart, a technician notices that the piston is pitted and the bore slightly rusted. What should be done?
    a. Crocus cloth the bore and replace the piston
    b. Bore out the caliper and install a new piston
    c. Crocus cloth the piston and replace the caliper
    d. Replace the caliper and the piston

15. Too much pedal effort is required to stop a vehicle with power disc brakes.

    Technician A says that this could be caused by air in the hydraulic system.

    Technician B says that this could be caused by a master cylinder not working properly.

    Who is right?
    a. A only
    b. B only
    c. Both A and B
    d. Neither A nor B

16. The vent on a master cylinder reservoir cover becomes plugged.

    Technician A says that this can cause the brakes to drag.

    Technician B says that this can cause a spongy brake pedal.

    Who is right?
    a. A only
    b. B only
    c. Both A and B
    d. Neither A nor B

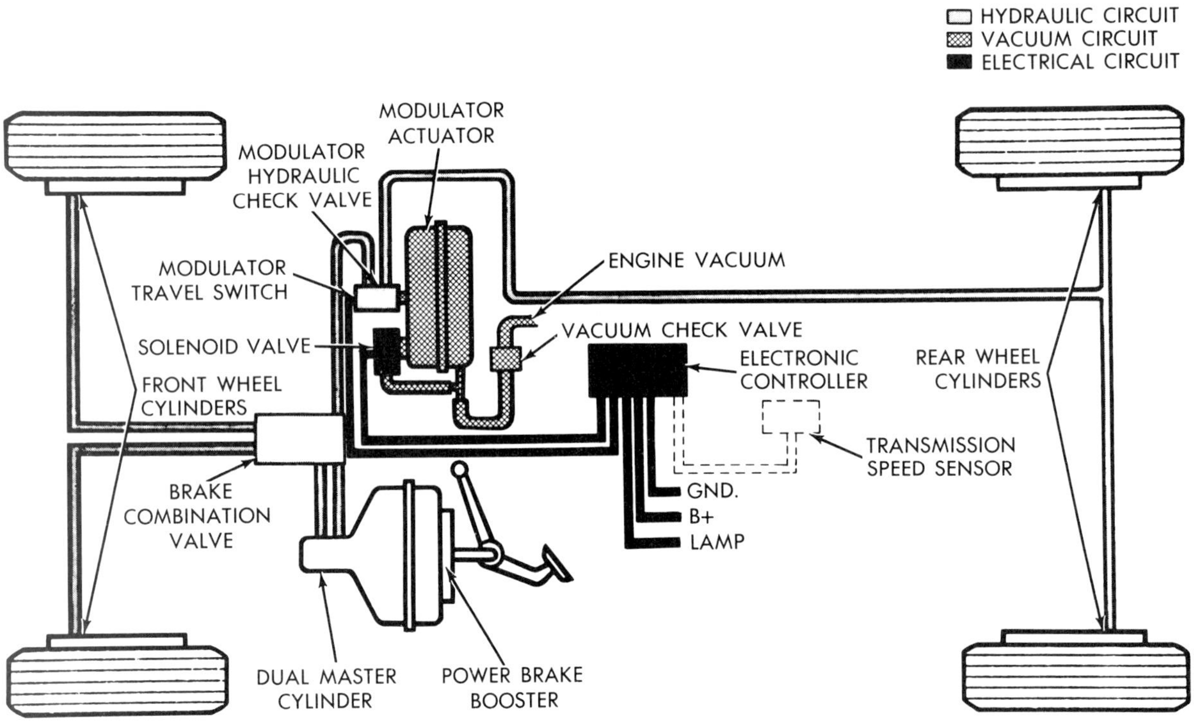

17. The battery is accidently hooked up backward in a car equipped with the above antilock brake system.

    Technician A says that this can damage the electronic controller.

    Technician B says that this can damage the power brake booster.

    Who is right?

    a. A only **c.** Both A and B
    b. B only **d.** Neither A nor B

18. A power brake system needs bleeding.

    Technician A says that the engine must be running.

    Technician B says that the bleeder screws must be removed.

    Who is right?

    a. A only **c.** Both A and B
    b. B only **d.** Neither A nor B

19. On a drum brake system, the pedal rises each time the brakes are applied during several miles of driving.

    Technician A says that this could be caused by the master cylinder piston assembly not returning to its stop.

    Technician B says that this could be caused by insufficient pedal free travel.

    Who is right?

    a. A only **c.** Both A and B
    b. B only **d.** Neither A nor B

20. On a vehicle with front disc brakes, violent cornering or rough roads may cause
    a. A drop in fluid pressure
    b. Excessive pedal travel
    c. Both a and b
    d. Neither a nor b

21. Which of the following pressure ranges is generally used when bleeding the brakes with a pressure tank?
    a. 2 to 8 psi
    b. 15 to 30 psi
    c. 40 to 65 psi
    d. 60 to 80 psi

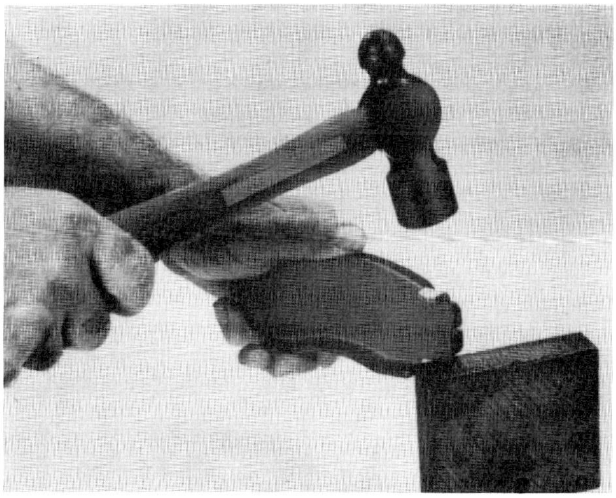

22. What is the technician doing in the above illustration?
    a. Aligning the wear indicator
    b. Adjusting the pad-to-rotor clearance
    c. Bending the inboard shoe retainer plate
    d. Getting rid of noise

23. To check rotor parallelism you would use a
    a. Dial indicator
    b. Micrometer
    c. Pair of dividers
    d. Special service tool (SST)

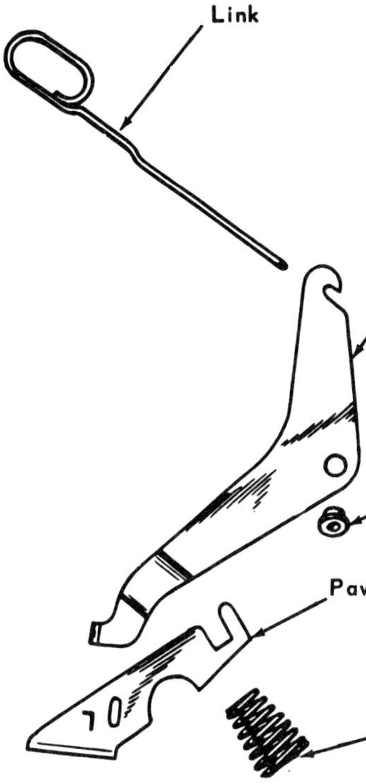

24. **Technician A** says that the above parts are positioned for a right rear brake assembly.

    Technician B says that the above parts are adjusted by bending.

    Who is right?
    a. A only
    b. B only
    c. Both A and B
    d. Neither A nor B

25. Some brake systems use a metering valve. When bleeding this type of system,
    a. The metering valve plunger must be pushed to the closed position
    b. The metering valve has no effect on bleeding brakes if a pressure tank is used
    c. The metering valve plunger must be held in the open position
    d. The metering valve must be removed

26. The above tool is being used to:
    a. Remove the grease seal
    b. Seat the inner bearing
    c. Pull the outer bearing
    d. Align the hub

27. When checking disc brake rotor runout
    a. Remove the caliper from the steering knuckle
    b. Use a micrometer
    c. Use a ruler
    d. Readjust wheel bearings when finished

28. The disc brake reservoir portion of a dual master cylinder is low on fluid.

    Technician A says that the front pads may be worn.

    Technician B says that the power brake booster may be faulty.

    Who is right?
    a. A only
    b. B only
    c. Both A and B
    d. Neither A nor B

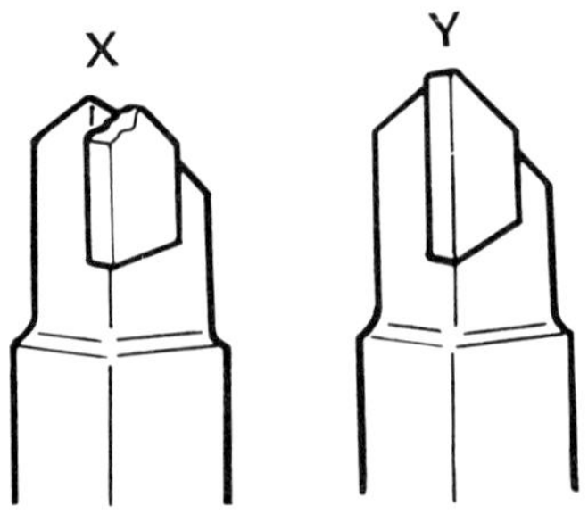

29. Refer to the above illustrations.

    Technician A says that tool bit X is sharp and ground to the correct shape.

    Technician B says that tool bit Y will produce a poor surface finish.

    Who is right?
    a. A only
    b. B only
    c. Both A and B
    d. Neither A nor B

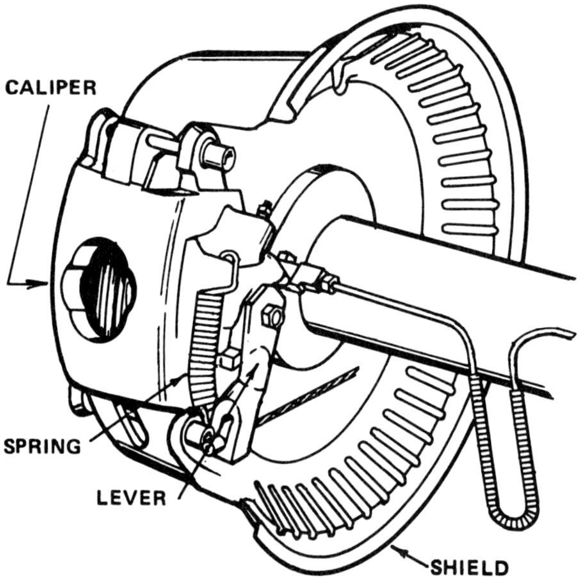

30. **Technician A** says that the above assembly houses an integral parking brake.

    Technician B says that the above assembly uses a floating caliper.

    Who is right?
    a. A only
    b. B only
    c. Both A and B
    d. Neither A nor B

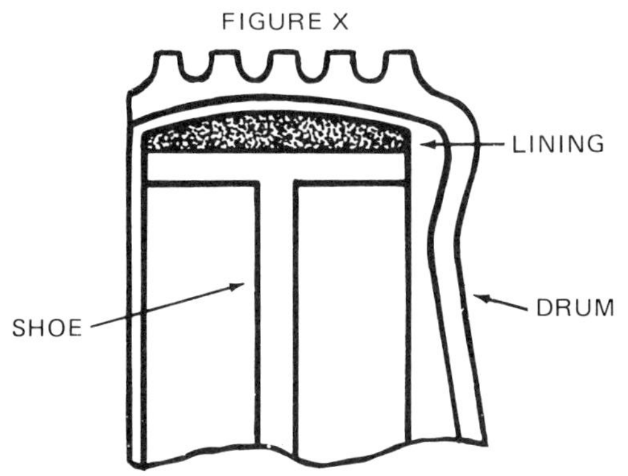

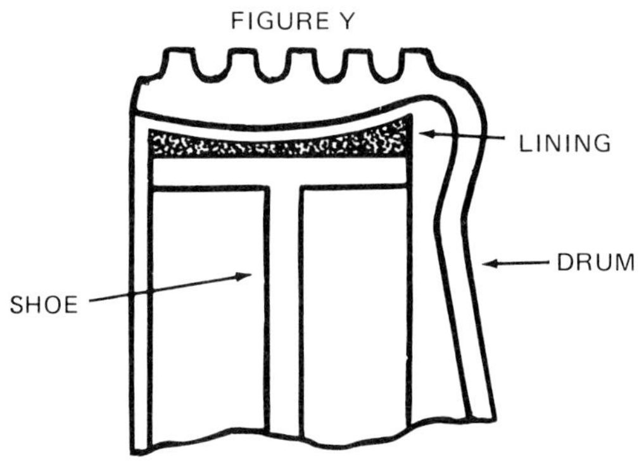

31. Which figure above shows a correct lining-to-drum fit?
    a. X only
    b. Y only
    c. Both X and Y
    d. Neither X nor Y

32. On a wet highway, what prevents the front brakes from locking up at light pedal pressure before the rear drum brakes develop enough stopping force?
    a. Bypass valve
    b. Residual valve
    c. Proportioning valve
    d. None of the above

33. Before installing new front disc pads, a technician removes some brake fluid from the master cylinder. This will
    a. Prevent the brake warning light from coming on
    b. Prevent fluid overflow
    c. Allow front and rear fluid pressure to equalize
    d. Permit gravity bleeding of the calipers

34. **Technician A** says that drum brake fade can be caused by glazed lining.

    Technician B says that drum brake fade can be caused by bell-mouthed drums.

    Who is right?
    a. A only
    b. B only
    c. Both A and B
    d. Neither A nor B

35. **Technician A** says that the above switch can be tested with a jumper wire.

    Technician B says that the above switch can be tested with an ohmmeter.

    Who is right?
    a. A only
    b. B only
    c. Both A and B
    d. Neither A nor B

36. **Technician A** says that brake drum hard spots can be removed by machining the drum with a carbide tool bit.

    Technician B says that brake drum hard spots can be removed by grinding. However, they may reappear when heat is applied.

    Who is right?
    a. A only
    b. B only
    c. Both A and B
    d. Neither A nor B

37. A honed wheel cylinder should not be more than _____ over the original diameter.
    a. 0.002″
    b. 0.005″
    c. 0.010″
    d. 0.0005″

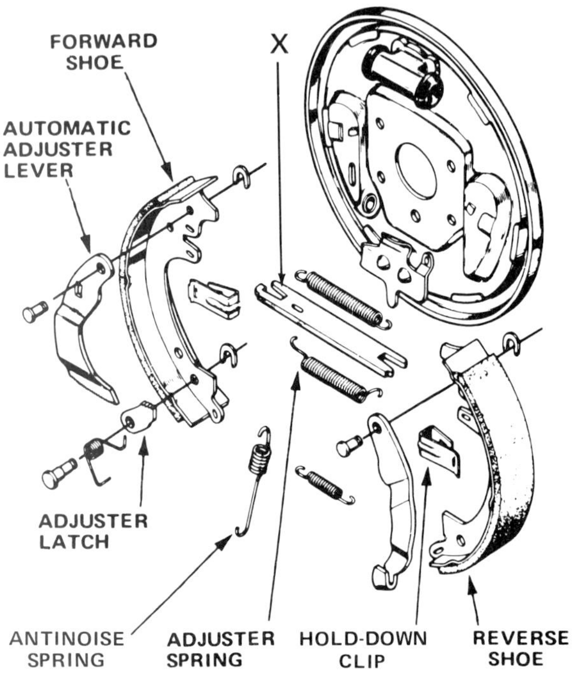

38. The purpose of part X above is to
    a. Center the shoes after each brake application
    b. Balance the force between the shoes during hard braking
    c. Both a and b
    d. Neither a nor b

39. A driver complains of a clicking noise when applying the brakes.

    Technician A says that this could be caused by grooved backing plate ledges.

    Technician B says that this could be caused by a loose anchor pin.

    Who is right?
    a. A only
    b. B only
    c. Both A and B
    d. Neither A nor B

40. After both front calipers have been rebuilt, the brake warning light stays on. What must the technician do?
    a. Reset the pressure differential valve
    b. Adjust pedal free play
    c. Replace the warning light switch
    d. Vacuum bleed the master cylinder

41. You are replacing a right front inner wheel bearing that has failed (for the third time in several months). Which of these could be the cause?
    a. Use of the wrong grease
    b. A worn cup
    c. A bad spindle
    d. Any of the above

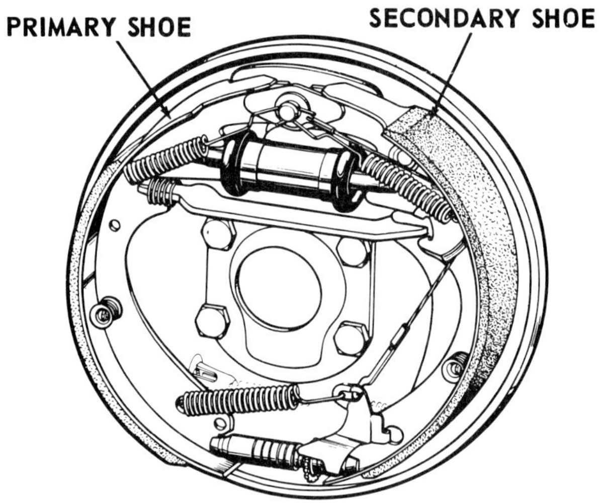

42. Refer to the above picture.

    Technician A says that a front brake assembly is shown.

    Technician B says that a right side brake assembly is shown.

    Who is right?
    a. A only
    b. B only
    c. Both A and B
    d. Neither A nor B

43. A car owner complains of a pull to the right when using the brakes.

    Technician A says that this could be caused by a faulty master cylinder.

    Technician B says that this could be caused by a faulty proportioning valve.

    Who is right?
    a. A only
    b. B only
    c. Both A and B
    d. Neither A nor B

44. When the brakes are applied hard, a car has a pull to the right along with a light vibration.

    Technician A says that this could be caused by incorrect tire pressure.

    Technician B says that this could be caused by collapsing strut rod bushings.

    Who is right?
    a. A only
    b. B only
    c. Both A and B
    d. Neither A nor B

46. A clean shop rag is used to wipe the old grease off an outer front wheel bearing. A lot of shiny silver particles are observed on the rag afterward.

    Technician A says that the bearing should be replaced.

    Technician B says that the cup should be replaced.

    Who is right?
    a. A only
    b. B only
    c. Both A and B
    d. Neither A nor B

47. **Technician A** says that the above part helps prevent water and dirt contamination.

    Technician B says that the above part eliminates the need for a residual pressure check valve.

    Who is right?
    a. A only
    b. B only
    c. Both A and B
    d. Neither A nor B

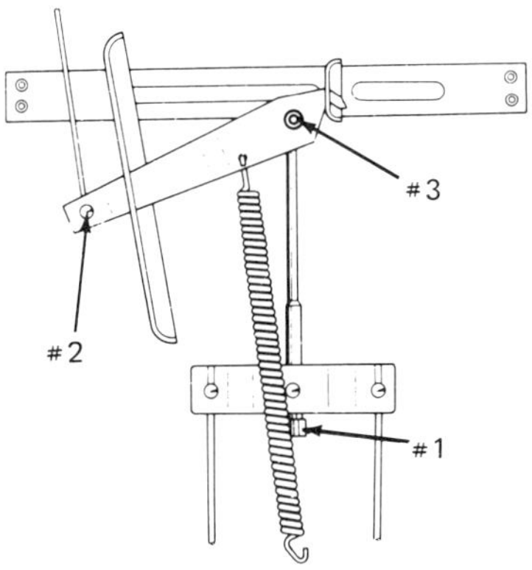

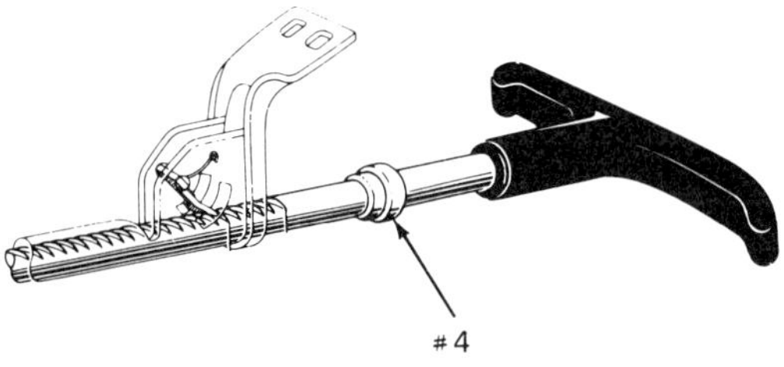

45. Excess slack in the above parking brake mechanism can be taken care of by an adjustment at point:
    a. #1
    b. #2
    c. #3
    d. #4

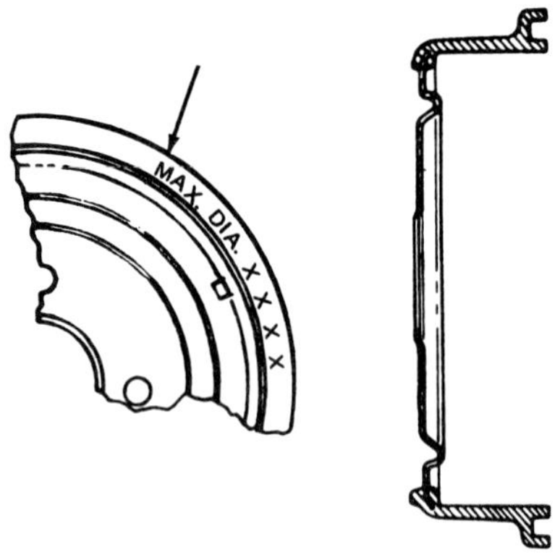

48. Refer to the above marking.

Technician A says that this shows the maximum outside drum diameter.

Technician B says that this shows the cutting dimension to use when installing new lining.

Who is right?
a. A only
b. B only
c. Both A and B
d. Neither A nor B

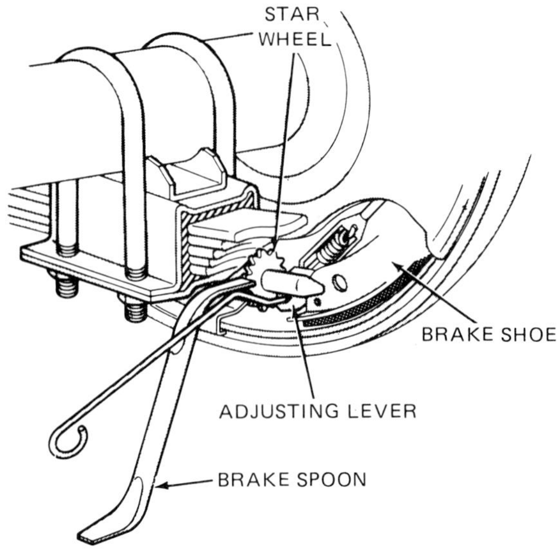

49. When making the above adjustment
a. Release the parking brake
b. Spin the tire
c. Both a and b
d. Neither a nor b

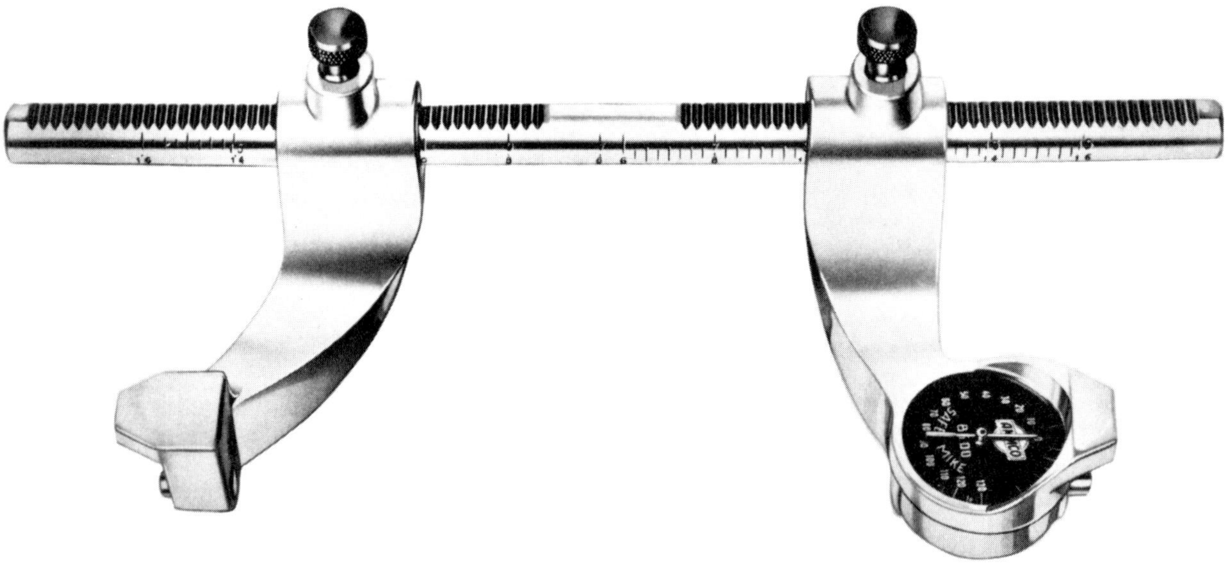

50. **Technician A** says that the above tool checks drum diameter.

Technician B says that the above tool checks drum out-of-roundness.

Who is right?
a. A only
b. B only
c. Both A and B
d. Neither A nor B

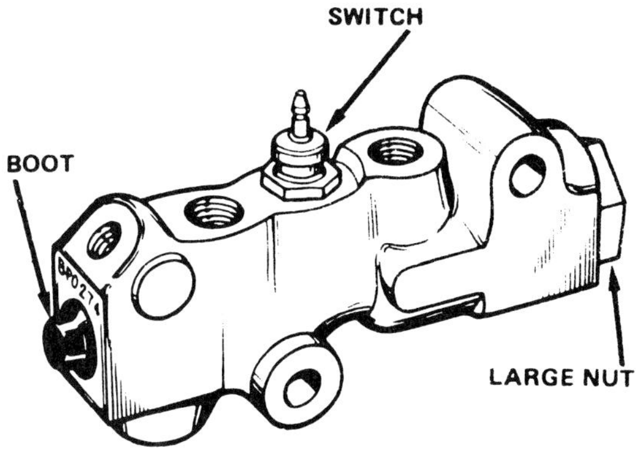

51. **Technician A** says that the above valve assembly delays pressure build-up to the front axle disc brakes.

    Technician B says that the above valve assembly turns on the brake warning lamp.

    Who is right?
    a. A only
    b. B only
    c. Both A and B
    d. Neither A nor B

52. Two technicians are discussing the replacement of a wheel bearing race (cup).

    Technician A says to drive in the race using a wooden dowel.

    Technician B says to drive in the race until it is flush with the top of the hub.

    Who is right?
    a. A only
    b. B only
    c. Both A and B
    d. Neither A nor B

53. The owner of a vehicle with front disc brakes says that the brake pedal pulsates during braking.

    Technician A says that this could be caused by uneven rotor thickness.

    Technician B says that this could be caused by a wheel bearing not being adjusted properly.

    Who is right?
    a. A only
    b. B only
    c. Both A and B
    d. Neither A nor B

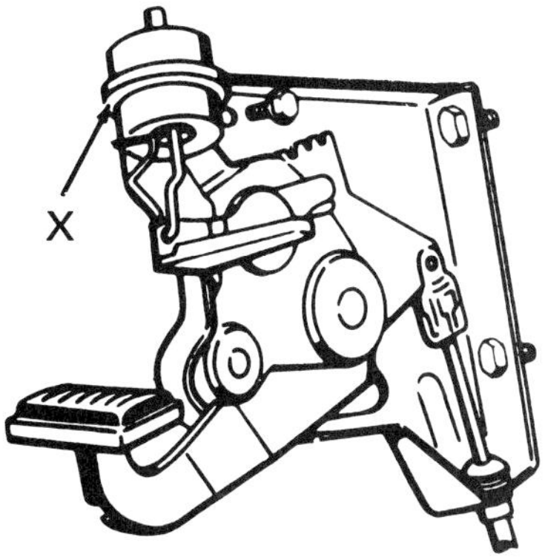

54. **Technician A** says that part X above releases the parking brake.

    Technician B says that part X above applies the parking brake.

    Who is right?
    a. A only
    b. B only
    c. Both A and B
    d. Neither A nor B

55. A set of right front brake linings has become badly contaminated with grease.

    Technician A says to inspect the wheel bearings.

    Technician B says to install a new set of linings on the right front and on the left front.

    Who is right?
    a. A only
    b. B only
    c. Both A and B
    d. Neither A nor B

56. The brake pedal is spongy on a car with a disc/drum system. This could be caused by
    a. Air in the system
    b. A missing master cylinder cover diaphragm
    c. Both a and b
    d. Neither a nor b

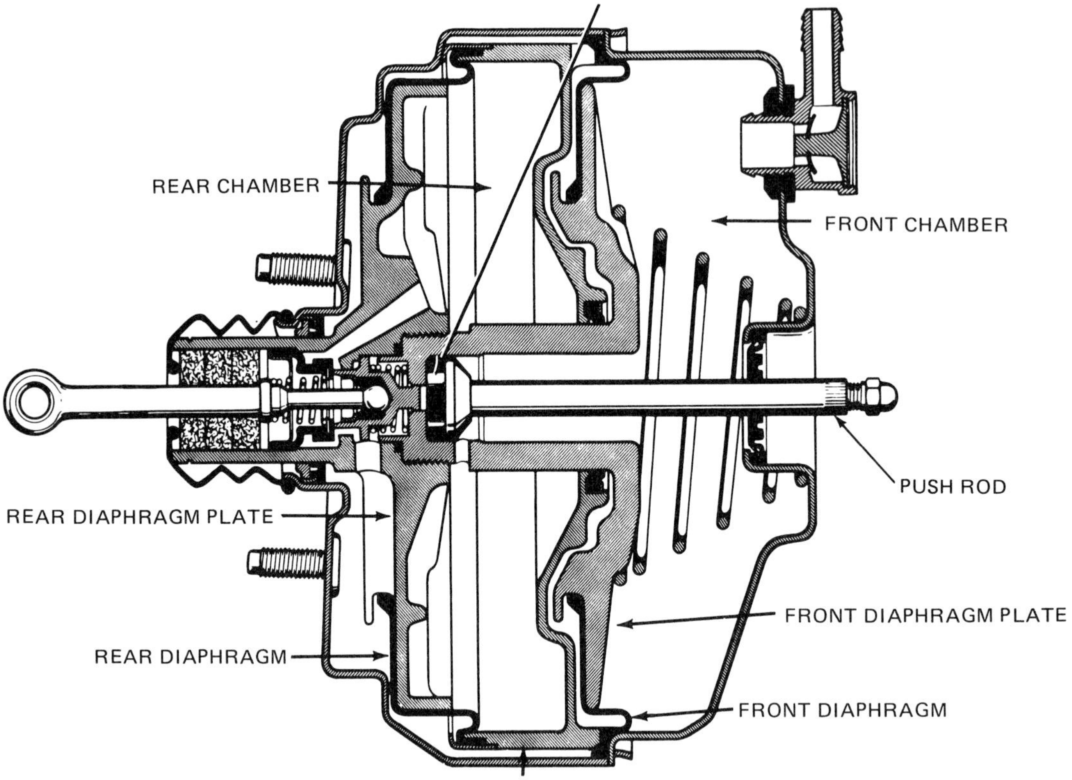

57. A technician is replacing a master cylinder. The push rod (see above) is accidentally pulled out, causing part X to fall into the bottom of the front chamber.

    Technician A says that this can cause sensitive brakes.

    Technician B says that this can cause excessive brake pedal travel.

    Who is right?
    a. A only **c.** Both A and B
    b. B only **d.** Neither A nor B

58. The power brake hose on a vehicle collapses when the engine is idling.

    Technician A says that the hose could be defective.

    Technician B says that the check valve could be defective.

    Who is right?
    a. A only **c.** Both A and B
    b. B only **d.** Neither A nor B

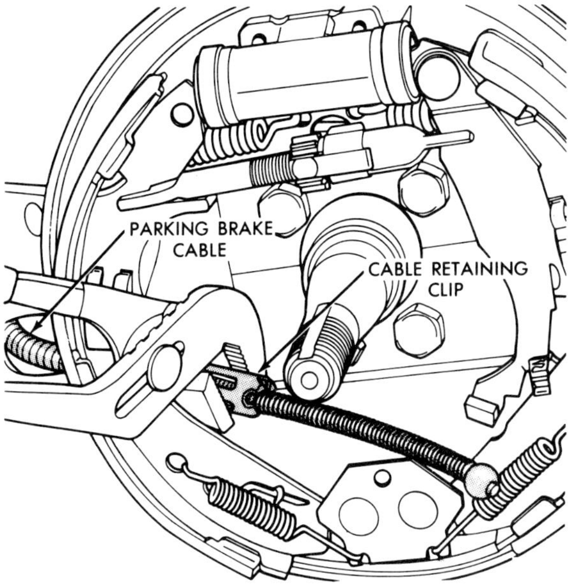

59. The tool shown above is being used to:
    a. Remove the brake cable
    b. Free up the linkage
    c. Install the brake cable
    d. Adjust parking brake tension

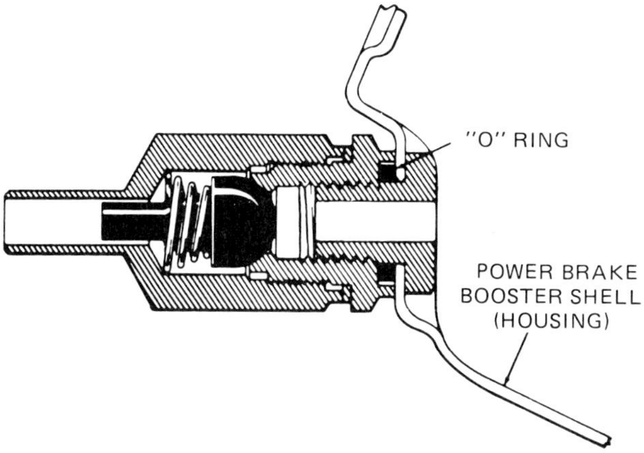

60. **Technician A** says that the above valve operates in one direction only.

    Technician B says that the above valve maintains vacuum in the power brake unit.

    Who is right?
    a. A only
    b. B only
    c. Both A and B
    d. Neither A nor B

Appendix A

SOURCES OF MANUALS

Information on the availability and price of factory service manuals may be obtained by contacting the specific car manufacturer listed below:

Alfa Romeo, S.P.A.
National Headquarters
250 Sylvan Avenue
Englewood Cliffs, NJ 07632

American Honda Motor Company, Inc.
100 West Alondra Blvd.
Gardena, CA 90247

American Isuzu Motors Inc.
Izuzu Motors Ltd.
2300 Pellissier Place
Whittier, CA 90601

American Motors Corporation
14250 Plymouth Road
Detroit, MI 48232

BMW of North America, Inc.
Bayerische Motoren Werke A.G.
BMW Plaza
Montvale, NJ 07645

Buick Motor Division
General Motors Corporation
Flint, MI 48550

Cadillac Motor Car Division
General Motors Corporation
2860 Clark Avenue
Detroit, MI 48202

Chrysler Corporation
P.O. Box 1919
Detroit, MI 48231

Dodge Division
Chrysler Corporation
P.O. Box 857
Detroit, MI 48231

FIAT, S.P.A.
Strada Torina 50
Orbassano, Italy 10043

Ford Motor Company
The American Road
Dearborn, MI 48121

Hyundai Motor America
7373 Hunt Avenue
P.O. Box 2669
Garden Grove, CA 92642-2669

Jaguar Rover Triumph, Inc.
600 Willow Tree Road
Leonia, NJ 07605

Lincoln-Mercury Division
Ford Motor Company
3000 Schaefer Road
Dearborn, MI 48216

Mazda Motor Corporation
Toyo Kogyo Company Ltd.
821 Brock Road South
Pickering, Ontario Canada L1W 3L6

Mercedes-Benz of North America, Inc.
Daimler-Benz A.G.
1 Mercedes Drive
Montvale, NJ 07645

Mitsubishi Motor Sales of America, Inc.
West Office Tower
10540 Talbert Avenue
Fountain Valley, CA 92728-9060

Nissan Motor Company, Ltd.
18501 S. Figueroa Street
Carson, CA 90248

Oldsmobile Division
General Motors Corporation
920 Townsend Street
Lansing, MI 48921

Peugeot Motors of America, Inc.
1 Peugeot Plaza
Lyndhurst, NJ 07071

Plymouth Division
Chrysler Corporation
P.O. Box 857
Detroit, MI 48231

Pontiac Motor Division
General Motors Corporation
1 Pontiac Plaza
Pontiac, MI 48053

Porsche-Audi
Worldwide Volkswagen Corporation
340 Rt. 59
West Nyack, NY 10994

Saab Scania of America
Saab-Scania Aktiebolag
Saab Drive
Orange, CT 06477

Subaru of America, Inc.
7040 Central Highway
Pennsauken, NJ 08109

Toyota Motor Sales, U.S.A., Inc.
19001 S. Western Avenue
Torrance, CA 90509

Volkswagen of America
Corporate Service Division
888 West Big Beaver Road
Troy, MI 48007

Volvo of America Corporation
AB Volvo
One Volvo Drive
Rockleigh, NJ 07647

SOURCES OF COMPREHENSIVE SHOP MANUALS

The following companies publish comprehensive shop manuals that contain commonly used specifications and the most frequently performed service operations.

Chilton Book Company
Chilton Way
Radnor, PA 19809

Mitchell Manuals Inc., A Cordura Company
P.O. Box 26260
San Diego, CA 92126

Motor Publications
555 West 57th Street
New York, NY 10019

Appendix B

PRE-BRAKE JOB INSPECTION CHECKLIST

For safe and efficient brake operation, every component in the system must be in good working condition. Periodic inspection is the best way to ensure that all components are functioning normally.

NOTE: Several states require yearly comprehensive brake inspections.

Completing an inspection checklist such as that on the following page will help you avoid missing any components.

COMMENT: This checklist can also serve as a written report for the customer.

REGISTERED OWNER ______________________ DATE ______________

ADDRESS ______________________ PHONE NO. ______________

MAKE __________ MODEL __________ MILEAGE __________ LICENSE NO. __________ YEAR __________

SPECIAL KEY REQUIRED FOR REMOVING HUBCAPS/WHEELS? YES ☐ NO ☐ DOES OWNER USE PARKING BRAKE? YES ☐ NO ☐

4 DRUM ☐ 4 DISC ☐ DISC/DRUM ☐ POWER BRAKES? YES ☐ NO ☐ VACUUM PUMP? YES ☐ NO ☐

OWNER COMPLAINTS ______________________

1. **CHECKS BEFORE ROAD TEST**	OK	BAD
STOPLAMP OPERATION WITH HEADLAMPS ON/OFF		
BRAKE WARNING LAMP OPERATION		
PARKING BRAKE LAMP OPERATION		
PARKING BRAKE ADJUSTMENT		

MASTER CYLINDER CHECKS

FLUID LEVEL		
FLUID CONTAMINATION		
UNDER HOOD FLUID LEAKS		
UNDER DASH FLUID LEAKS (NO POWER BRAKES)		
PEDAL HEIGHT STAYS WHEN FORCE APPLIED		
PEDAL FREE TRAVEL		

CHECK ONE		CHECK ONE		
LOW PEDAL HEIGHT		SPONGY PEDAL		
MED. PEDAL HEIGHT		FIRM PEDAL		
HIGH PEDAL HEIGHT				

POWER BRAKE UNIT CHECKS

VACUUM TYPE	OK	BAD	HYDRO TYPE	OK	BAD
VACUUM UNIT LEAKS			HYDRO UNIT LEAKS		
ENGINE VACUUM			P/S FLUID LEVEL		
VACUUM HOSE			P/S BELT TENSION		
UNIT CHECK VALVE			P/S BELT CONDITION		
RESERVE BRAKING			P/S FLUID LEAKS		
PEDAL DROP TEST			RESERVE BRAKING		

TIRE PRESSURE SPECS:

RECORD PRESSURE MEASURED

RF ______ LF ______ RR ______ LR ______

TIRE CONDITION

RF ______ LF ______ RR ______ LR ______

2. **ROAD TEST**	YES	NO
PULLING		
CLUNK NOISE		
SCRAPING NOISE		
SQUEAL NOISE		
GRABBING		
PREMATURE LOCKUP		
WHEEL BEARING NOISE		
VIBRATION		

	YES	NO
PEDAL PULSATION WHEN BRAKING		
STEERING WHEEL OSCILLATION WHEN BRAKING		
EXCESSIVE PEDAL EFFORT REQUIRED		
WARNING LAMP COMES ON WHEN BRAKING		
DIFFERENCE IN PEDAL HEIGHT AFTER CORNERING		
NOSE DIVE WHEN BRAKING		

3. **IN SHOP CHECKS ON HOIST**	YES	NO	RF	LF	RR	LR
BRAKE DRAG						
INTERMITTENT BRAKE DRAG						
BRAKE PEDAL LINKAGE BINDING						
WHEEL BEARING LOOSENESS						
MISSING OR DAMAGED LUGS						
SUSPENSION LOOSENESS						
Mark Wheels, and Remove:						
Caliper/Piston Stuck						
Mark Drums and Remove:						

ADDITIONAL COMMENTS:

	FRONT					REAR				
	RIGHT			LEFT		RIGHT			LEFT	
	SPEC	SAFE	UNSAFE	SAFE	UNSAFE	SPEC	SAFE	UNSAFE	SAFE	UNSAFE
MEASURE ROTOR THICKNESS OR DRUM DIAMETER										
MEASURE ROTOR THICKNESS VARIATION (PARALLELISM)										
MEASURE ROTOR RUNOUT (TIR)										
LINING THICKNESS/CONDITION										
LINES, HOSES, CONNECTIONS, VALVES										
FLUID LEAKS										
BROKEN BLEEDER SCREWS										
LEAKY GREASE SEALS										
SELF-ADJUSTER OPERATION										

	OK	BAD
PARKING BRAKE CABLE AND LINKAGE CONDITION		
HARDWARE/BACKING PLATES		

Appendix C

SHOE AND DRUM DIAMETER CHART

When replacing brake shoes, the diameter of the new shoe set *must* be the one specified by the manufacturer. Brake shoe diameter can be determined visually by comparing the old shoes with the new shoes, or by matching against the arcs drawn on the chart that follows.

NOTE: *Remember that the location of the lining on the shoe is also important. The secondary shoe normally has a full-length lining that is centered on the shoe.*

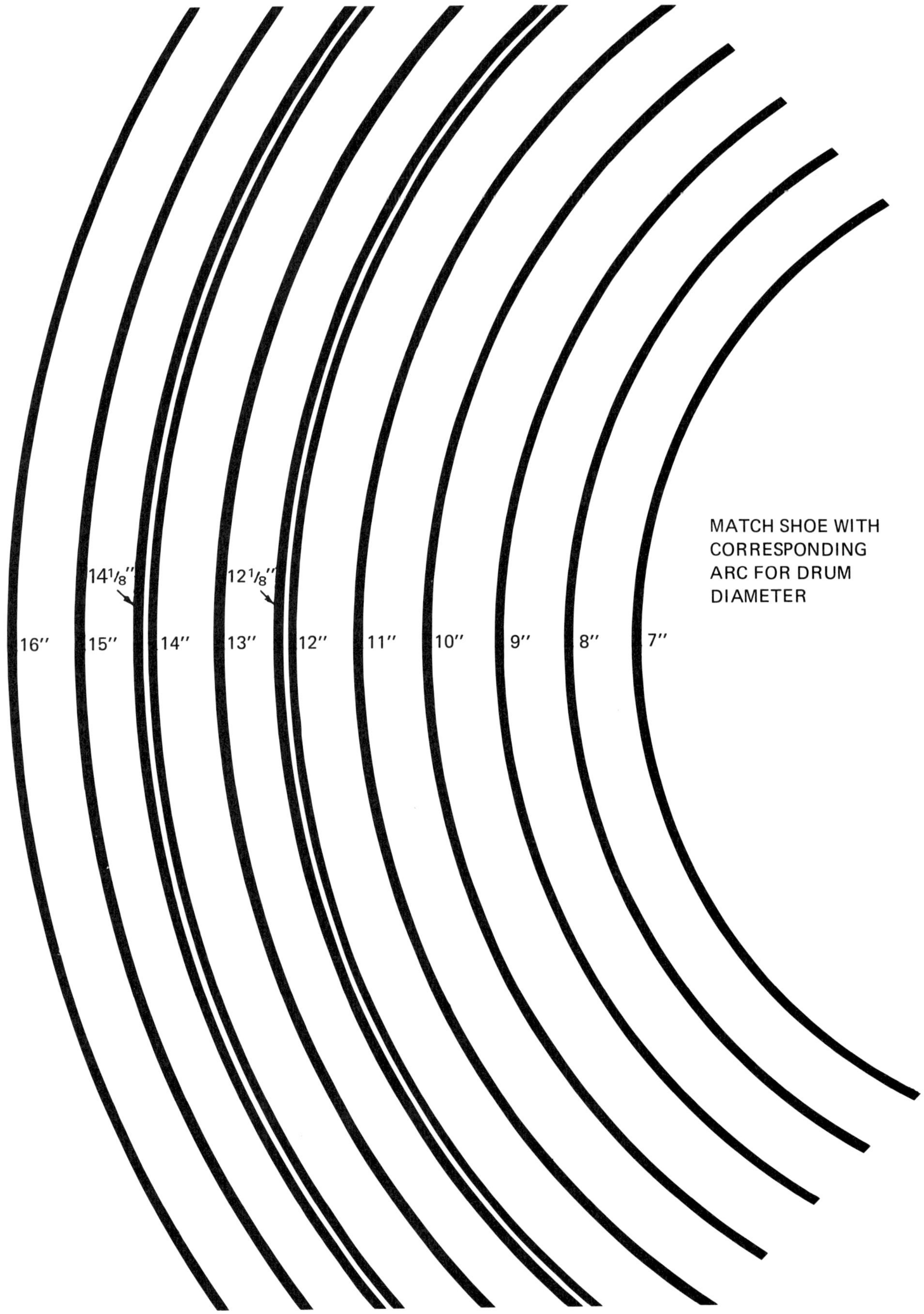
14 1/8″
12 1/8″
16″
15″
14″
13″
12″
11″
10″
9″
8″
7″
MATCH SHOE WITH
CORRESPONDING
ARC FOR DRUM
DIAMETER

Appendix D

TORQUE CHARTS

U.S. STANDARD

GRADE OF BOLT		S.A.E. 1 & 2	S.A.E. 5	S.A.E. 6	S.A.E. 8		
MINIMUM TENSILE STRENGTH		64,000 P.S.I.	105,000 P.S.I.	133,000 P.S.I.	150,000 P.S.I.		
GRADE MARKING ON HEAD ➔						SOCKET OR WRENCH SIZE	
U.S. STANDARD		**TORQUE (IN FOOT POUNDS)**				U.S. REGULAR	
BOLT DIAMETER	DEC. EQUIV.					BOLT HEAD	NUT
1/4	.250	5	7	10	10.5	3/8	7/16
5/16	.3125	9	14	19	22	1/2	9/16
3/8	.375	15	25	34	37	9/16	5/8
7/16	.4375	24	40	55	60	5/8	3/4
1/2	.500	37	60	85	92	3/4	13/16

THE FOLLOWING RULES APPLY TO THE CHART:

1. Consult manufacturers specific recommendations when available.
2. The chart may be used directly when any of the following lubricants are used: *NEVER-SEEZ COMPOUND, MOLYKOTE, FEL-PRO C-5, GRAPHITE AND OIL OR SIMILAR MIXTURES.*
3. Increase the torque by 20% when using engine oil or chassis grease as a lubricant.
4. Reduce torque by 20% when plated bolts are used.
5. Increase torque by 20% when multiple tapered tooth lockwashers are used.

CAUTION: Tightening into aluminum usually will require less torque.

METRIC STANDARD

GRADE OF BOLT		5D	8G	10K	12K		
MINIMUM TENSILE STRENGTH		71,160 P.S.I.	113,800 P.S.I.	142,000 P.S.I.	170,674 P.S.I.		
GRADE MARKING ON HEAD ➡		5D	8G	10K	12K	SOCKET OR WRENCH SIZE	
METRIC		TORQUE (IN FOOT POUNDS)				METRIC	
BOLT DIAMETER	U.S. DEC. EQUIV.					BOLT HEAD	NUT
6mm	.2362	5	6	8	10	10mm	10mm
8mm	.3150	10	16	22	27	14mm	14mm
10mm	.3937	19	31	40	49	17mm	17mm
12mm	.4720	34	54	70	86	19mm	19mm
14mm	.5512	55	89	117	137	22mm	22mm

THE FOLLOWING RULES APPLY TO THE CHART:

1. Consult manufacturers specific recommendations when available.
2. The chart may be used directly when any of the following lubricants are used: *NEVER-SEEZ COMPOUND, MOLYKOTE, FEL-PRO C-5, GRAPHITE AND OIL OR SIMILAR MIXTURES.*
3. Increase the torque by 20% when using engine oil or chassis grease as a lubricant.
4. Reduce torque by 20% when plated bolts are used.
5. Increase torque by 20% when multiple tapered tooth lockwashers are used.

CAUTION: Tightening into aluminum usually will require less torque.

Appendix E

INCHES AND MILLIMETERS CONVERSION CHART

Many dimensions in the automotive industry are specified in millimeter (mm) notation. As a result, today's technician must think in terms of metric measurement. However, for the present and the near future, he or she will also be using inch measurements. Until a full transition to metrics takes place, there will be times when it is necessary to convert from one system to the other. The chart on the next page will serve as an aid for making metric-inch conversions.

MILLIMETERS TO DECIMAL(inches)						DECIMAL(inches) TO METRIC					
MM	DECIMAL (Inches)	MM	DECIMAL (Inches)	MM	DECIMAL (Inches)	DECIMAL (Inches)	MM	DECIMAL (Inches)	MM	DECIMAL (Inches)	MM
1	.039	41	1.614	81	3.189	.001	.02	.320	8.12	.720	18.28
2	.079	42	1.654	82	3.228	.002	.05	.330	8.38	.730	18.54
3	.118	43	1.693	83	3.268	.003	.07	.340	8.63	.740	18.79
4	.157	44	1.732	84	3.307	.004	.10	.350	8.89	.750	19.05
5	.197	45	1.772	85	3.346	.005	.12	.360	9.14	.760	19.30
6	.236	46	1.811	86	3.386	.006	.15	.370	9.39	.770	19.55
7	.276	47	1.850	87	3.425	.007	.17	.380	9.65	.780	19.81
8	.315	48	1.890	88	3.465	.008	.20	.390	9.90	.790	20.06
9	.354	49	1.929	89	3.504	.009	.22	.400	10.16	.800	20.32
10	.394	50	1.969	90	3.543	.010	.25	.410	10.41	.810	20.57
11	.433	51	2.008	91	3.583	.020	.50	.420	10.66	.820	20.82
12	.472	52	2.047	92	3.622	.030	.76	.430	10.92	.830	21.08
13	.512	53	2.087	93	3.661	.040	1.01	.440	11.17	.840	21.33
14	.551	54	2.126	94	3.701	.050	1.27	.450	11.43	.850	21.59
15	.591	55	2.165	95	3.740	.060	1.52	.460	11.68	.860	21.84
16	.630	56	2.205	96	3.780	.070	1.77	.470	11.93	.870	22.09
17	.669	57	2.244	97	3.819	.080	2.03	.480	12.19	.880	22.35
18	.709	58	2.283	98	3.858	.090	2.28	.490	12.44	.890	22.60
19	.748	59	2.323	99	3.898	.100	2.54	.500	12.70	.900	22.86
20	.787	60	2.362	100	3.937	.110	2.79	.510	12.95	.910	23.11
21	.827	61	2.402	125	4.921	.120	3.04	.520	13.20	.920	23.36
22	.866	62	2.441	150	5.906	.130	3.30	.530	13.46	.930	23.62
23	.906	63	2.480	175	6.890	.140	3.55	.540	13.71	.940	23.87
24	.945	64	2.520			.150	3.81	.550	13.97	.950	24.13
25	.984	65	2.559			.160	4.06	.560	14.22	.960	24.38
26	1.024	66	2.598			.170	4.31	.570	14.47	.970	24.63
27	1.063	67	2.638	200	7.874	.180	4.57	.580	14.73	.980	24.89
28	1.102	68	2.677	225	8.858	.190	4.82	.590	14.98	.990	25.14
29	1.142	69	2.717	250	9.843	.200	5.08	.600	15.24	1.000	25.40
30	1.181	70	2.756	275	10.827	.210	5.33	.610	15.49		
31	1.220	71	2.795			.220	5.55	.620	15.74	2.000	50.80
32	1.260	72	2.835	300	11.811	.230	5.84	.630	16.00	3.000	76.20
33	1.299	73	2.874	325	12.795	.240	6.09	.640	16.25	4.000	101.60
34	1.339	74	2.913			.250	6.35	.650	16.51	5.000	127.00
35	1.378	75	2.953			.260	6.60	.660	16.76	10.000	254.00
36	1.417	76	2.992			.270	6.85	.670	17.01		
37	1.457	77	3.031			.280	7.11	.680	17.27		
38	1.496	78	3.071			.290	7.36	.690	17.52		
39	1.535	79	3.110			.300	7.62	.700	17.78		
40	1.575	80	3.150			.310	7.87	.710	18.03		

Appendix F

DRILL SIZES IN DECIMAL EQUIVALENTS

There are *fractional, number, letter,* and *metric* drill size series. The fractional divisions are in 1/64″ increments. The number, letter, and metric series have drill diameters that fall between the factional inch measures. Together, the four series make up a long-running series of decimal equivalents, as shown on the next page.

Inch	Decimal	Wire	M/M
1/64	.0156		
	.0157		.4
	.0160	78	
	.0165		.42
	.0173		.44
	.0177		.45
	.0180	77	
	.0181		.46
	.0189		.48
	.0197		.5
	.0200	76	
	.0210	75	
	.0217		.55
	.0225	74	
	.0236		.6
	.0240	73	
	.0250	72	
	.0256		.65
	.0260	71	
	.0276		.7
	.0280	70	
	.0292	69	
	0295		.75
	.0310	68	
1/32	.0312		
	.0315		.8
	.0320	67	
	.0330	66	
	.0335		.85
	.0350	65	
	.0354		.9
	.0360	64	
	.0370	63	
	.0374		.95
	.0380	62	
	.0390	61	
	.0394		1.
	.0400	60	
	.0410	59	
	.0413		1.05
	.0420	58	
	.0430	57	
	.0433		1.1
	.0453		1.15
	.0465	56	
3/64	.0469		
	.0472		1.2
	.0492		1.25
	.0512		1.3
	.0520	55	
	.0531		1.35
	.0550	54	
	.0551		1.4
	.0571		1.45
	.0591		1.5
	.0595	53	

Inch	Decimal	Wire	M/M
	.0610		1.55
1/16	.0625		
	.0630		1.6
	.0635	52	
	.0650		1.65
	.0669		1.7
	.0670	51	
	.0689		1.75
	.0700	50	
	.0709		1.8
	.0728		1.85
	.0730	49	
	.0748		1.9
	.0760	48	
	.0768		1.95
5/64	.0781		
	.0785	47	
	.0787		2.
	.0807		2.05
	.0810	46	
	.0820	45	
	.0827		2.1
	.0846		2.15
	.0860	44	
	.0866		2.2
	.0886		2.25
	.0890	43	
	.0906		2.3
	.0925		2.35
	.0935	42	
3/32	.0938		
	.0945		2.4
	.0960	41	
	.0965		2.45
	.0980	40	
	.0981		2.5
	.0995	39	
	.1015	38	
	.1024		2.6
	.1040	37	
	.1063		2.7
	.1065	36	
	.1083		2.75
7/64	.1094		
	.1100	35	
	.1102		2.8
	.1110	34	
	.1130	33	
	.1142		2.9
	.1160	32	
	.1181		3.
	.1200	31	
	.1220		3.1
1/8	.1250		
	.1260		3.2
	.1280		3.25

Inch	Decimal	Wire	M/M
	.1285	30	
	.1299		3.3
	.1339		3.4
	.1360	29	
	.1378		3.5
	.1405	28	
9/64	.1406		
	.1417		3.6
	.1440	27	
	.1457		3.7
	.1470	26	
	.1476		3.75
	.1495	25	
	.1496		3.8
	.1520	24	
	.1535		3.9
	.1540	23	
5/32	.1562		
	.1570	22	
	.1575		4.
	.1590	21	
	.1610	20	
	.1614		4.1
	.1654		4.2
	.1660	19	
	.1673		4.25
	.1693		4.3
	.1695	18	
11/64	.1719		
	.1730	17	
	.1732		4.4
	.1770	16	
	.1772		4.5
	.1800	15	
	.1811		4.6
	.1820	14	
	.1850	13	
	.1850		4.7
	.1870		4.75
3/16	.1875		
	.1890		4.8
	.1890	12	
	.1910	11	
	.1929		4.9
	.1935	10	
	.1960	9	
	.1969		5.
	.1990	8	
	.2008		5.1
	.2010	7	
13/64	.2031		
	.2040	6	
	.2047		5.2
	.2055	5	
	.2067		5.25
	.2087		5.3

Inch	Decimal	Wire/Letter	M/M
	.2090	4	
	.2126		5.4
	.2130	3	
	.2165		5.5
7/32	.2188		
	.2205		5.6
	.2210	2	
	.2244		5.7
	.2264		5.75
	.2280	1	
	.2283		5.8
	.2323		5.9
	.2340	A	
15/64	.2344		
	.2362		6.
	.2380	B	
	.2402		6.1
	.2420	C	
	.2441		6.2
	.2460	D	
	.2461		6.25
	.2480		6.3
1/4	.2500	E	
	.2520		6.4
	.2559		6.5
	.2570	F	
	.2598		6.6
	.2610	G	
	.2638		6.7
17/64	.2656		
	.2657		6.75
	.2660	H	
	.2677		6.8
	.2717		6.9
	.2720	I	
	.2756		7.
	.2770	J	
	.2795		7.1
	.2810	K	
9/32	.2812		
	.2835		7.2
	.2854		7.25
	.2874		7.3
	.2900	L	
	.2913		7.4
	.2950	M	
	.2953		7.5
19/64	.2969		
	.2992		7.6
	.3020	N	
	.3031		7.7
	.3051		7.75
	.3071		7.8
	.3110		7.9
5/16	.3125		
	.3150		8.

Inch	Decimal	Letter	M/M
	.3160	O	
	.3189		8.1
	.3228		8.2
	.3230	P	
	.3248		8.25
	.3268		8.3
21/64	.3281		
	.3307		8.4
	.3320	Q	
	.3346		8.5
	.3386		8.6
	.3390	R	
	.3425		8.7
11/32	.3438		
	.3445		8.75
	.3465		8.8
	.3480	S	
	.3504		8.9
	.3543		9.
	.3580	T	
	.3583		9.1
23/64	.3594		
	.3622		9.2
	.3642		9.25
	.3661		9.3
	.3680	U	
	.3701		9.4
	.3740		9.5
3/8	.3750		
	.3770	V	
	.3780		9.6
	.3819		9.7
	.3839		9.75
	.3858		9.8
	.3860	W	
	.3898		9.9
25/64	.3906		
	.3937		10.
	.3970	X	
	.4040	Y	
13/32	.4062		
	.4130	Z	
	.4134		10.5
27/64	.4219		
	.4331		11.
7/16	.4375		
	.4528		11.5
29/64	.4531		
15/32	.4688		
	.4724		12.
31/64	.4844		
	.4921		12.5
1/2	.5000		
	.5118		13.
33/64	.5156		
17/32	.5312		

Inch	Decimal	M/M
	.5315	13.5
35/64	.5469	
	.5512	14.
9/16	.5625	
	.5709	14.5
37/64	.5781	
	.5906	15.
19/32	.5938	
39/64	.6094	
	.6102	15.5
5/8	.6250	
	.6299	16.
41/64	.6406	
	.6496	16.5
21/32	.6562	
	.6693	17.
43/64	.6719	
11/16	.6875	
	.6890	17.5
45/64	.0731	
	.7087	18.
23/32	.7188	
	.7283	18.5
47/64	.7344	
	.7480	19.
3/4	.7500	
49/64	.7656	
	.7677	19.5
25/32	.7812	
	.7874	20.
51/64	.7969	
	.8071	20.5
13/16	.8125	
	.8268	21.
53/64	.8281	
27/32	.8438	
	.8465	21.5
55/64	.8594	
	.8661	22.
7/8	.8750	
	.8858	22.5
57/64	.8906	
	.9055	23.
29/32	.9062	
59/64	.9219	
	.9252	23.5
15/16	.9375	
	.9449	24.
61/64	.9531	
	.9646	24.5
31/32	.9688	
	.9843	25.
63/64	.9844	
1	1.0000	

Appendix G

MISCELLANEOUS CHARTS

Fractional/Metric Conversion

Eighths	Sixteenths	Thirty-Seconds	Milli-meters	Decimal
		1/32		.03125
			1	.03937
	1/16			.0625
			2	.07874
		3/32		.09375
			3	.11811
1/8				.125
		5/32		.15625
			4	.15748
	3/16			.1875
			5	.19685
		7/32		.21875
			6	.23622
1/4				.250
			7	.27559
		9/32		.28125
	5/16			.3125
			8	.31496
		11/32		.34375
			9	.35433
3/8				.375
			10	.3937
		13/32		.40625
			11	.43307
	7/16			.4375
		15/32		.46875
			12	.47244
1/2				.500
			13	.51181

Eighths	Sixteenths	Thirty-Seconds	Milli-meters	Decimal
		17/32		.53125
			14	.55118
	9/16			.5625
			15	.59055
		19/32		.59375
5/8				.625
			16	.62992
		21/32		.65625
			17	.66929
	11/16			.6875
			18	.70866
		23/32		.71875
			19	.74803
3/4				.750
		25/32		.78125
			20	.7874
	13/16			.8125
			21	.82677
		27/32		.84375
			22	.86614
7/8				.875
			23	.90551
		29/32		.90625
	15/16			.9375
			24	.94488
		31/32		.96875
			25	.98425
1″				1.000000

Fastener Type Identification

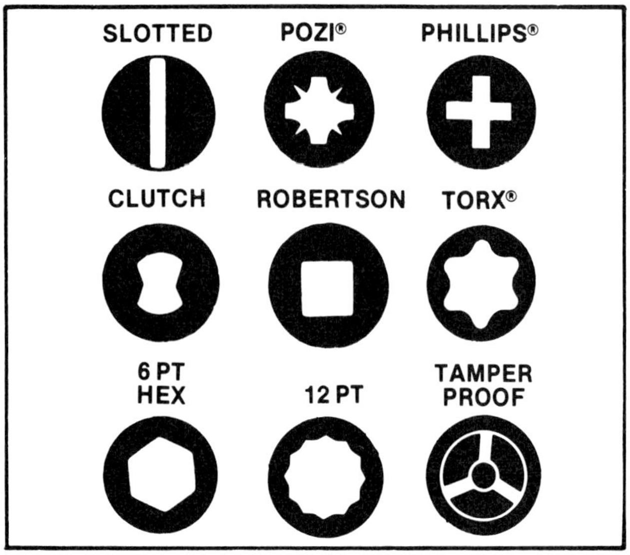

Hole Saw Operating Speeds

RECOMMENDED RPMS FOR OPERATING SAWS

Hole Saw Diameter		Mild Steel	Tool/ Stainless	Aluminum	Brass
Inch	M/M				
9/16	14.3	530	300	900	790
5/8	15.8	550	275	825	730
11/16	17.5	480	250	750	665
3/4	19.0	460	230	690	600
13/16	20.6	425	210	635	560
7/8	22.2	390	195	685	520
15/16	23.8	370	185	555	495
1	25.4	350	175	525	470
1 1/16	27.0	325	160	480	435
1 1/8	28.6	300	150	450	400
1 3/16	30.2	285	145	425	380
1 1/4	31.8	275	140	410	360
1 5/16	33.3	260	135	390	345
1 3/8	34.9	250	125	375	330
1 7/16	36.5	240	120	360	315
1 1/2	38.1	230	115	345	300
1 9/16	39.7	220	110	330	280
1 5/8	41.3	210	105	315	280
1 11/16	42.9	205	100	305	270
1 3/4	44.5	195	95	295	260
1 13/16	46.0	190	95	285	250
1 7/8	47.6	180	90	270	240
2	50.8	170	85	255	230
2 1/16	52.4	165	80	245	220
2 1/8	54.0	160	80	240	210
2 1/4	57.2	150	75	225	200
2 5/16	58.7	145	75	225	195
2 3/8	60.3	140	70	220	190
2 1/2	63.5	135	65	205	180
2 9/16	65.1	130	65	220	175
2 5/8	66.7	130	65	195	170
2 3/4	69.9	125	60	185	160
2 7/8	73.0	120	60	180	160
3	76.2	115	55	170	150
3 1/8	79.4	110	55	165	140
3 1/4	82.6	105	50	155	140
3 3/8	85.7	100	50	150	130
3 1/2	88.9	95	45	145	130
3 5/8	92.1	95	45	140	120
3 3/4	95.3	90	45	135	120
3 7/8	98.4	90	45	135	120
4	101.6	85	40	130	110
4 1/8	104.2	80	40	120	110
4 1/4	108.0	80	40	120	110
4 3/8	111.1	80	40	120	100
4 1/2	114.3	75	35	105	100
4 3/4	120.7	70	35	95	95
5	127.0	65	30	90	90
5 1/2	139.7	60	25	85	85
5 3/4	146.0	55	25	75	75
6	152.4	55	25	75	75

Drill Sizes for Tapping Threads

Screw Size	N. C. Pitch (Coarse)	Drill Size	Decimal Equiv.	N. F. Pitch (Fine)	Drill Size	Decimal Equiv.
No. 0	. . .	. . .	. . .	80	3/64	.0469
No. 1	64	#53	.0595	72	#53	.0595
No. 2	56	#50	.0700	64	#50	.7000
No. 3	48	#47	.0785	56	#45	.0820
No. 4	40	#43	.0890	48	#42	.0935
No. 5	40	#38	.1015	44	#37	.1040
No. 6	32	#36	.1065	40	#33	.1130
No. 8	32	#29	.1360	36	#29	.1360
No. 10	24	#25	.1495	32	#21	.1590
No. 12	24	#16	.1770	28	#14	.1820
1/4	20	#7	.2010	28	#3	.2130
5/16	18	F	.2570	24	I	.2720
3/8	16	5/16	.3125	24	Q	.3320
7/16	14	U	.3680	20	25/64	.3906
1/2	13	27/64	.4219	20	29/64	.4531
9/16	12	31/64	.4844	18	33/64	.5156
5/8	11	17/32	.5312	18	37/64	.5781
3/4	10	21/32	.6562	16	11/16	.6875
7/8	9	49/64	.7656	14	13/16	.8125
1	8	7/8	.8750	14	15/16	.9375
1 1/8	7	63/64	.9844			

ANSWERS

SECTION A—HYDRAULIC SYSTEM DIAGNOSIS AND REPAIR

1. c
2. b
3. d
4. c
5. a
6. b
7. a
8. b
9. b
10. c
11. a
12. d
13. d
14. d
15. b
16. d
17. d
18. d
19. d
20. d
21. c
22. d
23. a

SECTION B—DRUM BRAKE DIAGNOSIS AND REPAIR

1. c
2. c
3. c
4. d
5. c
6. d
7. d
8. d
9. a
10. b
11. a
12. b
13. a
14. b
15. c

SECTION C—DISC BRAKE DIAGNOSIS AND REPAIR

1. b
2. a
3. a
4. c
5. c
6. d
7. c
8. b
9. d
10. c
11. c
12. c
13. d
14. d

SECTION D—POWER ASSIST UNITS DIAGNOSIS AND REPAIR

1. c
2. b
3. d
4. c

SECTION E—MISCELLANEOUS DIAGNOSIS AND REPAIR

1. c
2. c
3. b
4. a
5. c
6. c
7. c

BOOK SUMMARY QUESTIONS

1. c
2. a
3. d
4. d
5. b
6. b
7. b
8. b
9. b
10. d
11. a
12. d
13. c
14. a
15. b
16. b
17. a
18. d
19. c
20. b
21. b
22. d
23. b
24. d
25. c
26. a
27. d
28. c
29. d
30. c
31. d
32. d

33. b
34. c
35. c
36. b
37. b
38. d
39. c
40. a
41. d
42. d
43. d
44. c
45. a
46. c
47. a
48. d
49. c
50. c
51. c
52. d
53. c
54. a
55. c
56. a
57. c
58. a
59. a
60. c

INDEX

REGISTERED OWNER ____________________ DATE __________

ADDRESS ____________________ PHONE NO. __________

MAKE __________ MODEL __________ MILEAGE __________ LICENSE NO. __________ YEAR __________

SPECIAL KEY REQUIRED FOR REMOVING HUBCAPS/WHEELS? YES ☐ NO ☐ DOES OWNER USE PARKING BRAKE? YES ☐ NO ☐

4 DRUM ☐ 4 DISC ☐ DISC/DRUM ☐ POWER BRAKES? YES ☐ NO ☐ VACUUM PUMP? YES ☐ NO ☐

OWNER COMPLAINTS ____________________

1. **CHECKS BEFORE ROAD TEST**	OK	BAD
STOPLAMP OPERATION WITH HEADLAMPS ON/OFF		
BRAKE WARNING LAMP OPERATION		
PARKING BRAKE LAMP OPERATION		
PARKING BRAKE ADJUSTMENT		

TIRE PRESSURE SPECS:

RECORD PRESSURE MEASURED

RF ______ LF ______ RR ______ LR ______

TIRE CONDITION

RF ______ LF ______ RR ______ LR ______

MASTER CYLINDER CHECKS

FLUID LEVEL		
FLUID CONTAMINATION		
UNDER HOOD FLUID LEAKS		
UNDER DASH FLUID LEAKS (NO POWER BRAKES)		
PEDAL HEIGHT STAYS WHEN FORCE APPLIED		
PEDAL FREE TRAVEL		

CHECK ONE		CHECK ONE		
LOW PEDAL HEIGHT		SPONGY PEDAL		
MED. PEDAL HEIGHT		FIRM PEDAL		
HIGH PEDAL HEIGHT				

POWER BRAKE UNIT CHECKS

VACUUM TYPE	OK	BAD	HYDRO TYPE	OK	BAD
VACUUM UNIT LEAKS			HYDRO UNIT LEAKS		
ENGINE VACUUM			P/S FLUID LEVEL		
VACUUM HOSE			P/S BELT TENSION		
UNIT CHECK VALVE			P/S BELT CONDITION		
RESERVE BRAKING			P/S FLUID LEAKS		
PEDAL DROP TEST			RESERVE BRAKING		

2. **ROAD TEST**	YES	NO
PULLING		
CLUNK NOISE		
SCRAPING NOISE		
SQUEAL NOISE		
GRABBING		
PREMATURE LOCKUP		
WHEEL BEARING NOISE		
VIBRATION		

	YES	NO
PEDAL PULSATION WHEN BRAKING		
STEERING WHEEL OSCILLATION WHEN BRAKING		
EXCESSIVE PEDAL EFFORT REQUIRED		
WARNING LAMP COMES ON WHEN BRAKING		
DIFFERENCE IN PEDAL HEIGHT AFTER CORNERING		
NOSE DIVE WHEN BRAKING		

3. **IN SHOP CHECKS ON HOIST**	YES	NO	RF	LF	RR	LR
BRAKE DRAG						
INTERMITTENT BRAKE DRAG						
BRAKE PEDAL LINKAGE BINDING						
WHEEL BEARING LOOSENESS						
MISSING OR DAMAGED LUGS						
SUSPENSION LOOSENESS						
Mark Wheels, and Remove:						
Caliper/Piston Stuck						
Mark Drums and Remove:						

ADDITIONAL COMMENTS:

	FRONT					REAR				
	RIGHT			LEFT		RIGHT			LEFT	
	SPEC	SAFE	UNSAFE	SAFE	UNSAFE	SPEC	SAFE	UNSAFE	SAFE	UNSAFE
MEASURE ROTOR THICKNESS OR DRUM DIAMETER										
MEASURE ROTOR THICKNESS VARIATION (PARALLELISM)										
MEASURE ROTOR RUNOUT (TIR)										
LINING THICKNESS/CONDITION										
LINES, HOSES, CONNECTIONS, VALVES										
FLUID LEAKS										
BROKEN BLEEDER SCREWS										
LEAKY GREASE SEALS										
SELF-ADJUSTER OPERATION										

	OK	BAD
PARKING BRAKE CABLE AND LINKAGE CONDITION		
HARDWARE/BACKING PLATES		

Notes

REGISTERED OWNER ______ DATE ______

ADDRESS ______ PHONE NO. ______

MAKE ______ MODEL ______ MILEAGE ______ LICENSE NO. ______ YEAR ______

SPECIAL KEY REQUIRED FOR REMOVING HUBCAPS/WHEELS? YES ☐ NO ☐ DOES OWNER USE PARKING BRAKE? YES ☐ NO ☐

4 DRUM ☐ 4 DISC ☐ DISC/DRUM ☐ POWER BRAKES? YES ☐ NO ☐ VACUUM PUMP? YES ☐ NO ☐

OWNER COMPLAINTS ______

1. CHECKS BEFORE ROAD TEST	OK	BAD
STOPLAMP OPERATION WITH HEADLAMPS ON/OFF		
BRAKE WARNING LAMP OPERATION		
PARKING BRAKE LAMP OPERATION		
PARKING BRAKE ADJUSTMENT		

TIRE PRESSURE SPECS:

RECORD PRESSURE MEASURED

RF ______ LF ______ RR ______ LR ______

TIRE CONDITION

RF ______ LF ______ RR ______ LR ______

MASTER CYLINDER CHECKS

FLUID LEVEL		
FLUID CONTAMINATION		
UNDER HOOD FLUID LEAKS		
UNDER DASH FLUID LEAKS (NO POWER BRAKES)		
PEDAL HEIGHT STAYS WHEN FORCE APPLIED		
PEDAL FREE TRAVEL		

CHECK ONE		CHECK ONE	
LOW PEDAL HEIGHT		SPONGY PEDAL	
MED. PEDAL HEIGHT		FIRM PEDAL	
HIGH PEDAL HEIGHT			

POWER BRAKE UNIT CHECKS

VACUUM TYPE	OK	BAD	HYDRO TYPE	OK	BAD
VACUUM UNIT LEAKS			HYDRO UNIT LEAKS		
ENGINE VACUUM			P/S FLUID LEVEL		
VACUUM HOSE			P/S BELT TENSION		
UNIT CHECK VALVE			P/S BELT CONDITION		
RESERVE BRAKING			P/S FLUID LEAKS		
PEDAL DROP TEST			RESERVE BRAKING		

2. ROAD TEST	YES	NO
PULLING		
CLUNK NOISE		
SCRAPING NOISE		
SQUEAL NOISE		
GRABBING		
PREMATURE LOCKUP		
WHEEL BEARING NOISE		
VIBRATION		

	YES	NO
PEDAL PULSATION WHEN BRAKING		
STEERING WHEEL OSCILLATION WHEN BRAKING		
EXCESSIVE PEDAL EFFORT REQUIRED		
WARNING LAMP COMES ON WHEN BRAKING		
DIFFERENCE IN PEDAL HEIGHT AFTER CORNERING		
NOSE DIVE WHEN BRAKING		

3. IN SHOP CHECKS ON HOIST	YES	NO	RF	LF	RR	LR
BRAKE DRAG						
INTERMITTENT BRAKE DRAG						
BRAKE PEDAL LINKAGE BINDING						
WHEEL BEARING LOOSENESS						
MISSING OR DAMAGED LUGS						
SUSPENSION LOOSENESS						
Mark Wheels, and Remove:						
Caliper/Piston Stuck						
Mark Drums and Remove:						

ADDITIONAL COMMENTS:

	FRONT					REAR				
	RIGHT			LEFT		RIGHT			LEFT	
	SPEC	SAFE	UNSAFE	SAFE	UNSAFE	SPEC	SAFE	UNSAFE	SAFE	UNSAFE
MEASURE ROTOR THICKNESS OR DRUM DIAMETER										
MEASURE ROTOR THICKNESS VARIATION (PARALLELISM)										
MEASURE ROTOR RUNOUT (TIR)										
LINING THICKNESS/CONDITION										
LINES, HOSES, CONNECTIONS, VALVES										
FLUID LEAKS										
BROKEN BLEEDER SCREWS										
LEAKY GREASE SEALS										
SELF-ADJUSTER OPERATION										

	OK	BAD
PARKING BRAKE CABLE AND LINKAGE CONDITION		
HARDWARE/BACKING PLATES		

Notes

REGISTERED OWNER ______________________ DATE ____________

ADDRESS ______________________ PHONE NO. ____________

MAKE ________ MODEL ________ MILEAGE ________ LICENSE NO. ________ YEAR ________

SPECIAL KEY REQUIRED FOR REMOVING HUBCAPS/WHEELS? YES ☐ NO ☐ DOES OWNER USE PARKING BRAKE? YES ☐ NO ☐

4 DRUM ☐ 4 DISC ☐ DISC/DRUM ☐ POWER BRAKES? YES ☐ NO ☐ VACUUM PUMP? YES ☐ NO ☐

OWNER COMPLAINTS ______________________

1. CHECKS BEFORE ROAD TEST	OK	BAD
STOPLAMP OPERATION WITH HEADLAMPS ON/OFF		
BRAKE WARNING LAMP OPERATION		
PARKING BRAKE LAMP OPERATION		
PARKING BRAKE ADJUSTMENT		

MASTER CYLINDER CHECKS

FLUID LEVEL		
FLUID CONTAMINATION		
UNDER HOOD FLUID LEAKS		
UNDER DASH FLUID LEAKS (NO POWER BRAKES)		
PEDAL HEIGHT STAYS WHEN FORCE APPLIED		
PEDAL FREE TRAVEL		

CHECK ONE		CHECK ONE		
LOW PEDAL HEIGHT		SPONGY PEDAL		
MED. PEDAL HEIGHT		FIRM PEDAL		
HIGH PEDAL HEIGHT				

POWER BRAKE UNIT CHECKS

VACUUM TYPE	OK	BAD	HYDRO TYPE	OK	BAD
VACUUM UNIT LEAKS			HYDRO UNIT LEAKS		
ENGINE VACUUM			P/S FLUID LEVEL		
VACUUM HOSE			P/S BELT TENSION		
UNIT CHECK VALVE			P/S BELT CONDITION		
RESERVE BRAKING			P/S FLUID LEAKS		
PEDAL DROP TEST			RESERVE BRAKING		

TIRE PRESSURE SPECS:

RECORD PRESSURE MEASURED

RF ______ LF ______ RR ______ LR ______

TIRE CONDITION

RF ______ LF ______ RR ______ LR ______

2. ROAD TEST	YES	NO
PULLING		
CLUNK NOISE		
SCRAPING NOISE		
SQUEAL NOISE		
GRABBING		
PREMATURE LOCKUP		
WHEEL BEARING NOISE		
VIBRATION		

	YES	NO
PEDAL PULSATION WHEN BRAKING		
STEERING WHEEL OSCILLATION WHEN BRAKING		
EXCESSIVE PEDAL EFFORT REQUIRED		
WARNING LAMP COMES ON WHEN BRAKING		
DIFFERENCE IN PEDAL HEIGHT AFTER CORNERING		
NOSE DIVE WHEN BRAKING		

3. IN SHOP CHECKS ON HOIST	YES	NO	RF	LF	RR	LR
BRAKE DRAG						
INTERMITTENT BRAKE DRAG						
BRAKE PEDAL LINKAGE BINDING						
WHEEL BEARING LOOSENESS						
MISSING OR DAMAGED LUGS						
SUSPENSION LOOSENESS						
Mark Wheels, and Remove:						
Caliper/Piston Stuck						
Mark Drums and Remove:						

ADDITIONAL COMMENTS:

	FRONT					REAR				
	RIGHT			LEFT		RIGHT			LEFT	
	SPEC	SAFE	UNSAFE	SAFE	UNSAFE	SPEC	SAFE	UNSAFE	SAFE	UNSAFE
MEASURE ROTOR THICKNESS OR DRUM DIAMETER										
MEASURE ROTOR THICKNESS VARIATION (PARALLELISM)										
MEASURE ROTOR RUNOUT (TIR)										
LINING THICKNESS/CONDITION										
LINES, HOSES, CONNECTIONS, VALVES										
FLUID LEAKS										
BROKEN BLEEDER SCREWS										
LEAKY GREASE SEALS										
SELF-ADJUSTER OPERATION										

	OK	BAD
PARKING BRAKE CABLE AND LINKAGE CONDITION		
HARDWARE/BACKING PLATES		

Notes

REGISTERED OWNER ______________________________ DATE ____________

ADDRESS ______________________________ PHONE NO. ____________

MAKE ____________ MODEL ____________ MILEAGE ____________ LICENSE NO. ____________ YEAR ____________

SPECIAL KEY REQUIRED FOR REMOVING HUBCAPS/WHEELS? YES ☐ NO ☐ DOES OWNER USE PARKING BRAKE? YES ☐ NO ☐

4 DRUM ☐ 4 DISC ☐ DISC/DRUM ☐ POWER BRAKES? YES ☐ NO ☐ VACUUM PUMP? YES ☐ NO ☐

OWNER COMPLAINTS ______________________________

1. **CHECKS BEFORE ROAD TEST**	OK	BAD
STOPLAMP OPERATION WITH HEADLAMPS ON/OFF		
BRAKE WARNING LAMP OPERATION		
PARKING BRAKE LAMP OPERATION		
PARKING BRAKE ADJUSTMENT		

TIRE PRESSURE SPECS:

RECORD PRESSURE MEASURED

RF ______ LF ______ RR ______ LR ______

TIRE CONDITION

RF ______ LF ______ RR ______ LR ______

MASTER CYLINDER CHECKS

FLUID LEVEL		
FLUID CONTAMINATION		
UNDER HOOD FLUID LEAKS		
UNDER DASH FLUID LEAKS (NO POWER BRAKES)		
PEDAL HEIGHT STAYS WHEN FORCE APPLIED		
PEDAL FREE TRAVEL		

CHECK ONE		CHECK ONE		
LOW PEDAL HEIGHT		SPONGY PEDAL		
MED. PEDAL HEIGHT		FIRM PEDAL		
HIGH PEDAL HEIGHT				

POWER BRAKE UNIT CHECKS

VACUUM TYPE	OK	BAD	HYDRO TYPE	OK	BAD
VACUUM UNIT LEAKS			HYDRO UNIT LEAKS		
ENGINE VACUUM			P/S FLUID LEVEL		
VACUUM HOSE			P/S BELT TENSION		
UNIT CHECK VALVE			P/S BELT CONDITION		
RESERVE BRAKING			P/S FLUID LEAKS		
PEDAL DROP TEST			RESERVE BRAKING		

2. **ROAD TEST**	YES	NO
PULLING		
CLUNK NOISE		
SCRAPING NOISE		
SQUEAL NOISE		
GRABBING		
PREMATURE LOCKUP		
WHEEL BEARING NOISE		
VIBRATION		

	YES	NO
PEDAL PULSATION WHEN BRAKING		
STEERING WHEEL OSCILLATION WHEN BRAKING		
EXCESSIVE PEDAL EFFORT REQUIRED		
WARNING LAMP COMES ON WHEN BRAKING		
DIFFERENCE IN PEDAL HEIGHT AFTER CORNERING		
NOSE DIVE WHEN BRAKING		

3. **IN SHOP CHECKS ON HOIST**	YES	NO	RF	LF	RR	LR
BRAKE DRAG						
INTERMITTENT BRAKE DRAG						
BRAKE PEDAL LINKAGE BINDING						
WHEEL BEARING LOOSENESS						
MISSING OR DAMAGED LUGS						
SUSPENSION LOOSENESS						
Mark Wheels, and Remove:						
Caliper/Piston Stuck						
Mark Drums and Remove:						

ADDITIONAL COMMENTS:

	FRONT					REAR				
	RIGHT			LEFT		RIGHT			LEFT	
	SPEC	SAFE	UNSAFE	SAFE	UNSAFE	SPEC	SAFE	UNSAFE	SAFE	UNSAFE
MEASURE ROTOR THICKNESS OR DRUM DIAMETER										
MEASURE ROTOR THICKNESS VARIATION (PARALLELISM)										
MEASURE ROTOR RUNOUT (TIR)										
LINING THICKNESS/CONDITION										
LINES, HOSES, CONNECTIONS, VALVES										
FLUID LEAKS										
BROKEN BLEEDER SCREWS										
LEAKY GREASE SEALS										
SELF-ADJUSTER OPERATION										

	OK	BAD
PARKING BRAKE CABLE AND LINKAGE CONDITION		
HARDWARE/BACKING PLATES		

Notes

REGISTERED OWNER ____________ DATE ____________

ADDRESS ____________ PHONE NO. ____________

MAKE ________ MODEL ________ MILEAGE ________ LICENSE NO. ________ YEAR ________

SPECIAL KEY REQUIRED FOR REMOVING HUBCAPS/WHEELS? YES ☐ NO ☐ DOES OWNER USE PARKING BRAKE? YES ☐ NO ☐

4 DRUM ☐ 4 DISC ☐ DISC/DRUM ☐ POWER BRAKES? YES ☐ NO ☐ VACUUM PUMP? YES ☐ NO ☐

OWNER COMPLAINTS ____________

1. **CHECKS BEFORE ROAD TEST**	OK	BAD
STOPLAMP OPERATION WITH HEADLAMPS ON/OFF		
BRAKE WARNING LAMP OPERATION		
PARKING BRAKE LAMP OPERATION		
PARKING BRAKE ADJUSTMENT		

TIRE PRESSURE SPECS:

RECORD PRESSURE MEASURED

RF ______ LF ______ RR ______ LR ______

TIRE CONDITION

RF ______ LF ______ RR ______ LR ______

MASTER CYLINDER CHECKS

FLUID LEVEL		
FLUID CONTAMINATION		
UNDER HOOD FLUID LEAKS		
UNDER DASH FLUID LEAKS (NO POWER BRAKES)		
PEDAL HEIGHT STAYS WHEN FORCE APPLIED		
PEDAL FREE TRAVEL		

CHECK ONE		CHECK ONE		
LOW PEDAL HEIGHT		SPONGY PEDAL		
MED. PEDAL HEIGHT		FIRM PEDAL		
HIGH PEDAL HEIGHT				

POWER BRAKE UNIT CHECKS

VACUUM TYPE	OK	BAD	HYDRO TYPE	OK	BAD
VACUUM UNIT LEAKS			HYDRO UNIT LEAKS		
ENGINE VACUUM			P/S FLUID LEVEL		
VACUUM HOSE			P/S BELT TENSION		
UNIT CHECK VALVE			P/S BELT CONDITION		
RESERVE BRAKING			P/S FLUID LEAKS		
PEDAL DROP TEST			RESERVE BRAKING		

2. **ROAD TEST**	YES	NO
PULLING		
CLUNK NOISE		
SCRAPING NOISE		
SQUEAL NOISE		
GRABBING		
PREMATURE LOCKUP		
WHEEL BEARING NOISE		
VIBRATION		

	YES	NO
PEDAL PULSATION WHEN BRAKING		
STEERING WHEEL OSCILLATION WHEN BRAKING		
EXCESSIVE PEDAL EFFORT REQUIRED		
WARNING LAMP COMES ON WHEN BRAKING		
DIFFERENCE IN PEDAL HEIGHT AFTER CORNERING		
NOSE DIVE WHEN BRAKING		

3. **IN SHOP CHECKS ON HOIST**	YES	NO	RF	LF	RR	LR
BRAKE DRAG						
INTERMITTENT BRAKE DRAG						
BRAKE PEDAL LINKAGE BINDING						
WHEEL BEARING LOOSENESS						
MISSING OR DAMAGED LUGS						
SUSPENSION LOOSENESS						
Mark Wheels, and Remove:						
Caliper/Piston Stuck						
Mark Drums and Remove:						

ADDITIONAL COMMENTS:

	FRONT					REAR				
	RIGHT			LEFT		RIGHT			LEFT	
	SPEC	SAFE	UNSAFE	SAFE	UNSAFE	SPEC	SAFE	UNSAFE	SAFE	UNSAFE
MEASURE ROTOR THICKNESS OR DRUM DIAMETER										
MEASURE ROTOR THICKNESS VARIATION (PARALLELISM)										
MEASURE ROTOR RUNOUT (TIR)										
LINING THICKNESS/CONDITION										
LINES, HOSES, CONNECTIONS, VALVES										
FLUID LEAKS										
BROKEN BLEEDER SCREWS										
LEAKY GREASE SEALS										
SELF-ADJUSTER OPERATION										

	OK	BAD
PARKING BRAKE CABLE AND LINKAGE CONDITION		
HARDWARE/BACKING PLATES		

Notes

REGISTERED OWNER ____________________ DATE __________

ADDRESS ____________________ PHONE NO. __________

MAKE __________ MODEL __________ MILEAGE __________ LICENSE NO. __________ YEAR __________

SPECIAL KEY REQUIRED FOR REMOVING HUBCAPS/WHEELS? YES ☐ NO ☐ DOES OWNER USE PARKING BRAKE? YES ☐ NO ☐

4 DRUM ☐ 4 DISC ☐ DISC/DRUM ☐ POWER BRAKES? YES ☐ NO ☐ VACUUM PUMP? YES ☐ NO ☐

OWNER COMPLAINTS ____________________

1. CHECKS BEFORE ROAD TEST	OK	BAD
STOPLAMP OPERATION WITH HEADLAMPS ON/OFF		
BRAKE WARNING LAMP OPERATION		
PARKING BRAKE LAMP OPERATION		
PARKING BRAKE ADJUSTMENT		

TIRE PRESSURE SPECS:

RECORD PRESSURE MEASURED

RF ______ LF ______ RR ______ LR ______

TIRE CONDITION

RF ______ LF ______ RR ______ LR ______

MASTER CYLINDER CHECKS

FLUID LEVEL		
FLUID CONTAMINATION		
UNDER HOOD FLUID LEAKS		
UNDER DASH FLUID LEAKS (NO POWER BRAKES)		
PEDAL HEIGHT STAYS WHEN FORCE APPLIED		
PEDAL FREE TRAVEL		

CHECK ONE		CHECK ONE		
LOW PEDAL HEIGHT		SPONGY PEDAL		
MED. PEDAL HEIGHT		FIRM PEDAL		
HIGH PEDAL HEIGHT				

POWER BRAKE UNIT CHECKS

VACUUM TYPE	OK	BAD	HYDRO TYPE	OK	BAD
VACUUM UNIT LEAKS			HYDRO UNIT LEAKS		
ENGINE VACUUM			P/S FLUID LEVEL		
VACUUM HOSE			P/S BELT TENSION		
UNIT CHECK VALVE			P/S BELT CONDITION		
RESERVE BRAKING			P/S FLUID LEAKS		
PEDAL DROP TEST			RESERVE BRAKING		

2. ROAD TEST	YES	NO
PULLING		
CLUNK NOISE		
SCRAPING NOISE		
SQUEAL NOISE		
GRABBING		
PREMATURE LOCKUP		
WHEEL BEARING NOISE		
VIBRATION		

	YES	NO
PEDAL PULSATION WHEN BRAKING		
STEERING WHEEL OSCILLATION WHEN BRAKING		
EXCESSIVE PEDAL EFFORT REQUIRED		
WARNING LAMP COMES ON WHEN BRAKING		
DIFFERENCE IN PEDAL HEIGHT AFTER CORNERING		
NOSE DIVE WHEN BRAKING		

3. IN SHOP CHECKS ON HOIST	YES	NO	RF	LF	RR	LR
BRAKE DRAG						
INTERMITTENT BRAKE DRAG						
BRAKE PEDAL LINKAGE BINDING						
WHEEL BEARING LOOSENESS						
MISSING OR DAMAGED LUGS						
SUSPENSION LOOSENESS						
Mark Wheels, and Remove:						
Caliper/Piston Stuck						
Mark Drums and Remove:						

ADDITIONAL COMMENTS:

	FRONT					REAR				
	RIGHT			LEFT		RIGHT			LEFT	
	SPEC	SAFE	UNSAFE	SAFE	UNSAFE	SPEC	SAFE	UNSAFE	SAFE	UNSAFE
MEASURE ROTOR THICKNESS OR DRUM DIAMETER										
MEASURE ROTOR THICKNESS VARIATION (PARALLELISM)										
MEASURE ROTOR RUNOUT (TIR)										
LINING THICKNESS/CONDITION										
LINES, HOSES, CONNECTIONS, VALVES										
FLUID LEAKS										
BROKEN BLEEDER SCREWS										
LEAKY GREASE SEALS										
SELF-ADJUSTER OPERATION										

	OK	BAD
PARKING BRAKE CABLE AND LINKAGE CONDITION		
HARDWARE/BACKING PLATES		

Notes

REGISTERED OWNER ____________ DATE ____________

ADDRESS ____________ PHONE NO. ____________

MAKE ________ MODEL ________ MILEAGE ________ LICENSE NO. ________ YEAR ________

SPECIAL KEY REQUIRED FOR REMOVING HUBCAPS/WHEELS? YES ☐ NO ☐ DOES OWNER USE PARKING BRAKE? YES ☐ NO ☐

4 DRUM ☐ 4 DISC ☐ DISC/DRUM ☐ POWER BRAKES? YES ☐ NO ☐ VACUUM PUMP? YES ☐ NO ☐

OWNER COMPLAINTS ____________

1. CHECKS BEFORE ROAD TEST	OK	BAD
STOPLAMP OPERATION WITH HEADLAMPS ON/OFF		
BRAKE WARNING LAMP OPERATION		
PARKING BRAKE LAMP OPERATION		
PARKING BRAKE ADJUSTMENT		

TIRE PRESSURE SPECS:

RECORD PRESSURE MEASURED

RF ______ LF ______ RR ______ LR ______

TIRE CONDITION

RF ______ LF ______ RR ______ LR ______

MASTER CYLINDER CHECKS

FLUID LEVEL		
FLUID CONTAMINATION		
UNDER HOOD FLUID LEAKS		
UNDER DASH FLUID LEAKS (NO POWER BRAKES)		
PEDAL HEIGHT STAYS WHEN FORCE APPLIED		
PEDAL FREE TRAVEL		

CHECK ONE		CHECK ONE		
LOW PEDAL HEIGHT		SPONGY PEDAL		
MED. PEDAL HEIGHT		FIRM PEDAL		
HIGH PEDAL HEIGHT				

POWER BRAKE UNIT CHECKS

VACUUM TYPE	OK	BAD	HYDRO TYPE	OK	BAD
VACUUM UNIT LEAKS			HYDRO UNIT LEAKS		
ENGINE VACUUM			P/S FLUID LEVEL		
VACUUM HOSE			P/S BELT TENSION		
UNIT CHECK VALVE			P/S BELT CONDITION		
RESERVE BRAKING			P/S FLUID LEAKS		
PEDAL DROP TEST			RESERVE BRAKING		

2. ROAD TEST	YES	NO
PULLING		
CLUNK NOISE		
SCRAPING NOISE		
SQUEAL NOISE		
GRABBING		
PREMATURE LOCKUP		
WHEEL BEARING NOISE		
VIBRATION		

	YES	NO
PEDAL PULSATION WHEN BRAKING		
STEERING WHEEL OSCILLATION WHEN BRAKING		
EXCESSIVE PEDAL EFFORT REQUIRED		
WARNING LAMP COMES ON WHEN BRAKING		
DIFFERENCE IN PEDAL HEIGHT AFTER CORNERING		
NOSE DIVE WHEN BRAKING		

3. IN SHOP CHECKS ON HOIST	YES	NO	RF	LF	RR	LR
BRAKE DRAG						
INTERMITTENT BRAKE DRAG						
BRAKE PEDAL LINKAGE BINDING						
WHEEL BEARING LOOSENESS						
MISSING OR DAMAGED LUGS						
SUSPENSION LOOSENESS						
Mark Wheels, and Remove:						
Caliper/Piston Stuck						
Mark Drums and Remove:						

ADDITIONAL COMMENTS:

	FRONT					REAR				
	RIGHT			LEFT		RIGHT			LEFT	
	SPEC	SAFE	UNSAFE	SAFE	UNSAFE	SPEC	SAFE	UNSAFE	SAFE	UNSAFE
MEASURE ROTOR THICKNESS OR DRUM DIAMETER										
MEASURE ROTOR THICKNESS VARIATION (PARALLELISM)										
MEASURE ROTOR RUNOUT (TIR)										
LINING THICKNESS/CONDITION										
LINES, HOSES, CONNECTIONS, VALVES										
FLUID LEAKS										
BROKEN BLEEDER SCREWS										
LEAKY GREASE SEALS										
SELF-ADJUSTER OPERATION										

	OK	BAD
PARKING BRAKE CABLE AND LINKAGE CONDITION		
HARDWARE/BACKING PLATES		

Notes

REGISTERED OWNER ______________________ DATE ______________

ADDRESS ______________________ PHONE NO. ______________

MAKE __________ MODEL __________ MILEAGE __________ LICENSE NO. __________ YEAR __________

SPECIAL KEY REQUIRED FOR REMOVING HUBCAPS/WHEELS? YES ☐ NO ☐ DOES OWNER USE PARKING BRAKE? YES ☐ NO ☐

4 DRUM ☐ 4 DISC ☐ DISC/DRUM ☐ POWER BRAKES? YES ☐ NO ☐ VACUUM PUMP? YES ☐ NO ☐

OWNER COMPLAINTS ______________________

1. **CHECKS BEFORE ROAD TEST**	OK	BAD
STOPLAMP OPERATION WITH HEADLAMPS ON/OFF		
BRAKE WARNING LAMP OPERATION		
PARKING BRAKE LAMP OPERATION		
PARKING BRAKE ADJUSTMENT		

TIRE PRESSURE SPECS:

RECORD PRESSURE MEASURED

RF ______ LF ______ RR ______ LR ______

TIRE CONDITION

RF ______ LF ______ RR ______ LR ______

MASTER CYLINDER CHECKS

FLUID LEVEL		
FLUID CONTAMINATION		
UNDER HOOD FLUID LEAKS		
UNDER DASH FLUID LEAKS (NO POWER BRAKES)		
PEDAL HEIGHT STAYS WHEN FORCE APPLIED		
PEDAL FREE TRAVEL		

CHECK ONE		CHECK ONE	
LOW PEDAL HEIGHT		SPONGY PEDAL	
MED. PEDAL HEIGHT		FIRM PEDAL	
HIGH PEDAL HEIGHT			

POWER BRAKE UNIT CHECKS

VACUUM TYPE	OK	BAD	HYDRO TYPE	OK	BAD
VACUUM UNIT LEAKS			HYDRO UNIT LEAKS		
ENGINE VACUUM			P/S FLUID LEVEL		
VACUUM HOSE			P/S BELT TENSION		
UNIT CHECK VALVE			P/S BELT CONDITION		
RESERVE BRAKING			P/S FLUID LEAKS		
PEDAL DROP TEST			RESERVE BRAKING		

2. **ROAD TEST**	YES	NO
PULLING		
CLUNK NOISE		
SCRAPING NOISE		
SQUEAL NOISE		
GRABBING		
PREMATURE LOCKUP		
WHEEL BEARING NOISE		
VIBRATION		

	YES	NO
PEDAL PULSATION WHEN BRAKING		
STEERING WHEEL OSCILLATION WHEN BRAKING		
EXCESSIVE PEDAL EFFORT REQUIRED		
WARNING LAMP COMES ON WHEN BRAKING		
DIFFERENCE IN PEDAL HEIGHT AFTER CORNERING		
NOSE DIVE WHEN BRAKING		

3. **IN SHOP CHECKS ON HOIST**	YES	NO	RF	LF	RR	LR
BRAKE DRAG						
INTERMITTENT BRAKE DRAG						
BRAKE PEDAL LINKAGE BINDING						
WHEEL BEARING LOOSENESS						
MISSING OR DAMAGED LUGS						
SUSPENSION LOOSENESS						
Mark Wheels, and Remove:						
Caliper/Piston Stuck						
Mark Drums and Remove:						

ADDITIONAL COMMENTS:

	FRONT					REAR				
	RIGHT			LEFT		RIGHT			LEFT	
	SPEC	SAFE	UNSAFE	SAFE	UNSAFE	SPEC	SAFE	UNSAFE	SAFE	UNSAFE
MEASURE ROTOR THICKNESS OR DRUM DIAMETER										
MEASURE ROTOR THICKNESS VARIATION (PARALLELISM)										
MEASURE ROTOR RUNOUT (TIR)										
LINING THICKNESS/CONDITION										
LINES, HOSES, CONNECTIONS, VALVES										
FLUID LEAKS										
BROKEN BLEEDER SCREWS										
LEAKY GREASE SEALS										
SELF-ADJUSTER OPERATION										

	OK	BAD
PARKING BRAKE CABLE AND LINKAGE CONDITION		
HARDWARE/BACKING PLATES		

Notes

REGISTERED OWNER ____________________ DATE ____________

ADDRESS ____________________ PHONE NO. ____________

MAKE ________ MODEL ________ MILEAGE ________ LICENSE NO. ________ YEAR ________

SPECIAL KEY REQUIRED FOR REMOVING HUBCAPS/WHEELS? YES ☐ NO ☐ DOES OWNER USE PARKING BRAKE? YES ☐ NO ☐

4 DRUM ☐ 4 DISC ☐ DISC/DRUM ☐ POWER BRAKES? YES ☐ NO ☐ VACUUM PUMP? YES ☐ NO ☐

OWNER COMPLAINTS ____________________

1. **CHECKS BEFORE ROAD TEST**	OK	BAD
STOPLAMP OPERATION WITH HEADLAMPS ON/OFF		
BRAKE WARNING LAMP OPERATION		
PARKING BRAKE LAMP OPERATION		
PARKING BRAKE ADJUSTMENT		

MASTER CYLINDER CHECKS

FLUID LEVEL		
FLUID CONTAMINATION		
UNDER HOOD FLUID LEAKS		
UNDER DASH FLUID LEAKS (NO POWER BRAKES)		
PEDAL HEIGHT STAYS WHEN FORCE APPLIED		
PEDAL FREE TRAVEL		

CHECK ONE		CHECK ONE	
LOW PEDAL HEIGHT		SPONGY PEDAL	
MED. PEDAL HEIGHT		FIRM PEDAL	
HIGH PEDAL HEIGHT			

POWER BRAKE UNIT CHECKS

VACUUM TYPE	OK	BAD	HYDRO TYPE	OK	BAD
VACUUM UNIT LEAKS			HYDRO UNIT LEAKS		
ENGINE VACUUM			P/S FLUID LEVEL		
VACUUM HOSE			P/S BELT TENSION		
UNIT CHECK VALVE			P/S BELT CONDITION		
RESERVE BRAKING			P/S FLUID LEAKS		
PEDAL DROP TEST			RESERVE BRAKING		

TIRE PRESSURE SPECS:

RECORD PRESSURE MEASURED

RF ______ LF ______ RR ______ LR ______

TIRE CONDITION

RF ______ LF ______ RR ______ LR ______

2. **ROAD TEST**	YES	NO
PULLING		
CLUNK NOISE		
SCRAPING NOISE		
SQUEAL NOISE		
GRABBING		
PREMATURE LOCKUP		
WHEEL BEARING NOISE		
VIBRATION		

	YES	NO
PEDAL PULSATION WHEN BRAKING		
STEERING WHEEL OSCILLATION WHEN BRAKING		
EXCESSIVE PEDAL EFFORT REQUIRED		
WARNING LAMP COMES ON WHEN BRAKING		
DIFFERENCE IN PEDAL HEIGHT AFTER CORNERING		
NOSE DIVE WHEN BRAKING		

3. **IN SHOP CHECKS ON HOIST**	YES	NO	RF	LF	RR	LR
BRAKE DRAG						
INTERMITTENT BRAKE DRAG						
BRAKE PEDAL LINKAGE BINDING						
WHEEL BEARING LOOSENESS						
MISSING OR DAMAGED LUGS						
SUSPENSION LOOSENESS						
Mark Wheels, and Remove:						
Caliper/Piston Stuck						
Mark Drums and Remove:						

ADDITIONAL COMMENTS:

	FRONT					REAR				
	RIGHT			LEFT		RIGHT			LEFT	
	SPEC	SAFE	UNSAFE	SAFE	UNSAFE	SPEC	SAFE	UNSAFE	SAFE	UNSAFE
MEASURE ROTOR THICKNESS OR DRUM DIAMETER										
MEASURE ROTOR THICKNESS VARIATION (PARALLELISM)										
MEASURE ROTOR RUNOUT (TIR)										
LINING THICKNESS/CONDITION										
LINES, HOSES, CONNECTIONS, VALVES										
FLUID LEAKS										
BROKEN BLEEDER SCREWS										
LEAKY GREASE SEALS										
SELF-ADJUSTER OPERATION										

	OK	BAD
PARKING BRAKE CABLE AND LINKAGE CONDITION		
HARDWARE/BACKING PLATES		

Notes

REGISTERED OWNER ____________________ DATE ________

ADDRESS ____________________ PHONE NO. ________

MAKE ________ MODEL ________ MILEAGE ________ LICENSE NO. ________ YEAR ________

SPECIAL KEY REQUIRED FOR REMOVING HUBCAPS/WHEELS? YES ☐ NO ☐ DOES OWNER USE PARKING BRAKE? YES ☐ NO ☐

4 DRUM ☐ 4 DISC ☐ DISC/DRUM ☐ POWER BRAKES? YES ☐ NO ☐ VACUUM PUMP? YES ☐ NO ☐

OWNER COMPLAINTS ____________________

1. **CHECKS BEFORE ROAD TEST**	OK	BAD
STOPLAMP OPERATION WITH HEADLAMPS ON/OFF		
BRAKE WARNING LAMP OPERATION		
PARKING BRAKE LAMP OPERATION		
PARKING BRAKE ADJUSTMENT		

TIRE PRESSURE SPECS:

RECORD PRESSURE MEASURED

RF ______ LF ______ RR ______ LR ______

TIRE CONDITION

RF ______ LF ______ RR ______ LR ______

MASTER CYLINDER CHECKS

FLUID LEVEL		
FLUID CONTAMINATION		
UNDER HOOD FLUID LEAKS		
UNDER DASH FLUID LEAKS (NO POWER BRAKES)		
PEDAL HEIGHT STAYS WHEN FORCE APPLIED		
PEDAL FREE TRAVEL		

CHECK ONE		CHECK ONE		
LOW PEDAL HEIGHT		SPONGY PEDAL		
MED. PEDAL HEIGHT		FIRM PEDAL		
HIGH PEDAL HEIGHT				

POWER BRAKE UNIT CHECKS

VACUUM TYPE	OK	BAD	HYDRO TYPE	OK	BAD
VACUUM UNIT LEAKS			HYDRO UNIT LEAKS		
ENGINE VACUUM			P/S FLUID LEVEL		
VACUUM HOSE			P/S BELT TENSION		
UNIT CHECK VALVE			P/S BELT CONDITION		
RESERVE BRAKING			P/S FLUID LEAKS		
PEDAL DROP TEST			RESERVE BRAKING		

2. **ROAD TEST**	YES	NO
PULLING		
CLUNK NOISE		
SCRAPING NOISE		
SQUEAL NOISE		
GRABBING		
PREMATURE LOCKUP		
WHEEL BEARING NOISE		
VIBRATION		

	YES	NO
PEDAL PULSATION WHEN BRAKING		
STEERING WHEEL OSCILLATION WHEN BRAKING		
EXCESSIVE PEDAL EFFORT REQUIRED		
WARNING LAMP COMES ON WHEN BRAKING		
DIFFERENCE IN PEDAL HEIGHT AFTER CORNERING		
NOSE DIVE WHEN BRAKING		

3. **IN SHOP CHECKS ON HOIST**	YES	NO	RF	LF	RR	LR
BRAKE DRAG						
INTERMITTENT BRAKE DRAG						
BRAKE PEDAL LINKAGE BINDING						
WHEEL BEARING LOOSENESS						
MISSING OR DAMAGED LUGS						
SUSPENSION LOOSENESS						
Mark Wheels, and Remove:						
Caliper/Piston Stuck						
Mark Drums and Remove:						

ADDITIONAL COMMENTS:

	FRONT					REAR				
	RIGHT			LEFT		RIGHT			LEFT	
	SPEC	SAFE	UNSAFE	SAFE	UNSAFE	SPEC	SAFE	UNSAFE	SAFE	UNSAFE
MEASURE ROTOR THICKNESS OR DRUM DIAMETER										
MEASURE ROTOR THICKNESS VARIATION (PARALLELISM)										
MEASURE ROTOR RUNOUT (TIR)										
LINING THICKNESS/CONDITION										
LINES, HOSES, CONNECTIONS, VALVES										
FLUID LEAKS										
BROKEN BLEEDER SCREWS										
LEAKY GREASE SEALS										
SELF-ADJUSTER OPERATION										

	OK	BAD
PARKING BRAKE CABLE AND LINKAGE CONDITION		
HARDWARE/BACKING PLATES		

Notes

REGISTERED OWNER ____________ DATE ____________

ADDRESS ____________ PHONE NO. ____________

MAKE ________ MODEL ________ MILEAGE ________ LICENSE NO. ________ YEAR ________

SPECIAL KEY REQUIRED FOR REMOVING HUBCAPS/WHEELS? YES ☐ NO ☐ DOES OWNER USE PARKING BRAKE? YES ☐ NO ☐

4 DRUM ☐ 4 DISC ☐ DISC/DRUM ☐ POWER BRAKES? YES ☐ NO ☐ VACUUM PUMP? YES ☐ NO ☐

OWNER COMPLAINTS ____________

1. **CHECKS BEFORE ROAD TEST**	OK	BAD
STOPLAMP OPERATION WITH HEADLAMPS ON/OFF		
BRAKE WARNING LAMP OPERATION		
PARKING BRAKE LAMP OPERATION		
PARKING BRAKE ADJUSTMENT		

MASTER CYLINDER CHECKS

FLUID LEVEL		
FLUID CONTAMINATION		
UNDER HOOD FLUID LEAKS		
UNDER DASH FLUID LEAKS (NO POWER BRAKES)		
PEDAL HEIGHT STAYS WHEN FORCE APPLIED		
PEDAL FREE TRAVEL		

CHECK ONE		CHECK ONE		
LOW PEDAL HEIGHT		SPONGY PEDAL		
MED. PEDAL HEIGHT		FIRM PEDAL		
HIGH PEDAL HEIGHT				

POWER BRAKE UNIT CHECKS

VACUUM TYPE	OK	BAD	HYDRO TYPE	OK	BAD
VACUUM UNIT LEAKS			HYDRO UNIT LEAKS		
ENGINE VACUUM			P/S FLUID LEVEL		
VACUUM HOSE			P/S BELT TENSION		
UNIT CHECK VALVE			P/S BELT CONDITION		
RESERVE BRAKING			P/S FLUID LEAKS		
PEDAL DROP TEST			RESERVE BRAKING		

TIRE PRESSURE SPECS:

RECORD PRESSURE MEASURED

RF ______ LF ______ RR ______ LR ______

TIRE CONDITION

RF ______ LF ______ RR ______ LR ______

2. **ROAD TEST**	YES	NO
PULLING		
CLUNK NOISE		
SCRAPING NOISE		
SQUEAL NOISE		
GRABBING		
PREMATURE LOCKUP		
WHEEL BEARING NOISE		
VIBRATION		

	YES	NO
PEDAL PULSATION WHEN BRAKING		
STEERING WHEEL OSCILLATION WHEN BRAKING		
EXCESSIVE PEDAL EFFORT REQUIRED		
WARNING LAMP COMES ON WHEN BRAKING		
DIFFERENCE IN PEDAL HEIGHT AFTER CORNERING		
NOSE DIVE WHEN BRAKING		

3. **IN SHOP CHECKS ON HOIST**	YES	NO	RF	LF	RR	LR
BRAKE DRAG						
INTERMITTENT BRAKE DRAG						
BRAKE PEDAL LINKAGE BINDING						
WHEEL BEARING LOOSENESS						
MISSING OR DAMAGED LUGS						
SUSPENSION LOOSENESS						
Mark Wheels, and Remove:						
Caliper/Piston Stuck						
Mark Drums and Remove:						

ADDITIONAL COMMENTS:

	FRONT					REAR				
	RIGHT			LEFT		RIGHT			LEFT	
	SPEC	SAFE	UNSAFE	SAFE	UNSAFE	SPEC	SAFE	UNSAFE	SAFE	UNSAFE
MEASURE ROTOR THICKNESS OR DRUM DIAMETER										
MEASURE ROTOR THICKNESS VARIATION (PARALLELISM)										
MEASURE ROTOR RUNOUT (TIR)										
LINING THICKNESS/CONDITION										
LINES, HOSES, CONNECTIONS, VALVES										
FLUID LEAKS										
BROKEN BLEEDER SCREWS										
LEAKY GREASE SEALS										
SELF-ADJUSTER OPERATION										

	OK	BAD
PARKING BRAKE CABLE AND LINKAGE CONDITION		
HARDWARE/BACKING PLATES		

Notes

REGISTERED OWNER ______ DATE ______

ADDRESS ______ PHONE NO. ______

MAKE ______ MODEL ______ MILEAGE ______ LICENSE NO. ______ YEAR ______

SPECIAL KEY REQUIRED FOR REMOVING HUBCAPS/WHEELS? YES ☐ NO ☐ DOES OWNER USE PARKING BRAKE? YES ☐ NO ☐

4 DRUM ☐ 4 DISC ☐ DISC/DRUM ☐ POWER BRAKES? YES ☐ NO ☐ VACUUM PUMP? YES ☐ NO ☐

OWNER COMPLAINTS ______

1. **CHECKS BEFORE ROAD TEST**	OK	BAD
STOPLAMP OPERATION WITH HEADLAMPS ON/OFF		
BRAKE WARNING LAMP OPERATION		
PARKING BRAKE LAMP OPERATION		
PARKING BRAKE ADJUSTMENT		

MASTER CYLINDER CHECKS

FLUID LEVEL		
FLUID CONTAMINATION		
UNDER HOOD FLUID LEAKS		
UNDER DASH FLUID LEAKS (NO POWER BRAKES)		
PEDAL HEIGHT STAYS WHEN FORCE APPLIED		
PEDAL FREE TRAVEL		

CHECK ONE		CHECK ONE	
LOW PEDAL HEIGHT		SPONGY PEDAL	
MED. PEDAL HEIGHT		FIRM PEDAL	
HIGH PEDAL HEIGHT			

POWER BRAKE UNIT CHECKS

VACUUM TYPE	OK	BAD	HYDRO TYPE	OK	BAD
VACUUM UNIT LEAKS			HYDRO UNIT LEAKS		
ENGINE VACUUM			P/S FLUID LEVEL		
VACUUM HOSE			P/S BELT TENSION		
UNIT CHECK VALVE			P/S BELT CONDITION		
RESERVE BRAKING			P/S FLUID LEAKS		
PEDAL DROP TEST			RESERVE BRAKING		

TIRE PRESSURE SPECS:

RECORD PRESSURE MEASURED

RF ______ LF ______ RR ______ LR ______

TIRE CONDITION

RF ______ LF ______ RR ______ LR ______

2. **ROAD TEST**	YES	NO
PULLING		
CLUNK NOISE		
SCRAPING NOISE		
SQUEAL NOISE		
GRABBING		
PREMATURE LOCKUP		
WHEEL BEARING NOISE		
VIBRATION		

	YES	NO
PEDAL PULSATION WHEN BRAKING		
STEERING WHEEL OSCILLATION WHEN BRAKING		
EXCESSIVE PEDAL EFFORT REQUIRED		
WARNING LAMP COMES ON WHEN BRAKING		
DIFFERENCE IN PEDAL HEIGHT AFTER CORNERING		
NOSE DIVE WHEN BRAKING		

3. **IN SHOP CHECKS ON HOIST**	YES	NO	RF	LF	RR	LR
BRAKE DRAG						
INTERMITTENT BRAKE DRAG						
BRAKE PEDAL LINKAGE BINDING						
WHEEL BEARING LOOSENESS						
MISSING OR DAMAGED LUGS						
SUSPENSION LOOSENESS						
Mark Wheels, and Remove:						
Caliper/Piston Stuck						
Mark Drums and Remove:						

ADDITIONAL COMMENTS:

	FRONT					REAR				
	RIGHT			LEFT		RIGHT			LEFT	
	SPEC	SAFE	UNSAFE	SAFE	UNSAFE	SPEC	SAFE	UNSAFE	SAFE	UNSAFE
MEASURE ROTOR THICKNESS OR DRUM DIAMETER										
MEASURE ROTOR THICKNESS VARIATION (PARALLELISM)										
MEASURE ROTOR RUNOUT (TIR)										
LINING THICKNESS/CONDITION										
LINES, HOSES, CONNECTIONS, VALVES										
FLUID LEAKS										
BROKEN BLEEDER SCREWS										
LEAKY GREASE SEALS										
SELF-ADJUSTER OPERATION										

	OK	BAD
PARKING BRAKE CABLE AND LINKAGE CONDITION		
HARDWARE/BACKING PLATES		

Notes

REGISTERED OWNER ______________________ DATE __________

ADDRESS ______________________ PHONE NO. __________

MAKE ________ MODEL ________ MILEAGE ________ LICENSE NO. ________ YEAR ________

SPECIAL KEY REQUIRED FOR REMOVING HUBCAPS/WHEELS? YES ☐ NO ☐ DOES OWNER USE PARKING BRAKE? YES ☐ NO ☐

4 DRUM ☐ 4 DISC ☐ DISC/DRUM ☐ POWER BRAKES? YES ☐ NO ☐ VACUUM PUMP? YES ☐ NO ☐

OWNER COMPLAINTS ______________________

1. **CHECKS BEFORE ROAD TEST**	OK	BAD
STOPLAMP OPERATION WITH HEADLAMPS ON/OFF		
BRAKE WARNING LAMP OPERATION		
PARKING BRAKE LAMP OPERATION		
PARKING BRAKE ADJUSTMENT		

TIRE PRESSURE SPECS:

RECORD PRESSURE MEASURED

RF ______ LF ______ RR ______ LR ______

TIRE CONDITION

RF ______ LF ______ RR ______ LR ______

MASTER CYLINDER CHECKS

FLUID LEVEL		
FLUID CONTAMINATION		
UNDER HOOD FLUID LEAKS		
UNDER DASH FLUID LEAKS (NO POWER BRAKES)		
PEDAL HEIGHT STAYS WHEN FORCE APPLIED		
PEDAL FREE TRAVEL		

CHECK ONE		CHECK ONE		
LOW PEDAL HEIGHT		SPONGY PEDAL		
MED. PEDAL HEIGHT		FIRM PEDAL		
HIGH PEDAL HEIGHT				

POWER BRAKE UNIT CHECKS

VACUUM TYPE	OK	BAD	HYDRO TYPE	OK	BAD
VACUUM UNIT LEAKS			HYDRO UNIT LEAKS		
ENGINE VACUUM			P/S FLUID LEVEL		
VACUUM HOSE			P/S BELT TENSION		
UNIT CHECK VALVE			P/S BELT CONDITION		
RESERVE BRAKING			P/S FLUID LEAKS		
PEDAL DROP TEST			RESERVE BRAKING		

2. **ROAD TEST**	YES	NO
PULLING		
CLUNK NOISE		
SCRAPING NOISE		
SQUEAL NOISE		
GRABBING		
PREMATURE LOCKUP		
WHEEL BEARING NOISE		
VIBRATION		

	YES	NO
PEDAL PULSATION WHEN BRAKING		
STEERING WHEEL OSCILLATION WHEN BRAKING		
EXCESSIVE PEDAL EFFORT REQUIRED		
WARNING LAMP COMES ON WHEN BRAKING		
DIFFERENCE IN PEDAL HEIGHT AFTER CORNERING		
NOSE DIVE WHEN BRAKING		

3. **IN SHOP CHECKS ON HOIST**	YES	NO	RF	LF	RR	LR
BRAKE DRAG						
INTERMITTENT BRAKE DRAG						
BRAKE PEDAL LINKAGE BINDING						
WHEEL BEARING LOOSENESS						
MISSING OR DAMAGED LUGS						
SUSPENSION LOOSENESS						
Mark Wheels, and Remove:						
Caliper/Piston Stuck						
Mark Drums and Remove:						

ADDITIONAL COMMENTS:

	FRONT					REAR				
	RIGHT			LEFT		RIGHT			LEFT	
	SPEC	SAFE	UNSAFE	SAFE	UNSAFE	SPEC	SAFE	UNSAFE	SAFE	UNSAFE
MEASURE ROTOR THICKNESS OR DRUM DIAMETER										
MEASURE ROTOR THICKNESS VARIATION (PARALLELISM)										
MEASURE ROTOR RUNOUT (TIR)										
LINING THICKNESS/CONDITION										
LINES, HOSES, CONNECTIONS, VALVES										
FLUID LEAKS										
BROKEN BLEEDER SCREWS										
LEAKY GREASE SEALS										
SELF-ADJUSTER OPERATION										

	OK	BAD
PARKING BRAKE CABLE AND LINKAGE CONDITION		
HARDWARE/BACKING PLATES		

Notes

REGISTERED OWNER ______ DATE ______

ADDRESS ______ PHONE NO. ______

MAKE ______ MODEL ______ MILEAGE ______ LICENSE NO. ______ YEAR ______

SPECIAL KEY REQUIRED FOR REMOVING HUBCAPS/WHEELS? YES ☐ NO ☐ DOES OWNER USE PARKING BRAKE? YES ☐ NO ☐

4 DRUM ☐ 4 DISC ☐ DISC/DRUM ☐ POWER BRAKES? YES ☐ NO ☐ VACUUM PUMP? YES ☐ NO ☐

OWNER COMPLAINTS ______

1. CHECKS BEFORE ROAD TEST	OK	BAD
STOPLAMP OPERATION WITH HEADLAMPS ON/OFF		
BRAKE WARNING LAMP OPERATION		
PARKING BRAKE LAMP OPERATION		
PARKING BRAKE ADJUSTMENT		

MASTER CYLINDER CHECKS

FLUID LEVEL		
FLUID CONTAMINATION		
UNDER HOOD FLUID LEAKS		
UNDER DASH FLUID LEAKS (NO POWER BRAKES)		
PEDAL HEIGHT STAYS WHEN FORCE APPLIED		
PEDAL FREE TRAVEL		

CHECK ONE		CHECK ONE	
LOW PEDAL HEIGHT		SPONGY PEDAL	
MED. PEDAL HEIGHT		FIRM PEDAL	
HIGH PEDAL HEIGHT			

POWER BRAKE UNIT CHECKS

VACUUM TYPE	OK	BAD	HYDRO TYPE	OK	BAD
VACUUM UNIT LEAKS			HYDRO UNIT LEAKS		
ENGINE VACUUM			P/S FLUID LEVEL		
VACUUM HOSE			P/S BELT TENSION		
UNIT CHECK VALVE			P/S BELT CONDITION		
RESERVE BRAKING			P/S FLUID LEAKS		
PEDAL DROP TEST			RESERVE BRAKING		

TIRE PRESSURE SPECS:

RECORD PRESSURE MEASURED

RF ______ LF ______ RR ______ LR ______

TIRE CONDITION

RF ______ LF ______ RR ______ LR ______

2. ROAD TEST	YES	NO
PULLING		
CLUNK NOISE		
SCRAPING NOISE		
SQUEAL NOISE		
GRABBING		
PREMATURE LOCKUP		
WHEEL BEARING NOISE		
VIBRATION		

	YES	NO
PEDAL PULSATION WHEN BRAKING		
STEERING WHEEL OSCILLATION WHEN BRAKING		
EXCESSIVE PEDAL EFFORT REQUIRED		
WARNING LAMP COMES ON WHEN BRAKING		
DIFFERENCE IN PEDAL HEIGHT AFTER CORNERING		
NOSE DIVE WHEN BRAKING		

3. IN SHOP CHECKS ON HOIST	YES	NO	RF	LF	RR	LR
BRAKE DRAG						
INTERMITTENT BRAKE DRAG						
BRAKE PEDAL LINKAGE BINDING						
WHEEL BEARING LOOSENESS						
MISSING OR DAMAGED LUGS						
SUSPENSION LOOSENESS						
Mark Wheels, and Remove:						
Caliper/Piston Stuck						
Mark Drums and Remove:						

ADDITIONAL COMMENTS:

	FRONT					REAR				
	RIGHT			LEFT		RIGHT			LEFT	
	SPEC	SAFE	UNSAFE	SAFE	UNSAFE	SPEC	SAFE	UNSAFE	SAFE	UNSAFE
MEASURE ROTOR THICKNESS OR DRUM DIAMETER										
MEASURE ROTOR THICKNESS VARIATION (PARALLELISM)										
MEASURE ROTOR RUNOUT (TIR)										
LINING THICKNESS/CONDITION										
LINES, HOSES, CONNECTIONS, VALVES										
FLUID LEAKS										
BROKEN BLEEDER SCREWS										
LEAKY GREASE SEALS										
SELF-ADJUSTER OPERATION										

	OK	BAD
PARKING BRAKE CABLE AND LINKAGE CONDITION		
HARDWARE/BACKING PLATES		

Notes

REGISTERED OWNER ______ DATE ______

ADDRESS ______ PHONE NO. ______

MAKE ______ MODEL ______ MILEAGE ______ LICENSE NO. ______ YEAR ______

SPECIAL KEY REQUIRED FOR REMOVING HUBCAPS/WHEELS? YES ☐ NO ☐ DOES OWNER USE PARKING BRAKE? YES ☐ NO ☐

4 DRUM ☐ 4 DISC ☐ DISC/DRUM ☐ POWER BRAKES? YES ☐ NO ☐ VACUUM PUMP? YES ☐ NO ☐

OWNER COMPLAINTS ______

1. **CHECKS BEFORE ROAD TEST**	OK	BAD
STOPLAMP OPERATION WITH HEADLAMPS ON/OFF		
BRAKE WARNING LAMP OPERATION		
PARKING BRAKE LAMP OPERATION		
PARKING BRAKE ADJUSTMENT		

TIRE PRESSURE SPECS:

RECORD PRESSURE MEASURED

RF ______ LF ______ RR ______ LR ______

TIRE CONDITION

RF ______ LF ______ RR ______ LR ______

MASTER CYLINDER CHECKS

FLUID LEVEL		
FLUID CONTAMINATION		
UNDER HOOD FLUID LEAKS		
UNDER DASH FLUID LEAKS (NO POWER BRAKES)		
PEDAL HEIGHT STAYS WHEN FORCE APPLIED		
PEDAL FREE TRAVEL		

CHECK ONE		CHECK ONE		
LOW PEDAL HEIGHT		SPONGY PEDAL		
MED. PEDAL HEIGHT		FIRM PEDAL		
HIGH PEDAL HEIGHT				

POWER BRAKE UNIT CHECKS

VACUUM TYPE	OK	BAD	HYDRO TYPE	OK	BAD
VACUUM UNIT LEAKS			HYDRO UNIT LEAKS		
ENGINE VACUUM			P/S FLUID LEVEL		
VACUUM HOSE			P/S BELT TENSION		
UNIT CHECK VALVE			P/S BELT CONDITION		
RESERVE BRAKING			P/S FLUID LEAKS		
PEDAL DROP TEST			RESERVE BRAKING		

2. **ROAD TEST**	YES	NO
PULLING		
CLUNK NOISE		
SCRAPING NOISE		
SQUEAL NOISE		
GRABBING		
PREMATURE LOCKUP		
WHEEL BEARING NOISE		
VIBRATION		

	YES	NO
PEDAL PULSATION WHEN BRAKING		
STEERING WHEEL OSCILLATION WHEN BRAKING		
EXCESSIVE PEDAL EFFORT REQUIRED		
WARNING LAMP COMES ON WHEN BRAKING		
DIFFERENCE IN PEDAL HEIGHT AFTER CORNERING		
NOSE DIVE WHEN BRAKING		

3. **IN SHOP CHECKS ON HOIST**	YES	NO	RF	LF	RR	LR
BRAKE DRAG						
INTERMITTENT BRAKE DRAG						
BRAKE PEDAL LINKAGE BINDING						
WHEEL BEARING LOOSENESS						
MISSING OR DAMAGED LUGS						
SUSPENSION LOOSENESS						
Mark Wheels, and Remove:						
Caliper/Piston Stuck						
Mark Drums and Remove:						

ADDITIONAL COMMENTS:

	FRONT					REAR				
	RIGHT			LEFT		RIGHT			LEFT	
	SPEC	SAFE	UNSAFE	SAFE	UNSAFE	SPEC	SAFE	UNSAFE	SAFE	UNSAFE
MEASURE ROTOR THICKNESS OR DRUM DIAMETER										
MEASURE ROTOR THICKNESS VARIATION (PARALLELISM)										
MEASURE ROTOR RUNOUT (TIR)										
LINING THICKNESS/CONDITION										
LINES, HOSES, CONNECTIONS, VALVES										
FLUID LEAKS										
BROKEN BLEEDER SCREWS										
LEAKY GREASE SEALS										
SELF-ADJUSTER OPERATION										

	OK	BAD
PARKING BRAKE CABLE AND LINKAGE CONDITION		
HARDWARE/BACKING PLATES		

Notes

REGISTERED OWNER ______ DATE ______

ADDRESS ______ PHONE NO. ______

MAKE ______ MODEL ______ MILEAGE ______ LICENSE NO. ______ YEAR ______

SPECIAL KEY REQUIRED FOR REMOVING HUBCAPS/WHEELS? YES ☐ NO ☐ DOES OWNER USE PARKING BRAKE? YES ☐ NO ☐

4 DRUM ☐ 4 DISC ☐ DISC/DRUM ☐ POWER BRAKES? YES ☐ NO ☐ VACUUM PUMP? YES ☐ NO ☐

OWNER COMPLAINTS ______

1. **CHECKS BEFORE ROAD TEST**	OK	BAD
STOPLAMP OPERATION WITH HEADLAMPS ON/OFF		
BRAKE WARNING LAMP OPERATION		
PARKING BRAKE LAMP OPERATION		
PARKING BRAKE ADJUSTMENT		

TIRE PRESSURE SPECS:

RECORD PRESSURE MEASURED

RF ______ LF ______ RR ______ LR ______

TIRE CONDITION

RF ______ LF ______ RR ______ LR ______

MASTER CYLINDER CHECKS

FLUID LEVEL		
FLUID CONTAMINATION		
UNDER HOOD FLUID LEAKS		
UNDER DASH FLUID LEAKS (NO POWER BRAKES)		
PEDAL HEIGHT STAYS WHEN FORCE APPLIED		
PEDAL FREE TRAVEL		

CHECK ONE		CHECK ONE	
LOW PEDAL HEIGHT		SPONGY PEDAL	
MED. PEDAL HEIGHT		FIRM PEDAL	
HIGH PEDAL HEIGHT			

POWER BRAKE UNIT CHECKS

VACUUM TYPE	OK	BAD	HYDRO TYPE	OK	BAD
VACUUM UNIT LEAKS			HYDRO UNIT LEAKS		
ENGINE VACUUM			P/S FLUID LEVEL		
VACUUM HOSE			P/S BELT TENSION		
UNIT CHECK VALVE			P/S BELT CONDITION		
RESERVE BRAKING			P/S FLUID LEAKS		
PEDAL DROP TEST			RESERVE BRAKING		

2. **ROAD TEST**	YES	NO
PULLING		
CLUNK NOISE		
SCRAPING NOISE		
SQUEAL NOISE		
GRABBING		
PREMATURE LOCKUP		
WHEEL BEARING NOISE		
VIBRATION		

	YES	NO
PEDAL PULSATION WHEN BRAKING		
STEERING WHEEL OSCILLATION WHEN BRAKING		
EXCESSIVE PEDAL EFFORT REQUIRED		
WARNING LAMP COMES ON WHEN BRAKING		
DIFFERENCE IN PEDAL HEIGHT AFTER CORNERING		
NOSE DIVE WHEN BRAKING		

3. **IN SHOP CHECKS ON HOIST**	YES	NO	RF	LF	RR	LR
BRAKE DRAG						
INTERMITTENT BRAKE DRAG						
BRAKE PEDAL LINKAGE BINDING						
WHEEL BEARING LOOSENESS						
MISSING OR DAMAGED LUGS						
SUSPENSION LOOSENESS						
Mark Wheels, and Remove:						
Caliper/Piston Stuck						
Mark Drums and Remove:						

ADDITIONAL COMMENTS:

	FRONT					REAR				
	RIGHT			LEFT		RIGHT			LEFT	
	SPEC	SAFE	UNSAFE	SAFE	UNSAFE	SPEC	SAFE	UNSAFE	SAFE	UNSAFE
MEASURE ROTOR THICKNESS OR DRUM DIAMETER										
MEASURE ROTOR THICKNESS VARIATION (PARALLELISM)										
MEASURE ROTOR RUNOUT (TIR)										
LINING THICKNESS/CONDITION										
LINES, HOSES, CONNECTIONS, VALVES										
FLUID LEAKS										
BROKEN BLEEDER SCREWS										
LEAKY GREASE SEALS										
SELF-ADJUSTER OPERATION										

	OK	BAD
PARKING BRAKE CABLE AND LINKAGE CONDITION		
HARDWARE/BACKING PLATES		

Notes

REGISTERED OWNER ____________________ DATE ____________

ADDRESS ____________________ PHONE NO. ____________

MAKE ________ MODEL ________ MILEAGE ________ LICENSE NO. ________ YEAR ________

SPECIAL KEY REQUIRED FOR REMOVING HUBCAPS/WHEELS? YES ☐ NO ☐ DOES OWNER USE PARKING BRAKE? YES ☐ NO ☐

4 DRUM ☐ 4 DISC ☐ DISC/DRUM ☐ POWER BRAKES? YES ☐ NO ☐ VACUUM PUMP? YES ☐ NO ☐

OWNER COMPLAINTS ____________________

1. **CHECKS BEFORE ROAD TEST**	OK	BAD
STOPLAMP OPERATION WITH HEADLAMPS ON/OFF		
BRAKE WARNING LAMP OPERATION		
PARKING BRAKE LAMP OPERATION		
PARKING BRAKE ADJUSTMENT		

TIRE PRESSURE SPECS:

RECORD PRESSURE MEASURED

RF ______ LF ______ RR ______ LR ______

TIRE CONDITION

RF ______ LF ______ RR ______ LR ______

MASTER CYLINDER CHECKS

FLUID LEVEL		
FLUID CONTAMINATION		
UNDER HOOD FLUID LEAKS		
UNDER DASH FLUID LEAKS (NO POWER BRAKES)		
PEDAL HEIGHT STAYS WHEN FORCE APPLIED		
PEDAL FREE TRAVEL		

CHECK ONE		CHECK ONE		
LOW PEDAL HEIGHT		SPONGY PEDAL		
MED. PEDAL HEIGHT		FIRM PEDAL		
HIGH PEDAL HEIGHT				

POWER BRAKE UNIT CHECKS

VACUUM TYPE	OK	BAD	HYDRO TYPE	OK	BAD
VACUUM UNIT LEAKS			HYDRO UNIT LEAKS		
ENGINE VACUUM			P/S FLUID LEVEL		
VACUUM HOSE			P/S BELT TENSION		
UNIT CHECK VALVE			P/S BELT CONDITION		
RESERVE BRAKING			P/S FLUID LEAKS		
PEDAL DROP TEST			RESERVE BRAKING		

2. **ROAD TEST**	YES	NO
PULLING		
CLUNK NOISE		
SCRAPING NOISE		
SQUEAL NOISE		
GRABBING		
PREMATURE LOCKUP		
WHEEL BEARING NOISE		
VIBRATION		

	YES	NO
PEDAL PULSATION WHEN BRAKING		
STEERING WHEEL OSCILLATION WHEN BRAKING		
EXCESSIVE PEDAL EFFORT REQUIRED		
WARNING LAMP COMES ON WHEN BRAKING		
DIFFERENCE IN PEDAL HEIGHT AFTER CORNERING		
NOSE DIVE WHEN BRAKING		

3. **IN SHOP CHECKS ON HOIST**	YES	NO	RF	LF	RR	LR
BRAKE DRAG						
INTERMITTENT BRAKE DRAG						
BRAKE PEDAL LINKAGE BINDING						
WHEEL BEARING LOOSENESS						
MISSING OR DAMAGED LUGS						
SUSPENSION LOOSENESS						
Mark Wheels, and Remove:						
Caliper/Piston Stuck						
Mark Drums and Remove:						

ADDITIONAL COMMENTS:

	FRONT					REAR				
	RIGHT			LEFT		RIGHT			LEFT	
	SPEC	SAFE	UNSAFE	SAFE	UNSAFE	SPEC	SAFE	UNSAFE	SAFE	UNSAFE
MEASURE ROTOR THICKNESS OR DRUM DIAMETER										
MEASURE ROTOR THICKNESS VARIATION (PARALLELISM)										
MEASURE ROTOR RUNOUT (TIR)										
LINING THICKNESS/CONDITION										
LINES, HOSES, CONNECTIONS, VALVES										
FLUID LEAKS										
BROKEN BLEEDER SCREWS										
LEAKY GREASE SEALS										
SELF-ADJUSTER OPERATION										

	OK	BAD
PARKING BRAKE CABLE AND LINKAGE CONDITION		
HARDWARE/BACKING PLATES		

Notes

REGISTERED OWNER ______________________ DATE ______________

ADDRESS ______________________ PHONE NO. ______________

MAKE ____________ MODEL ____________ MILEAGE ____________ LICENSE NO. ____________ YEAR ____________

SPECIAL KEY REQUIRED FOR REMOVING HUBCAPS/WHEELS? YES ☐ NO ☐ DOES OWNER USE PARKING BRAKE? YES ☐ NO ☐

4 DRUM ☐ 4 DISC ☐ DISC/DRUM ☐ POWER BRAKES? YES ☐ NO ☐ VACUUM PUMP? YES ☐ NO ☐

OWNER COMPLAINTS ______________________

____

1. **CHECKS BEFORE ROAD TEST**	OK	BAD
STOPLAMP OPERATION WITH HEADLAMPS ON/OFF		
BRAKE WARNING LAMP OPERATION		
PARKING BRAKE LAMP OPERATION		
PARKING BRAKE ADJUSTMENT		

TIRE PRESSURE SPECS:

RECORD PRESSURE MEASURED

RF ______ LF ______ RR ______ LR ______

TIRE CONDITION

RF ______ LF ______ RR ______ LR ______

MASTER CYLINDER CHECKS

FLUID LEVEL		
FLUID CONTAMINATION		
UNDER HOOD FLUID LEAKS		
UNDER DASH FLUID LEAKS (NO POWER BRAKES)		
PEDAL HEIGHT STAYS WHEN FORCE APPLIED		
PEDAL FREE TRAVEL		

CHECK ONE		CHECK ONE		
LOW PEDAL HEIGHT		SPONGY PEDAL		
MED. PEDAL HEIGHT		FIRM PEDAL		
HIGH PEDAL HEIGHT				

POWER BRAKE UNIT CHECKS

VACUUM TYPE	OK	BAD	HYDRO TYPE	OK	BAD
VACUUM UNIT LEAKS			HYDRO UNIT LEAKS		
ENGINE VACUUM			P/S FLUID LEVEL		
VACUUM HOSE			P/S BELT TENSION		
UNIT CHECK VALVE			P/S BELT CONDITION		
RESERVE BRAKING			P/S FLUID LEAKS		
PEDAL DROP TEST			RESERVE BRAKING		

2. **ROAD TEST**	YES	NO
PULLING		
CLUNK NOISE		
SCRAPING NOISE		
SQUEAL NOISE		
GRABBING		
PREMATURE LOCKUP		
WHEEL BEARING NOISE		
VIBRATION		

	YES	NO
PEDAL PULSATION WHEN BRAKING		
STEERING WHEEL OSCILLATION WHEN BRAKING		
EXCESSIVE PEDAL EFFORT REQUIRED		
WARNING LAMP COMES ON WHEN BRAKING		
DIFFERENCE IN PEDAL HEIGHT AFTER CORNERING		
NOSE DIVE WHEN BRAKING		

3. **IN SHOP CHECKS ON HOIST**	YES	NO	RF	LF	RR	LR
BRAKE DRAG						
INTERMITTENT BRAKE DRAG						
BRAKE PEDAL LINKAGE BINDING						
WHEEL BEARING LOOSENESS						
MISSING OR DAMAGED LUGS						
SUSPENSION LOOSENESS						
Mark Wheels, and Remove:						
Caliper/Piston Stuck						
Mark Drums and Remove:						

ADDITIONAL COMMENTS:

	FRONT					REAR				
	RIGHT			LEFT		RIGHT			LEFT	
	SPEC	SAFE	UNSAFE	SAFE	UNSAFE	SPEC	SAFE	UNSAFE	SAFE	UNSAFE
MEASURE ROTOR THICKNESS OR DRUM DIAMETER										
MEASURE ROTOR THICKNESS VARIATION (PARALLELISM)										
MEASURE ROTOR RUNOUT (TIR)										
LINING THICKNESS/CONDITION										
LINES, HOSES, CONNECTIONS, VALVES										
FLUID LEAKS										
BROKEN BLEEDER SCREWS										
LEAKY GREASE SEALS										
SELF-ADJUSTER OPERATION										

	OK	BAD
PARKING BRAKE CABLE AND LINKAGE CONDITION		
HARDWARE/BACKING PLATES		

Notes